nature

The Living Record of Science

《自然》百年科学经典

英汉对照版（平装本）

第九卷（上）

总顾问：李政道（Tsung-Dao Lee）

英方主编：Sir John Maddox
Sir Philip Campbell　中方主编：路甬祥

IX

1998-2001

外语教学与研究出版社　·　麦克米伦教育　·　自然科研

FOREIGN LANGUAGE TEACHING AND RESEARCH PRESS · MACMILLAN EDUCATION · NATURE RESEARCH

北京 BEIJING

图书在版编目 (CIP) 数据

《自然》百年科学经典. 第九卷. 上, 1998—2001：英汉对照 ／（英）约翰·马多克斯 (John Maddox)，（英）菲利普·坎贝尔 (Philip Campbell)，路甬祥主编. —— 北京 ：外语教学与研究出版社，2019.12
ISBN 978-7-5213-1472-4

Ⅰ . ①自… Ⅱ . ①约… ②菲… ③路… Ⅲ . ①自然科学－文集－英、汉 Ⅳ . ①N53

中国版本图书馆 CIP 数据核字 (2020) 第 014448 号

地图审图号：GS (2018) 5219 号

出 版 人　徐建忠
项目统筹　章思英
项目负责　刘晓楠　黄小斌
责任编辑　王丽霞
责任校对　黄小斌
封面设计　孙莉明　曹志远
版式设计　孙莉明
出版发行　外语教学与研究出版社
社　　址　北京市西三环北路 19 号（100089）
网　　址　http://www.fltrp.com
印　　刷　北京华联印刷有限公司
开　　本　787×1092　1/16
印　　张　25
版　　次　2020 年 4 月第 1 版 2020 年 4 月第 1 次印刷
书　　号　ISBN 978-7-5213-1472-4
定　　价　168.00 元

购书咨询：（010）88819926　电子邮箱: club@fltrp.com
外研书店: https://waiyants.tmall.com
凡印刷、装订质量问题，请联系我社印制部
联系电话：（010）61207896　电子邮箱: zhijian@fltrp.com
凡侵权、盗版书籍线索，请联系我社法律事务部
举报电话：（010）88817519　电子邮箱: banquan@fltrp.com
物料号：314720001

记载人类文明
沟通世界文化
www.fltrp.com

《自然》百年科学经典（英汉对照版）

总顾问：李政道（Tsung-Dao Lee）

英方主编：Sir John Maddox 中方主编：路甬祥

Sir Philip Campbell

编审委员会

编译委员会

本卷翻译工作组稿人（以姓氏笔画为序）

王丽霞	王晓蕾	王耀杨	刘 明	刘晓楠	关秀清	何 铭
沈乃澂	周家斌	郭红锋	黄小斌	蔡 迪	蔡则怡	

本卷翻译人员（以姓氏笔画为序）

王海纳	王耀杨	毛晨晖	卢 皓	吕 静	任 奕	刘项琨
刘振明	刘皓芳	齐红艳	李 梅	杨 晶	肖 莉	余 恒
沈乃澂	张玉光	金世超	周家斌	姜 克	高如丽	崔 宁
董培智	蔡则怡					

本卷校对人员（以姓氏笔画为序）

马 昊	王晓蕾	王德孚	牛慧冲	龙娉娉	卢 皓	田晓阳
吉 祥	任 奕	刘本琼	刘立云	刘项琨	刘琰璐	许静静
阮玉辉	李 龙	李 平	李 景	李 婷	李志军	李若男
李照涛	李霄霞	杨 晶	杨学良	吴 茜	邱珍琳	邱彩玉
何 敏	张亚盟	张茜楠	张美月	张瑶楠	陈 秀	陈贝贝
陈思原	周 晔	周少贞	郑旭峰	郑婧澜	赵凤轩	胡海霞
娄 研	洪雅强	贺舒雅	顾海成	徐 玲	黄小斌	第文龙
蒋世仰	韩少卿	焦晓林	蔡则怡	裴 琳	潘卫东	薛 陕
Eric Leher (澳)						

Contents
目录

Volume IX

(1998-2001)

Discovery of a Supernova Explosion at Half the Age of the Universe

S. Perlmutter *et al.*

Editor's Note

In the early 1990s it became possible to use type Ia supernovae as "standard candles" to determine astronomical distances. Using this approach, Saul Perlmutter and coworkers here report a supernova at a redshift of 0.83 that is fainter than expected. Although they initially interpreted the faintness as evidence that the universe has a lower average density than was thought, it was soon realized that the best explanation is that the universe is expanding at an accelerating rate. This is now the accepted view, although the reason is unclear and represents one of the central puzzles in contemporary cosmology. One interpretation is that the universe is pervaded by "dark energy" that creates a repulsive force, counteracting gravitational attraction.

The ultimate fate of the Universe, infinite expansion or a big crunch, can be determined by using the redshifts and distances of very distant supernovae to monitor changes in the expansion rate. We can now find[1] large numbers of these distant supernovae, and measure their redshifts and apparent brightnesses; moreover, recent studies of nearby type Ia supernovae have shown how to determine their intrinsic luminosities[2-4]—and therefore with their apparent brightnesses obtain their distances. The > 50 distant supernovae discovered so far provide a record of changes in the expansion rate over the past several billion years[5-7]. However, it is necessary to extend this expansion history still farther away (hence further back in time) in order to begin to distinguish the causes of the expansion-rate changes—such as the slowing caused by the gravitational attraction of the Universe's mass density, and the possibly counteracting effect of the cosmological constant[8]. Here we report the most distant spectroscopically confirmed supernova. Spectra and photometry from the largest telescopes on the ground and in space show that this ancient supernova is strikingly similar to nearby, recent type Ia supernovae. When combined with previous measurements of nearer supernovae[2,5], these new measurements suggest that we may live in a low-mass-density universe.

SN1997ap was discovered by the Supernova Cosmology Project collaboration on 5 March 1997 UT, during a two-night search at the Cerro Tololo Interamerican Observatory (CTIO) 4-m telescope that yielded 16 new supernovae. The search technique finds such sets of high-redshift supernovae on the rising part of their light curves and guarantees the date of discovery, thus allowing follow-up photometry and spectroscopy of the transient supernovae to be scheduled[1]. The supernova light curves were followed

在宇宙年龄一半处发现的超新星爆发

珀尔马特等

编者按

在 20 世纪 90 年代的早期，使用 Ia 型超新星作为"标准烛光"测定天文学距离已成为可能。使用这种方法，索尔·珀尔马特以及他的合作者们在这里报道了在红移 0.83 处比预期的暗淡的一颗超新星。尽管他们最初把这颗超新星的暗淡解释为宇宙的平均密度比普遍认为的要低的证据，但是很快他们意识到最好的解释是宇宙的膨胀在加速。现在这个观点已经被普遍认可，尽管原因尚不知晓，这也是当代宇宙学最主要的难题之一。一种解释是宇宙中充斥着暗能量，产生排斥力，与引力相抗衡。

宇宙的命运最终是无限膨胀还是大挤压，可以通过测量遥远超新星的红移和距离，进而监测宇宙膨胀速率的变化来确定。我们现在发现了大量遥远的超新星 [1]，并且测定了它们的红移和视亮度；而对较近 Ia 型超新星的研究已经找到了测定本征光度 [2-4] 的方法，通过它们的视亮度就能得到距离。至今已发现的 50 多个远距离 Ia 型超新星记录了过去几十亿年间宇宙膨胀速率的变化 [5-7]。不过我们还需要进一步追溯更远（即时间上更早）的宇宙膨胀历史，从而找出宇宙膨胀速率变化的原因——是在宇宙质量密度的引力影响下变慢，或是由宇宙学常数的反作用而加速 [8]。我们在此报道一颗已获光谱确认的最遥远的超新星。来自地面和空间中最大望远镜的光谱和测光数据表明这颗古老的超新星和较近的 Ia 型超新星非常相似。结合以前对较近距离的 Ia 型超新星的观测数据 [2,5]，这些新测量表明我们可能生活在一个质量密度偏低的宇宙之中。

SN1997ap 于世界时（UT）1997 年 3 月 5 日被超新星宇宙学项目团队发现。位于智利的托洛洛山美洲天文台（CTIO）4 米望远镜在两个晚上的搜寻中总共发现了 16 颗新的超新星。搜寻技术能够在这些高红移超新星处于光变曲线上升阶段的时候发现它们，这就保证了发现的日期。因此我们可以安排对这些暂现的超新星进行进一步的测光和光谱观测 [1]。利用 CTIO、WIYN、ESO 3.6 米和 INT 的望远镜，我们按

with scheduled R-, I- and some B-band photometry at the CTIO, WIYN, ESO 3.6-m, and INT telescopes, and with spectroscopy at the ESO 3.6-m and Keck II telescopes. (Here WIYN is the Wisconsin, Indiana, Yale, NOAO Telescope, ESO is the European Southern Observatory, and INT is the Isaac Newton Telescope.) In addition, SN1997ap was followed with scheduled photometry on the Hubble Space Telescope (HST).

Figure 1 shows the spectrum of SN1997ap, obtained on 14 March 1997 UT with a 1.5-h integration on the Keck II 10-m telescope. There is negligible ($\leqslant 5\%$) host-galaxy light contaminating the supernova spectrum, as measured from the ground- and space-based images. When fitted to a time series of well-measured nearby type Ia supernova spectra[9], the spectrum of SN1997ap is most consistent with a "normal" type Ia supernova at redshift $z = 0.83$ observed 2 ± 2 supernova-restframe days (~4 observer's days) before the supernova's maximum light in the rest-frame B band. It is a poor match to the "abnormal" type Ia supernovae, such as the brighter SN1991T or the fainter SN1986G. For comparison, the spectra of low-redshift, "normal" type Ia supernovae are shown in Fig. 1 with wavelengths redshifted as they would appear at $z = 0.83$. These spectra show the time evolution from 7 days before, to 2 days after, maximum light.

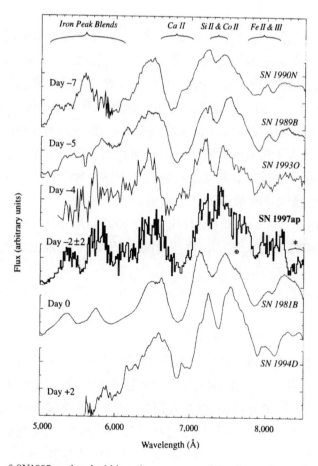

Fig. 1. Spectrum of SN1997ap placed within a time sequence of five "normal" type Ia supernovae. The

计划在 R、I 及 B 波段对超新星进行了测光从而得到光变曲线，同时还用 ESO 3.6 米和凯克 II 望远镜拍摄了光谱。（这里的 WIYN 是指威斯康星大学–印第安纳大学–耶鲁大学–美国国家光学天文台望远镜；ESO 是指欧洲南方天文台；INT 是指艾萨克·牛顿望远镜。）此外，我们还安排哈勃空间望远镜（HST）对 SN1997ap 进行测光。

图 1 是 SN1997ap 的光谱，它是由凯克 II 10 米天文望远镜于 1997 年 3 月 14 日 UT 持续 1.5 小时的观测数据积分而成。地面以及空间望远镜测量到的图像显示，寄主星系对超新星光谱的污染可以忽略不计（≤5%）。我们把 SN1997ap 的光谱与一系列较近且已充分测量过的 Ia 型超新星光谱做时间序列拟合，发现它的光谱与红移 $z = 0.83$、在 B 波段极大亮度前 2±2 日（超新星静止系，约为 4 个观测者日）观测的"正常"Ia 型超新星最为一致。SN1997ap 的光谱与其他"非正常"Ia 型超新星（例如较亮的 SN1991T 或较暗的 SN1986G）的光谱并不匹配。图 1 中为了方便比较，低红移的"正常"Ia 型超新星的光谱被红移到 $z = 0.83$ 处。这些光谱显示了超新星从最大亮度的前 7 天到后 2 天的时间演化。

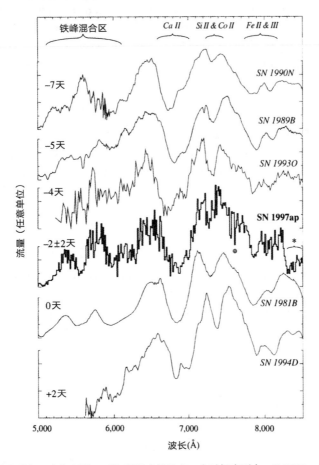

图 1. SN1997ap 的光谱与五个"正常"Ia 型超新星光谱放在一个时间序列中。SN1997ap 的光谱以 12.5 Å

data for SN1997ap have been binned by 12.5 Å ; the time series of spectra of the other supernovae[17-21] (the spectrum of SN1993O was provided courtesy of the Calán/Tololo Supernova Survey) are given as they would appear redshifted to $z = 0.83$. The spectra show the evolution of spectral features between 7 rest-frame days before, and 2 days after, rest-frame B-band maximum light. SN1997ap matches best at 2 ± 2 days before maximum light. The symbol $\oplus$ indicates an atmospheric absorption line and * indicates a region affected by night-sky line subtraction residuals. The redshift of $z = 0.83 \pm 0.005$ was determined from the supernova spectrum itself, as there were no host galaxy lines detected.

Figure 2 shows the photometry data for SN1997ap, with significantly smaller error bars for the HST observations (Fig. 2a) than for the ground-based observations (Fig. 2b and c). The width of the light curve of a type Ia supernova has been shown to be an excellent indicator of its intrinsic luminosity, both at low redshift[2-4] and at high redshift[5]: the broader and slower the light curve, the brighter the supernova is at maximum. We characterize this width by fitting the photometry data to a "normal" type Ia supernova template light curve that has its time axis stretched or compressed by a linear factor, called the "stretch factor"[1,5]; a "normal" supernova such as SN1989B, SN1993O or SN1981B in Fig. 1 thus has a stretch factor of $s \approx 1$. To fit the photometry data for SN1997ap, we use template U- and B-band light curves that have first been $1 + z$ time-dilated and wavelength-shifted ("K-corrected") to the R- and I-bands as they would appear at $z = 0.83$ (see ref. 5 and P.N. *et al.*, manuscript in preparation). The best-fit stretch factor for all the photometry of Fig. 2 indicates that SN1997ap is a "normal" type Ia supernova: $s = 1.03 \pm 0.05$ when fitted for a date of maximum at 16.3 March 1997 UT (the error-weighted average of the best-fit dates from the light curve, 15.3 ± 1.6 March 1997 UT, and from the spectrum, 18 ± 3 March 1997 UT).

It is interesting to note that we could alternatively fit the $1 + z$ time dilation of the event while holding the stretch factor constant at $s = 1.0^{+0.05}_{-0.14}$ (the best fit value from the spectral features obtained in ref. 10). We find that the event lasted $1 + z = 1.86^{+0.31}_{-0.09}$ times longer than a nearby $s = 1$ supernova, providing the strongest confirmation yet of the cosmological nature of redshift[9,11,12].

The best-fit peak magnitudes for SN1997ap are $I = 23.20 \pm 0.07$ and $R = 24.10 \pm 0.09$. (All magnitudes quoted or plotted here are transformed to the standard Cousins[13] R and I bands.) These peak magnitudes are relatively insensitive to the details of the fit: if the date of maximum is left unconstrained or set to the date indicated by the best-match spectrum, or if the ground- and space-based data are fitted alone, the peak magnitudes still agree well within errors.

为区间合并；其他超新星的光谱 [17-21](SN1993O 的光谱数据由 Calán/Tololo 超新星巡天提供）都被红移到 $z = 0.83$ 处。这些光谱反映了静止系 B 波段光强达到峰值的 7 天前至 2 天后之间的光谱特征变化。SN1997ap 的数据与最大亮度前 2±2 天的光谱数据最吻合。符号 ⊕ 表示大气吸收线，* 表示受夜天光抵扣残余影响的区域。红移 $z = 0.83 \pm 0.005$ 得自超新星光谱，我们并没有检测到寄主星系的谱线。

图 2 显示的是超新星 SN1997ap 的测光数据。哈勃空间望远镜的观测（图 2a）误差显著小于地面望远镜的观测（图 2b 和 2c）。Ia 型超新星光变曲线的宽度被证明是"本征光度"的绝佳表征，无论是在低红移 [2-4] 还是在高红移 [5] 处：超新星的光变曲线越宽、变化越慢，那么最大亮度就越高。我们通过将测光数据与"正常"Ia 型超新星的光变曲线模板进行拟合来得到这个宽度，其中模板的时间轴由一个称为"伸展因子"[1,5] 的线性参数进行"拉伸"或"压缩"。图 1 中的"正常"超新星 SN1989B、SN1993O 及 SN1981B 的伸展因子 s 约等于 1。为了拟合 SN1997ap 的测光数据，我们使用 U 和 B 波段的光变曲线作为模板，将其经过 $1+z$ 倍的时间拉伸和波长平移（即"K 修正"）之后移动到 R 和 I 波段，就像它们在红移 0.83 处一样（见参考文献 5 和纽金特等人正在撰写的文章）。利用图 2 中所有测光数据对延展因子进行拟合得到 $s = 1.03 \pm 0.05$，这表明 SN1997ap 是一个"正常"的 Ia 型超新星：拟合的亮度极大值日期为 1997 年 3 月 16.3 日 UT（根据光变曲线得到的误差加权平均值为 1997 年 3 月 15.3±1.6 日 UT，根据光谱得到的结果为 1997 年 3 月 18±3 日 UT）。

值得一提的是，我们也可以保持伸展因子 $s = 1.0^{+0.05}_{-0.14}$ 不变（这个最佳拟合值来自文献 10 中的光谱数据），然后对事件的时间膨胀因子 $1+z$ 进行拟合。我们发现 SN1997ap 爆发事件的持续时间是一颗较近超新星（$s = 1$）的 $1+z = 1.86^{+0.31}_{-0.09}$ 倍，这一点为红移的宇宙学属性提供了迄今最强的确认 [9,11,12]。

超新星 SN1997ap 的峰值星等最佳拟合值为 $I = 23.20 \pm 0.07$ 以及 $R = 24.10 \pm 0.09$。（在本文中，所有提到和绘出的星等都已转换为标准的库森 [13]R 和 I 波段星等。）这些峰值星等对拟合的细节并不敏感：如果我们不限制最大值日期或是以最佳匹配光谱来设定最大值日期，或者是只用地面观测数据或空间观测数据来单独拟合，拟合的星等峰值仍然在误差以内。

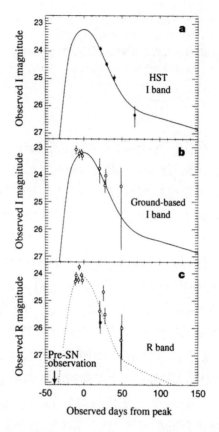

Fig. 2. Photometry points for SN1997ap. **a**, As observed by the HST in the F814W filter; **b**, as observed with ground-based telescopes in the Harris I filter; and **c**, as observed with the ground-based telescopes in the Harris R filter (open circles) and the HST in the F675W filter (filled circle); with all magnitudes corrected to the Cousins I or R systems[13]. The solid line shown in both **a** and **b** is the simultaneous best fit to the ground- and space-based data to the K-corrected, $(1+z)$ time-dilated Leibundgut B-band type Ia supernova template light curve[22], and the dotted line in **c** is the best fit to a K-corrected, time-dilated U-band type Ia supernova template light curve. The ground-based data was reduced and calibrated following the techniques of ref. 5, but with no host-galaxy light subtraction necessary. The HST data was calibrated and corrected for charge-transfer inefficiency following the prescriptions of refs 23, 24. K-corrections were calculated as in ref. 25, modified for the HST filter system. Correlated zero-point errors are accounted for in the simultaneous fit of the light curve. The errors in the calibration, charge-transfer inefficiency correction and K-corrections for the HST data are much smaller (~4% total) than the contributions from the photon noise. No corrections were applied to the HST data for a possible ~4% error in the zero points (P. Stetson, personal communication) or for nonlinearities in the WFPC2 response[26], which might bring the faintest of the HST points into tighter correspondence with the best-fit light curve in **a** and **c**. Note that the individual fits to the data in **a** and **b** agree within their error bars, providing a first-order cross-check of the HST calibration.

The ground-based data show no evidence of host-galaxy light, but the higher-resolution HST imaging shows a marginal detection (after co-adding all four dates of observation) of a possible $I = 25.2 \pm 0.3$ host galaxy 1 arcsec from the supernova. This light does not contaminate the supernova photometry from the HST and it contributes negligibly to the ground-based photometry. The projected separation is ~6 kpc (for $\Omega_M = 1$, $\Omega_\Lambda = 0$ and $h_0 = 0.65$, the dimensionless cosmological parameters describing the mass density,

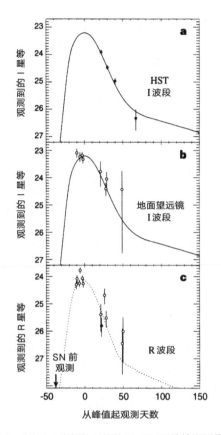

图 2. SN1997ap 的测光数据点。(**a**)哈勃空间望远镜使用 F814W 滤镜观测的结果；(**b**)地面望远镜使用哈里斯 I 波段滤镜的观测结果；(**c**)地面望远镜使用哈里斯 R 波段滤镜所观测的结果(空心点)和哈勃空间望远镜使用 F675W 滤镜所观测的结果(实心点)；所有星等已被转换到库森 I 或 R 波段[13]。图 **a** 和图 **b** 中的实线是同时将地面和空间的观测数据对经 K 修正和 $1+z$ 倍时间延展的 Ia 型超新星 Leibundgut B 波段光变曲线模板[22]拟合的最佳结果；图 **c** 中的点线是对经 K 修正和 $1+z$ 倍时间延展的 Ia 型超新星 U 波段光变曲线模板拟合的最佳结果。地面观测的数据使用参考文献 5 中的方法进行处理和校准，但没有扣除寄主星系背景光的必要。哈勃空间望远镜的观测数据根据参考文献 23 和 24 中的方法进行了电荷迁移低效率的处理和修正。K 修正使用了参考文献 25 中的方法计算，并根据哈勃空间望远镜的滤镜系统做了调整。在对光变曲线的同步拟合中，我们考虑了零点误差。在哈勃空间望远镜数据的校正、电荷迁移低效率修正和 K 修正中的误差(共约 4%)远小于光子噪声的贡献。对于哈勃空间望远镜数据潜在的约 4% 的零点误差(斯特森，个人交流)和 WFPC2 响应的非线性[26]没有做任何修正。这可能会使哈勃空间望远镜最暗的数据点与图 **a** 和图 **c** 中的最佳拟合曲线吻合得更好。注意图 **a** 和图 **b** 的每个数据点都在各自误差范围内与拟合结果一致，这提供了对哈勃空间望远镜校准的一阶交叉检验。

　　地面望远镜观测的数据没有显示寄主星系的光存在的证据，但是在分辨率更高的哈勃空间望远镜所拍摄的图像中有微弱的迹象(在叠加全部四次观测之后)表明，在距离该超新星 1 角秒的地方可能存在一个 $I = 25.2 \pm 0.3$ 的寄主星系。这些光并没有污染到哈勃空间望远镜的超新星测光结果，同时对地面望远镜测光的影响也微不足道。寄主星系到 SN1997ap 的投影距离约为 6 kpc(条件为 $\Omega_M = 1$、$\Omega_\Lambda = 0$ 和

vacuum energy density and Hubble constant, respectively) and the corresponding B-band rest-frame magnitude is $M_B \approx -17$ and its surface brightness is $\mu_B \approx 21$ mag arcsec^{-2}, consistent with properties of local spiral galaxies. We note that the analysis will need a final measurement of any host-galaxy light after the supernova has faded, in the unlikely event that there is a very small knot of host-galaxy light directly under the HST image of SN1997ap.

We compare the K-corrected $R-I$ observed difference of peak magnitudes (measured at the peak of each band, not the same day) to the $U-B$ colour found for "normal" low-redshift type Ia supernovae. We find that the rest-frame colour of SN1997ap [$(U-B)_{SN1997ap} = -0.28 \pm 0.11$] is consistent with an unreddened "normal" type Ia supernova colour, $(U-B)_{normal} = -0.32 \pm 0.12$ (see ref. 14 and also P.N. *et al.*, manuscript in preparation). In this region of the sky, there is also no evidence for Galactic reddening[15]. Given the considerable projected distance from the putative host galaxy, the supernova colour, and the lack of galaxy contamination in the supernova spectra, we proceed with an analysis under the hypothesis that the supernova suffers negligible host-galaxy extinction, but with the following caveat.

Although correcting for $E(U-B) \approx 0.04$ of reddening would shift the magnitude by only one standard deviation, $A_B = 4.8E(U-B) = 0.19 \pm 0.78$, the uncertainty in this correction would then be the most significant source of uncertainty for this one supernova. This is because of the large uncertainty in the $(U-B)_{SN1997ap}$ measurement, and the sparse low-redshift U-band reference data. HST J-band observations are currently planned for future $z > 0.8$ supernovae, to allow a comparison with the restframe $B-V$ colour, a much better indicator of reddening for type Ia supernovae. Such data will thus provide an important improvement in extinction correction uncertainties for future supernovae and eliminate the need for assumptions regarding host-galaxy extinction. In the following analysis, we also do not correct the lower-redshift supernovae for possible host-galaxy extinction, so any similar distribution of extinction would partly compensate for this possible bias in the cosmological measurements.

The significance of type Ia supernovae at $z = 0.83$ for measurements of the cosmological parameters is illustrated on the Hubble diagram of Fig. 3. To compare with low-redshift magnitudes, we plot SN1997ap at an effective rest-frame B-band magnitude of $B = 24.50 \pm 0.15$, derived, as in ref. 5, by adding a K-correction and increasing the error bar by the uncertainty due to the (small) width-luminosity correction and by the intrinsic dispersion remaining after this correction. By studying type Ia supernovae at twice the redshift of our first previous sample at $z \approx 0.4$, we can look for a correspondingly larger magnitude difference between the cosmologies considered. At the redshift of SN1997ap, a flat $\Omega_M = 1$ universe is separated from a flat $\Omega_M = 0.1$ universe by almost one magnitude, as opposed to half a magnitude at $z \approx 0.4$. For comparison, the uncertainty in the peak magnitude of SN1997ap is only 0.15 mag, while the intrinsic dispersion amongst stretch-calibrated type Ia supernovae is ~0.17 mag (ref. 5). Thus, at such redshifts even individual type Ia supernovae become powerful tools for discriminating amongst various world models, provided observations are obtained, such as those presented here, where the

$h_0 = 0.65$，这些无量纲宇宙学参数分别表示质量密度、真空能密度和哈勃常数），对应的 B 波段静止系星等为 $M_B \approx -17$，它的面亮度 $\mu_B \approx 21 \ mag \cdot arcsec^{-2}$，这个属性与较近的旋涡星系的一致。需要指出的是，通常需要在超新星逐渐暗淡后，才能最终测量寄主星系的亮度，以免哈勃空间望远镜拍摄的 SN1997ap 图像正下方正好有个寄主星系的非常小的亮结，不过这种情况不太可能发生。

我们将 K 修正之后的 $R-I$ 峰值星等差（分别观测每个波段的亮度峰值，并非在同一天）和低红移"正常"Ia 型超新星的 $U-B$ 颜色结果作比较。发现 SN1997ap 的静止系颜色 $[(U-B)_{SN1997ap} = -0.28 \pm 0.11]$ 与未红化的"正常"Ia 型超新星的颜色 $[(U-B)_{normal} = -0.32 \pm 0.12]$（见参考文献 14 以及纽金特等人正在撰写的文章）相一致。在这片天区中，我们也没有发现"银河红化"[15]的证据。考虑到假定的寄主星系相当远的投影距离、超新星的颜色，以及光谱并没有受到寄主星系光的污染，我们后面的分析都假定寄主星系对该超新星的消光可以忽略。不过下面几点需要注意。

尽管 $E(U-B) \approx 0.04$ 的红化修正只会让星等偏移一个标准差，$A_B = 4.8E(U-B) = 0.19 \pm 0.78$，这个修正的不确定性是这颗超新星最主要的误差来源。这是因为 $(U-B)_{SN1997ap}$ 的测量有很大的不确定性，而且还缺乏低红移处的 U 波段参考数据。哈勃空间望远镜计划对未来 $z > 0.8$ 的超新星进行 J 波段的观测，以便能够对比静止系 $B-V$ 颜色，它可以作为 Ia 型超新星更好的红化表征。这些数据可以帮助减小未来超新星消光修正的不确定性，并消除有关寄主星系消光的假定。在接下来的分析中，我们也没有对较低红移超新星进行可能的寄主星系的消光修正，所以任何相似的消光分布对宇宙学测量可能造成的偏差会部分地抵消。

红移 $z = 0.83$ 的 Ia 型超新星对宇宙参数测量的意义在图 3 的哈勃图中已经标示出。为了和低红移星等进行对比，我们将 SN1997ap 的数据画在有效静止系下 $B = 24.50 \pm 0.15$ 处，这个值的推导和文献 5 中一样，经过了 K 修正，并考虑了由（小的）宽度–光度修正的不确定性和修正之后遗留的内禀弥散度所引起的误差棒增加。通过研究我们此前 $z \approx 0.4$ 的第一个样本红移的两倍处的超新星，我们可以寻找不同宇宙学模型之间相应更大的星等差。在 SN1997ap 的红移处，$\Omega_M = 1$ 的平坦宇宙与 $\Omega_M = 0.1$ 的平坦宇宙模型相差将近一个星等，而在红移值 $z \approx 0.4$ 处，仅相差半个星等。相比之下，SN1997ap 的峰值星等的不确定性仅有 0.15 星等，然而 Ia 型超新星在延展修正后的内禀弥散度约为 0.17 星等（参考文献 5）。因此，在这样的红移处，只要观测能够提供本文这样的测光误差低于内禀弥散度的数据，即使是单个 Ia

photometric errors are below the intrinsic dispersion of type Ia supernova.

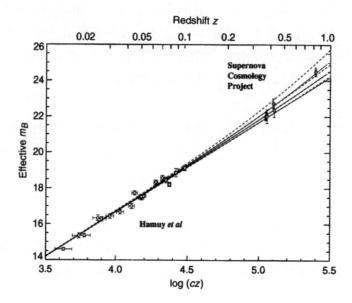

Fig. 3. SN1997ap at $z = 0.83$ plotted on the Hubble diagram from ref. 5. Also plotted are the 5 of the first 7 high-redshift supernovae that could be width-luminosity corrected, and the 18 of the lower-redshift supernovae from the Calán/Tololo Supernova Survey that were observed earlier then 5 d after maximum light[3]. Magnitudes have been K-corrected, and also corrected for the width-luminosity relation. The inner error bar on the SN1997ap point corresponds to the photometry error alone, while the outer error bar includes the intrinsic dispersion of type Ia supernovae after stretch correction (see text). The solid curves are theoretical m_B for $(\Omega_M, \Omega_\Lambda) = (0, 0)$ on top, $(1, 0)$ in middle and $(2, 0)$ on bottom. The dotted curves are for the flat-universe case, with $(\Omega_M, \Omega_\Lambda) = (0, 1)$ on top, $(0.5, 0.5)$, $(1, 0)$ and $(1.5, -0.5)$ on bottom.

By combining such data spanning a large range of redshift, it is also possible to distinguish between the effects of mass density Ω_M and cosmological constant Λ on the Hubble diagram[8]. The blue contours of Fig. 4 show the allowed confidence region on the Ω_Λ ($\equiv \Lambda/(3H_0^2)$) versus Ω_M plane for the $z \approx 0.4$ supernovae[5]. The yellow contours show the confidence region from SN1997ap by itself, demonstrating the change in slope of the confidence region at higher redshift. The red contours show the result of the combined fit, which yields a closed confidence region in the Ω_M–Ω_Λ plane. This fit corresponds to a value of $\Omega_M = 0.6 \pm 0.2$ if we constrain the result to a flat universe ($\Omega_\Lambda + \Omega_M = 1$), or $\Omega_M = 0.2 \pm 0.4$ if we constrain the result to a $\Lambda = 0$ universe. These results are preliminary evidence for a relatively low-mass-density universe. The addition of SN1997ap to the previous sample of lower-redshift supernovae decreases the best-fit Ω_M by approximately 1 standard deviation compared to the earlier results[5].

型超新星也能成为区分各种宇宙模型的有力工具。

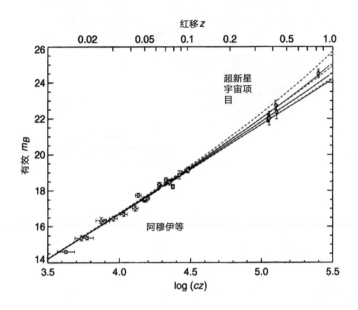

图 3. 红移 $z = 0.83$ 的 SN1997ap 标注在参考文献 5 中的哈勃图上。图上还画出首次能做宽度–光度修正的 7 颗高红移超新星中的 5 个，和 18 颗由 Calán/Tololo 超新星巡天观测到的较低红移超新星，它们是在亮度极大前和后五天的区间内被观测到的 [3]。星等已经经过 K 修正和宽度–光度关系修正。SN1997ap 数据点上的内侧误差棒只表示测光误差，而外侧误差棒则包含了经过伸展修正后的 Ia 型超新星内禀弥散度（见正文）。实线对应不同宇宙模型的理论 m_B：上，$(\Omega_M, \Omega_\Lambda) = (0, 0)$；中，$(\Omega_M, \Omega_\Lambda) = (1, 0)$；下，$(\Omega_M, \Omega_\Lambda) = (2, 0)$。虚线对应平坦宇宙模型：上，$(\Omega_M, \Omega_\Lambda) = (0, 1)$；下，$(\Omega_M, \Omega_\Lambda) = (0.5, 0.5)$、$(1, 0)$ 和 $(1.5, -0.5)$。

综合这些大范围红移的数据，我们还有可能在哈勃图中区分质量密度 Ω_M 和宇宙学常数 Λ 的影响 [8]。图 4 中的蓝色等值线显示了 $z \approx 0.4$ 的超新星所允许的 Ω_Λ （ $\equiv \Lambda / (3H_0^2)$ ）与 Ω_M 平面的置信区域 [5]。黄色等值线显示的是 SN1997ap 独立限制的 Ω_Λ 和 Ω_M 置信区域，表明在较高红移处置信区域的斜率有所变化。红色等值线显示的是综合所有超新星进行拟合的结果，得到了一个 Ω_M–Ω_Λ 的封闭置信区域。如果我们限定宇宙是平坦的（$\Omega_\Lambda + \Omega_M = 1$），拟合可以得到 $\Omega_M = 0.6 \pm 0.2$；如果我们限定于 $\Lambda = 0$ 的宇宙，可以得到 $\Omega_M = 0.2 \pm 0.4$。这些结果是宇宙质量密度偏低的初步证据。将超新星 SN1997ap 的数据加入以往较低红移的超新星样本中，我们所得的最佳拟合值比以前的结果减少了大约一个标准差 [5]。

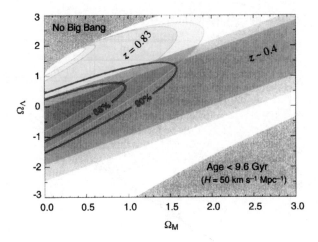

Fig. 4. Contour plot of the best fit confidence regions in the Ω_Λ versus Ω_M plane for SN1997ap and the five supernovae at $z \approx 0.4$ (see ref. 5). The 68% (1σ) and 90% confidence regions for (blue shading) the supernovae at $z \approx 0.4$, (yellow shading) SN1997ap at $z = 0.83$ by itself, and (red contours) all of these supernovae taken together. The two labelled corners of the plot are ruled out because they imply: (upper left corner) a "bouncing" universe with no Big Bang[27], or (lower right corner) a universe younger than the oldest heavy elements, $t_0 < 9.6$ Gyr (ref. 28), for any value of $H_0 \geqslant 50$ km s^{-1} Mpc^{-1}.

Our data for SN1997ap demonstrate: (1) that type Ia supernovae at $z > 0.8$ exist; (2) that they can be compared spectroscopically with nearby supernovae to determine supernova ages and luminosities and check for indications of supernova evolution; and (3) that calibrated peak magnitudes with precision better than the intrinsic dispersion of type Ia supernovae can be obtained at these high redshifts. The width of the confidence regions in Fig. 4 and the size of the corresponding projected measurement uncertainties show that with additional type Ia supernovae having data of quality comparable to that of SN1997ap, a simultaneous measurement of Ω_Λ and Ω_M is now possible. It is important to note that this measurement is based on only one supernova at the highest ($z > 0.8$) redshifts, and that a larger sample size is required to find a statistical peak and identify any "outliers". In particular, SN1997ap was discovered near the search detection threshold and thus may be drawn from the brighter tail of a distribution ("Malmquist bias"). There is similar potential bias in the lower-redshift supernovae of the Calán/Tololo Survey[2], making it unclear which direction such a bias would change Ω_M.

Several more supernovae at comparably high redshift have already been discovered by the Supernova Cosmology Project collaboration, including SN1996cl, also at $z = 0.83$. SN1996cl can be identified as a very probable type Ia supernova, as a serendipitous HST observation (M. Donahue et al., personal communication) shows its host galaxy to be an elliptical or S0. Its magnitude and colour, although much more poorly constrained by photometry data, agree within uncertainty with those of SN1997ap. The next most distant spectroscopically confirmed type Ia supernovae are at $z = 0.75$ and $z = 0.73$ (ref. 16; these supernovae are awaiting final calibration data). In the redshift range $z = 0.3–0.7$, we have discovered over 30 additional spectroscopically confirmed type Ia supernovae, and followed them with two-filter photometry. (The first sample of supernovae with $z \approx 0.4$

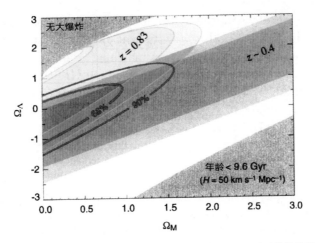

图 4. 由 SN1997ap 和 5 个 $z \approx 0.4$ 的超新星（参考文献 5）限制的 Ω_M-Ω_Λ 平面的最佳拟合置信区域等值线图。图中画出了 68%（1 个标准差）和 90% 的置信区域：蓝色为五个 $z \approx 0.4$ 超新星的结果，黄色为 $z = 0.83$ 的 SN1997ap 单独限制的结果，红色等值线为所有超新星的结果。图中有标注的两角是被排除的区域：左上角表示一个没有大爆炸的反弹宇宙 [27]，右下角表示了一个比最老的重元素年轻的宇宙，在哈勃常数 $H_0 \geqslant 50$ km · s^{-1} · Mpc^{-1} 时，$t_0 < 96$ 亿年（参考文献 28）。

我们的 SN1997ap 观测数据说明：(1) 存在红移值 $z > 0.8$ 的 Ia 型超新星；(2) 与较近超新星的光谱做对比可以确定超新星的年龄和光度，并检查超新星演化的迹象；(3) 在高红移处，可以获得比 Ia 型超新星内禀弥散度更加精确的校准后峰值星等。图 4 中置信区域的宽度与其对应的投影测量的不确定性的大小表明，通过增加与 SN1997ap 数据质量相当的 Ia 型超新星数据，现在是可能同时测量 Ω_Λ 和 Ω_M 的。需要特别指出的是：这一测量只依赖于一颗最高红移超新星（$z > 0.8$），需要更大的超新星数据样本来得到统计峰值并排除虚假的源。特别是，超新星 SN1997ap 的发现接近搜寻探测极限，可能来自亮度分布的明亮尾段（所谓"马姆奎斯特偏差"）。在 Calán/Tololo 巡天观测的较低红移超新星中存在相似的潜在偏差 [2]，我们不清楚这些偏差对 Ω_M 的测量的影响。

超新星宇宙学观测项目已经发现了其他几颗相对高红移的超新星。其中有 SN1996cl，红移值也是 $z = 0.83$，它很可能是一颗 Ia 型超新星。哈勃空间望远镜偶然发现它的寄主星系可能是一个椭圆星系或者是 S0 星系（多纳休等，个人交流）。SN1996cl 的星等和颜色，尽管测光数据限定得不太好，但与 SN1997ap 的数据在误差范围内是一致的。距离上居次的已由光谱确定的两颗 Ia 型超新星的红移值分别为 $z = 0.75$ 和 $z = 0.73$（参考文献 16；正等待这两颗超新星的最后校准数据）。在红移值 $z = 0.3$ 到 0.7 区间，我们已经发现了 30 多颗额外的经光谱确认的 Ia 型超新星，并继以双色测光。（第一组 $z \approx 0.4$ 的超新星样本不是全部经过光谱确认和双色测光的 [5]）

15

were not all spectroscopically confirmed and observed with two-filter photometry[5].) These new supernovae will improve both the statistical and systematic uncertainties in our measurement of Ω_M and Ω_Λ in combination. A matching sample of $\geqslant 6$ type Ia supernovae at $z > 0.7$ is to be observed in two filters in Hubble Space Telescope observations due to start on 5 January 1998. SN1997ap demonstrates the efficacy of these complementary higher-redshift measurements in separating the contribution of Ω_M and Ω_Λ to the total mass-energy density of the Universe.

(**391**, 51-54; 1998)

S. Perlmutter[1,2], G. Aldering[1], M. Della Valle[3], S. Deustua[1,4], R. S. Ellis[5], S. Fabbro[1,6,7], A. Fruchter[8], G. Goldhaber[1,2], A. Goobar[9], D. E. Groom[1], I. M. Hook[1,10], A. G. Kim[1,11], M. Y. Kim[1], R. A. Knop[1], C. Lidman[12], R. G. McMahon[5], P. Nugent[1], R. Pain[1,6], N. Panagia[13], C. R. Pennypacker[1,4], P. Ruiz-Lapuente[14], B. Schaefer[15] & N. Walton[16]

[1] E. O. Lawrence Berkeley National Laboratory, 1 Cyclotron Road, MS 50-232, Berkeley, California 94720, USA

[2] Center for Particle Astrophysics, University of California, Berkeley, California 94720, USA

[3] Dipartimento di Astronomia, Universita' di Padova, Vicolo Osservatorio 5, 35122, Padova, Italy

[4] Space Sciences Laboratory, University of California, Berkeley, California 94720, USA

[5] Institute of Astronomy, Madingley Road, Cambridge CB3 0HA, UK

[6] LPNHE, Universites Paris VI & VII, T33 Rdc, 4, Place Jussieu, 75252 Paris Cedex 05, France

[7] Observatoire de Strasbourg, 11, Rue de l'Universite, 67000 Strasbourg, France

[8] Space Telescope Science Institute, 3700 San Martin Drive, Baltimore, Maryland 21218, USA

[9] Physics Department, Stockholm University, Box 6730, S-11385 Stockholm, Sweden

[10] European Souther Observatory, Karl-Schwarzschild-Strasse 2, D-85748 Garching bei Munchen, Germany

[11] Physique Corpusculaire et Cosmologie, Collège de France, 11, Place Marcelin-Berthelot, 75231 Paris, France

[12] European Southern Observatory, Alonso de Cordova, 3107, Vitacura, Casilla 19001, Santiago, 19, Chile

[13] Space Telescope Science Institute, 3700 San Martin Drive, Baltimore, Maryland 21218, USA; affiliated with the Astrophysics Division, Space Science Department of ESA

[14] Department of Astronomy, Faculty of Physics, University of Barcelona, Diagonal 647, E-08028 Barcelona, Spain

[15] Department of Physics, Yale University, 260 Whitney Avenue, JWG 463, New Haven, Connecticut 06520, USA

[16] Isaac Newton Group, Apartado 321, 38780 Santa Cruz de La Palma, The Canary Islands, Spain

Received 7 October; accepted 18 November 1997.

References:

1. Perlmutter, S. *et al.* in *Thermonuclear Supernovae* (eds Ruiz-Lapuente, P. *et al.*) 749-763 (Kluwer, Dordrecht, 1997).

2. Phillips, M. M. The absolute magnitudes of Type Ia supernovae. *Astrophys. J.* **413**, L105-L108 (1993).

3. Hamuy, M. *et al.* The absolute luminosities of the Calán/Tololo Type Ia supernovae. *Astron. J.* **112**, 2391-2397 (1996).

4. Riess, A. G., Press, W. H. & Kirshner, R. P. Using Type Ia supernova light curve shapes to measure the Hubble constant. *Astrophys. J.* **438**, L17-L20 (1995).

5. Perlmutter, S. *et al.* Measurements of the cosmological parameters Ω and Λ from the first seven supernovae at $z \geqslant 0.35$. *Astrophys. J.* **483**, 565-581 (1997).

6. Perlmutter, S. *et al.* *IAU Circ.* No. 6621 (1997).

7. Schmidt, B. *et al.* *IAU Circ.* No. 6646 (1997).

8. Goobar, A. & Perlmutter, S. Feasibility of measuring the cosmological constant Λ and mass density Ω using Type Ia supernovae. *Astrophys. J.* **450**, 14-18 (1995).

9. Riess, A. G. *et al.* Time dilation from spectral feature age measurements of Type Ia supernovae. *Astron. J.* **114**, 722-729 (1997).

10. Nugent, P. *et al.* Evidence for a spectroscopic sequence among Type Ia supernovae. *Astrophys. J.* **455**, L147-L150 (1993).

11. Goldhaber, G. *et al.* in *Thermonuclear Supernovae* (eds Ruiz-Lapuente, P. *et al.*) 777-784 (Kluwer, Dordrecht, 1997).

12. Leibundgut, B. *et al.* Time dilation in the light curve of the distant Type Ia supernova SN 1995K. *Astrophys. J.* **466**, L21-L44 (1996).

13. Bessell, M. S. UBVRI passbands. *Publ. Astron. Soc. Pacif.* **102**, 1181-1199 (1990).

14. Branch, D., Nugent, P. & Fisher, A. in *Thermonuclear Supernovae* (eds Ruiz-Lapuente, P. *et al.*) 715-734 (Kluwer, Dordrecht, 1997).

15. Burstein, D. & Heiles, C. Reddenings derived from H I and galaxy counts—accuracy and maps. *Astron. J.* **87**, 1165-1189 (1982).

16. Perlmutter, S. *et al.* *IAU Circ.* No. 6540 (1997).

这些新的超新星有利于在把 Ω_{M} 和 Ω_{Λ} 结合在一起测量时减小统计误差和系统误差。哈勃空间望远镜从 1998 年 1 月 5 日起将开始使用两片滤镜观测至少 6 颗 $z > 0.7$ 的 Ia 型超新星匹配样本。SN1997ap 表明，我们可以有效地使用额外的较高红移超新星的数据区分 Ω_{M} 和 Ω_{Λ} 在宇宙总质能密度中的贡献。

（余恒 翻译；陈阳 审稿）

17. Leibundgut, B. *et al.* Premaximum observations of the type Ia SN 1990N. *Astrophys. J.* **371**, L23-L26 (1991).

18. Wells, L. A. *et al.* The type Ia supernova 1989B in NGC 3627 (M66). *Astron. J.* **108**, 2233-2250 (1994).

19. Branch, D. *et al.* The type I supernova 1981b in NGC 4536: the first 100 days. *Astrophys. J.* **270**, 123-139 (1983).

20. Patat, F. *et al.* The Type Ia supernova 1994D in NGC 4526: the early phases. *Mon. Not. R. Astron. Soc.* **278**, 111-124 (1996).

21. Cappellaro, E., Turatto, M. & Fernley, J. in *IUE—ULDA Access Guide No. 6: Supernovae* (eds Cappellaro, E., Turatto, M. & Fernley, J.) 1-180 (ESA, Noordwijk, 1995).

22. Leibundgut, B., Tammann, G., Cadonau, R. & Cerrito, D. Supernova studies. VII. An atlas of light curves of supernovae type I. *Astron. Astrophys. Suppl. Ser.* **89**, 537-579 (1991).

23. Holtzman, J. *et al.* The photometric performance and calibration of WFPC2. *Publ. Astron. Soc. Pacif.* **107**, 1065-1093 (1995).

24. Whitmore, B. & Heyer, I. *New Results on Charge Transfer Efficiency and Constraints on Flat-Field Accuracy* (Instrument Sci. Rep. WFPC2 97-08, Space Telescope Science Institute, Baltimore, 1997).

25. Kim, A., Goobar, A. & Perlmutter, S. A generalized K-corrections for Type Ia supernovae: comparing R-band photometry beyond z = 0.2 with B, V, and R-band nearby photometry. *Publ. Astron. Soc. Pacif.* **108**, 190-201 (1996).

26. Stiavelli, M. & Mutchler, M. *WFPC2 Electronics Verification* (Instrument Sci. Rep. WFPC2 97-07, Space Telescope Science Institute, Baltimore, 1997).

27. Carrol, S., Press, W. & Turner, E. The cosmological constant. *Annu. Rev. Astron. Astrophys.* **30**, 499-542 (1992).

28. Schramm, D. in *Astrophysical Ages and Dating Methods* (eds Vangioni-Flam, E. *et al.*) 365-384 (Editions Frontières, Gif sur Yvette, 1990).

Acknowledgements. The authors are members of the Supernova Cosmology Project. We thank CTIO, Keck, HST, WIYN, ESO and the ORM–La Palma observatories for a generous allocation of time, and the support of dedicated staff in pursuit of this project; D. Harmer, P. Smith and D. Willmarth for their help as WIYN queue observers; and G. Bernstein and A. Tyson for developing and supporting the Big Throughput Camera which was instrumental in the discovery of this supernova.

Correspondence and requests for materials should be addressed to S.P. (e-mail: saul@lbl.gov).

An Exceptionally Well-preserved Theropod Dinosaur from the Yixian Formation of China

Pei-ji Chen *et al.*

Editor's Note

The first sign that the palaeontological world was about to be turned upside down at the 1996 meeting of the Society of Vertebrate Paleontology in New York came when Pei-Ji Chen showed snapshots of a dinosaur from the hitherto obscure Yixian Formation of Liaoning Province, PRC. A dinosaur—with feathers. The relationship between dinosaurs and birds had long been suspected. Indeed, *Archaeopteryx*, the earliest known bird, looked like a feathered dinosaur. However, there had been no hard evidence for feathers in non-flying dinosaurs. To be sure, Chen's *Sinosauropteryx* had more of a fringe of feather-like fibrils than true feathers, but more was to come, and soon. Fossils of feathered dinosaurs began to pour from Liaoning to capture the world's imagination.

Two spectacular fossilized dinosaur skeletons were recently discovered in Liaoning in northeastern China. Here we describe the two nearly complete skeletons of a small theropod that represent a species closely related to *Compsognathus*. *Sinosauropteryx* has the longest tail of any known theropod, and a three-fingered hand dominated by the first finger, which is longer and thicker than either of the bones of the forearm. Both specimens have interesting integumentary structures that could provide information about the origin of feathers. The larger individual also has stomach contents, and a pair of eggs in the abdomen.

THE Jehol biota[1] was widely distributed in eastern Asia during latest Jurassic and Early Cretaceous times. These freshwater and terrestrial fossils include macroplants, palynomorphs, charophytes, flagellates, conchostracans, ostracods, shrimps, insects, bivalves, gastropods, fish, turtles, lizards, pterosaurs, crocodiles, dinosaurs, birds and mammals. In recent years, the Jehol biota has become famous as an abundant source of remains of early birds[2,3]. Dinosaurs are less common in the lacustrine beds, but the specimens described here consist of two nearly complete skeletons of a small theropod discovered by farmers in Liaoning. The skeletons are from the basal part of the Yixian Formation, from the same horizon as the fossil birds *Confuciusornis* and *Liaoningornis*[3]. Both are remarkably well preserved, and include fossilized integument, organ pigmentation and abdominal contents. One of the two was split into part and counterpart, and the sections were deposited in two different institutions. One side (in the National Geological Museum of China, Beijing) became the holotype of *Sinosauropteryx prima*, a supposed bird[4].

在中国义县组发现的一具保存异常完美的兽脚类恐龙

陈丕基等

编者按

当陈丕基在 1996 年组约举办的古脊椎动物学会会议上展示了来自中国辽宁省义县组（年代还没有研究透彻）的一只恐龙的照片时，古生物学世界被震撼了。一只带羽毛的恐龙！恐龙与鸟类之间的关联一直被怀疑。的确，始祖鸟——已知最早的鸟类——看上去非常像一只带羽毛的恐龙。然而，之前并未发现不会飞的恐龙长有羽毛的确凿证据。诚然，陈丕基命名的中华龙鸟具有的更像是羽毛状的纤维，而不是真的羽毛，但很快就有更多的发现。带羽毛恐龙化石开始在辽宁大量出现，吸引了全世界的目光。

最近在中国东北部的辽宁省发现了两具非常吸引人的恐龙骨架化石。在本文中我们描述了这两具近乎完整的小型兽脚类恐龙骨架，它们代表了一种与美颌龙具有密切关系的恐龙。中华龙鸟是已知的兽脚类恐龙中尾巴最长的，前肢具有三指，由第一指主导运动，第一指要比前肢的任何骨骼都长且粗壮。两件标本都保存了吸引人的皮肤衍生物结构，这些结构能够为羽毛的起源提供信息。较大个体的腹腔中还保存着胃容物以及一对蛋化石。

热河生物群[1]广泛地分布在东亚地区的侏罗纪末到早白垩世地层中，这里发现的淡水陆相化石包括大型植物、孢粉类、轮藻类、鞭毛虫、叶肢介、介形类、虾、昆虫、双壳类、腹足类、鱼类、龟鳖类、蜥蜴、翼龙、鳄类、恐龙、鸟和哺乳动物。在最近几年里，热河生物群成了著名的早期鸟类化石的丰富产地[2,3]。恐龙化石在湖相地层较少，但是本文记述的两具接近完整的小型兽脚类恐龙骨架是由在辽宁的农民发现的。这两具标本发现自义县组的基部，与孔子鸟和辽宁鸟[3]层位相同。这两件标本都保存完好，并且包括了石化的皮肤衍生物结构、器官色素沉着和腹容物。两者中有一件标本分离成为正体和负体两块，这两块标本保存在两个不同的机构，一面作为原始中华龙鸟（曾被误认为是鸟类[4]）的正模标本（保存于中国地质博物馆，北京），而负体则和第二件较大的标本一起保存于中国科学院南京地质古

The counterpart and the second larger specimen are in the collections of the Nanjing Institute of Geology and Palaeontology.

The Yixian Formation is mainly composed of andesites, andesite-breccia, agglomerates and basalts, but has four fossil-bearing sedimentary intercalations that are rich in tuffaceous materials. The Jianshangou (formerly Jianshan[5,6]) intercalated bed (60 m thick) is the basal part of this volcanic sedimentary formation, and is made up of greyish–white, greyish–yellow and greyish–black sandstones, siltstones, mudstones and shales. These sediments are rich in fossils of mixed Jurassic–Cretaceous character. The primitive nature of the fossil birds of the Jianshangou fossil group has led to suggestions that the beds could be as early as Tithonian in age[2]. But although *Confuciusornis* and the other birds[3] are more advanced than *Archaeopteryx* in a number of significant features, we can only conclude that the beds that the fossils came from are probably younger than the Solnhofen Lithographic Limestones (Early Tithonian). The presence of *Psittacosaurus* in the same beds is more consistent with an Early Cretaceous age[7], as are the palynomorphs[8] and a recent radiometric date of the formation[9], but other radiometric dating attempts have indicated older ages[10].

Dinosauria OWEN 1842

Theropoda MARSH 1881

Coelurosauria VON HUENE 1914

Compsognathidae MARSH 1882

Sinosauropteryx prima JI and JI 1996

Holotype. Part (National Geological Museum of China, GMV 2123) and counterpart (Nanjing Institute of Geology and Palaeontology, NIGP 127586) slabs of a complete skeleton.

Referred specimen. Nanjing Institute of Geology and Palaeontology NIGP 127587. Nearly complete skeleton, lacking only the distal half of the tail.

Locality and horizon. Jianshangou-Sihetun area of Beipiao, Liaoning, People's Republic of China. Yixian Formation, Jehol Group, Upper Jurassic or Lower Cretaceous (Fig. 1; Fig. 1 has been omitted in this edited version).

Diagnosis. Compsognathid with longest tail known for any theropod (64 caudals). Skull 15% longer than femur, and forelimb (humerus plus radius) only 30% length of leg (femur plus tibia), in contrast with *Compsognathus* where skull is same length as femur, and forelimb length is 40% leg length. Within the Compsognathidae, forelimb length (compared to femur length) is shorter in *Sinosauropteryx* (61–65%) than it is in *Compsognathus* (90–99%). In contrast with all other theropods, ungual phalanx II–2 is longer than the radius. Haemal spines simple and spatulate, whereas those of *Compsognathus* taper distally.

生物研究所。

义县组的主要岩性组成为安山岩、安山角砾岩、集块岩和玄武岩，但是有四个富含凝灰质的含化石沉积岩夹层。尖山沟（以前称"尖山"[5,6]）夹层（60 米厚）位于火山岩沉积构造的基部，组成层序是：灰白色、灰黄色和灰黑色的砂岩、粉砂岩、泥岩和页岩。这段沉积富含混合侏罗纪–白垩纪特征的化石。尖山沟化石群含有的鸟类化石的原始特征表明该层位可能与提塘阶年代相当[2]。尽管孔子鸟和其他鸟类[3]在许多重要特征上比始祖鸟更加进步，我们仅能推断该化石层位可能要比德国索伦霍芬石印灰岩的层位（早提塘阶）靠上。发现于相同层位的鹦鹉嘴龙生存时代为早白垩世[7]，这与孢粉测年[8]和最近对该地层进行的放射性同位素测年[9]的结果相一致，但是其他的放射性测年显示该地层的年代更古老[10]。

恐龙总目 Dinosauria OWEN 1842
兽脚亚目 Theropoda MARSH 1881
虚骨龙类 Coelurosauria VON HUENE 1914
美颌龙科 Compsognathidae MARSH 1882
原始中华龙鸟 *Sinosauropteryx prima* JI and JI 1996

正模标本 正体（中国地质博物馆，标本编号 GMV 2123）和负体（南京地质古生物研究所，NIGP 127586）完整骨骼的板状化石。

归入标本 南京地质古生物研究所，NIGP 127587。近乎完整的骨骼，仅缺少尾部的远端一半。

产地与层位 中国辽宁省北票市尖山沟–四合屯地区。上侏罗统或下白垩统，热河群，义县组（图 1；该图显示中华龙鸟和孔子鸟产地，此版本中省略）。

鉴别特征 一种美颌龙科恐龙，具有已知兽脚类中最长的尾（64 个尾椎）。头骨比股骨长 15%，前肢（肱骨 + 桡骨）仅是后肢（股骨 + 胫骨）长度的 30%，而美颌龙头骨与股骨等长，前肢长是后肢长的 40%。在美颌龙科成员中，中华龙鸟的前肢与股骨的长度比例为 61% ~ 65%，短于美颌龙（90% ~ 99%）。不同于其他兽脚类恐龙，中华龙鸟的第二指爪比桡骨长。脉弧简单且呈片状，而美颌龙的则是锥形。

Description

Sinosauropteryx is comparable in size and morphology to known specimens of *Compsognathus*[11,12] from Germany and France. The smaller Chinese specimen (Fig. 2) is 0.68 m long (snout to end of tail) and has a femur length of 53.2 mm, whereas the second specimen (Fig. 3) has a femur length of 86.4 mm. The former is smaller than the type specimen of *Compsognathus longipes* (femur length about 67 mm) and the latter is smaller than the second specimen of *Compsognathus* from Canjuers (France), which has a femur length of 110 mm and an estimated length of 1.25 m. Although size and body proportions indicate that the smaller specimen was younger when it died, well-ossified limb joints and tarsals suggest that it was approaching maturity.

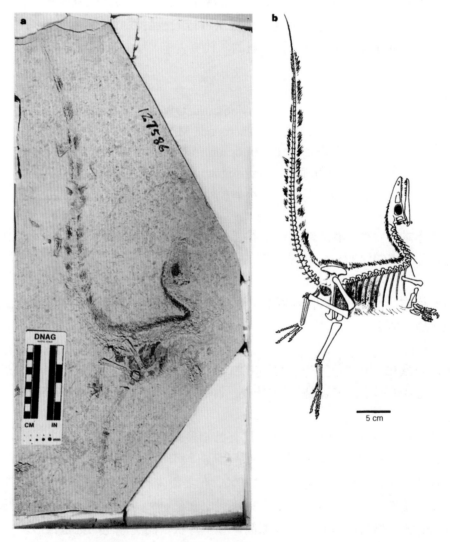

Fig. 2. *Sinosauropteryx prima* Ji and Ji. **a**, NIGP 127586, the counterpart of holotype (GMV 2123). **b**, Skeletal reconstruction of NIGP 127586. The integumentary structures are along the dorsal side and tail and dark pigmentation in the abdominal region might be some soft tissues of viscera.

描　述

中华龙鸟在大小和形态上同德国、法国已知的美颌龙标本[11,12]相近。中国标本中较小的一块（图 2）为 0.68 米长（从吻到尾巴末端），股骨长为 53.2 毫米，而第二件标本（图 3）的股骨长为 86.4 毫米。前者比模式种长足美颌龙（股骨长约 67 毫米）小，而后者比另一件来自法国 Canjuers 的美颌龙（股骨长为 110 毫米，估计全长 1.25 米）小。尽管大小和身体比例反映出小的标本死时比较年轻，但它死亡后保存下来的完全骨化的肢骨关节和跗骨表明它已经接近成熟。

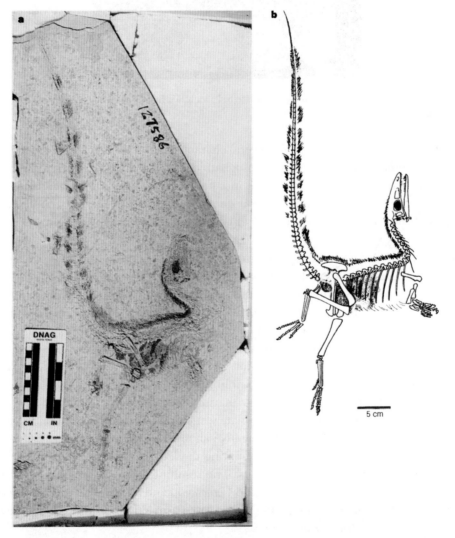

图 2. 原始中华龙鸟。**a**，NIGP 127586，是正模（GMV 2123）的负体。**b**，标本 NIGP 127586 的骨骼复原。皮肤衍生物结构沿着背部到尾部，腹部区域的黑色沉着可能是一些内脏软体组织残留。

25

Fig. 3. *Sinosauropteryx prima* Ji and Ji, NIGP 127587, an adult individual from the same locality as holotype.

Sinosauropteryx and *Compsognathus* share several characteristics that indicate close relationship. These can be used to diagnose the Compsognathidae and include unserrated premaxillary but serrated maxillary teeth, a powerful manual phalanx I–1 (shaft diameter is greater than that of the radius), fan-shaped neural spines on the dorsal vertebrae, limited anterior expansion of the pubic boot and a prominent obturator process of the ischium.

Other characteristics were used to diagnose *Compsognathus*, including the presence of a relatively large skull and short forelimbs. In *Compsognathus*, skull length is 30% of that of the presacral vertebral column, whereas this same ratio is 40% in the new specimen NIGP 127586 and 36% in NIGP 127587. Unfortunately, relative skull length is highly variable in theropods. Comparing skull length with femur length, which is less variable than vertebral length, most theropods have skulls 100–119% the length of the femur[13]. The *Compsognathus* skulls are 99–100% and *Sinosauropteryx* skulls are 113–117%. Compsognathids have short forelimbs[11], 40% of the length of the hindlimb in *Compsognathus*. In *Sinosauropteryx*, the lengths of humerus plus radius divided by the sum of femur and tibia lengths produces a figure of less than 30% (Table 1). Unfortunately, such ratios are dependent on the absolute size of the animal, mostly because of negative allometry experienced by the tibia during growth or interspecific size increase. Comparing the lengths of humerus plus radius with femur length produces more useful results. The resulting figures fall within the range of most theropods (60–110%). The abelisaurid *Carnosaurus* and all tyrannosaurs have relatively shorter arms. Within the Compsognathidae, however, arm length is shorter in *Sinosauropteryx* (61–65%) than it is in *Compsognathus* (90–99%).

26

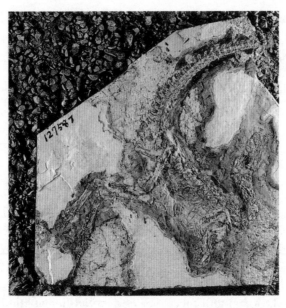

图 3. 原始中华龙鸟，NIGP 127587，一个成年个体，与正模标本来自同一产地

中华龙鸟和美颌龙具有几个共同的特征，显示出它们之间存在密切的关系。这些特征通常能够用于鉴别美颌龙科，包括无锯齿的前颌骨齿和有锯齿的上颌骨齿，强有力的第一指第一骨（直径要比桡骨大），背椎上扇形的神经棘，耻骨靴状突前端有限的扩展以及坐骨明显的闭孔突。

其他鉴定美颌龙的特征包括相对较大的头骨和短小的前肢。美颌龙头骨的长度占荐前椎总长的 30%，然而在新标本 NIGP 127586 和 NIGP 127587 中，该比率分别为 40% 和 36%。不巧的是，相对头骨长度在兽脚类恐龙中是一个非常易变的变量。头骨和股骨的长度的比值相对于与椎体的长度比值在兽脚类恐龙中变化较小，大部分兽脚类恐龙的头骨是股骨的长度的 100%～119%[13]。其中美颌龙的该值为 99%～100%，中华龙鸟为 113%～117%。美颌龙科具有较短的前肢[11]，其中美颌龙前肢长度是后肢的 40%。在中华龙鸟中，肱骨加上桡骨的长度与股骨和胫骨的总和相比小于 30%（表 1）。不过，这个比率取决于动物的绝对大小，主要是由生长或种间大小增长的过程中胫骨的异速生长造成的。肱骨加上桡骨的长度与股骨长度的比值更加可靠。该比例在大多数兽脚类的范围内（60%～110%）。在阿贝力龙类和暴龙类中，前肢相对较短。在美颌龙类中，中华龙鸟的前肢长度比值（61%～65%）要比美颌龙的（90%～99%）小。

Table 1. Comparison of size and proportions of *Sinosauropteryx* and *Compsognathus*

Species	Specimen	Skull	Humerus	Radius	Femur	Tibia	Skull/femur	Arm/leg
Compsognathus sp.	BSP ASI	70	39	24.7	71	87.6	0.99	0.40
Compsognathus sp.	MNHN	110	67	42	110		1.00	
Sinosauropteryx primus	NIGP 127586	62.5	20.3	12.4	53.2	61	1.17	0.29
Sinosauropteryx primus	NIGP 127587	97.2	35.5	21	86.4	97	1.13	0.31

Length measurements are given in millimetres. Data about *Compsognathus* are from ref. 11.

Both specimens of *Sinosauropteryx* have 10 cervical and 13 dorsal vertebrae. The posterior cervical vertebrae have biconcave centra. We could not determine the number of sacrals. The tail is extremely long. In the smaller specimen it is almost double the snout–vent length, and there are 59 caudal vertebrae exposed with an estimated five more than have been lost from the middle of the tail of NIGP 127586 (but present in GMV 2123). Only the first 23 vertebrae are preserved in the larger specimen, but this section is longer than the summed lengths of the cervical, dorsal and sacral vertebrae. Neither of the European specimens has a complete tail, but in both cases the tail was clearly longer than the body. When vertebral lengths are normalized (divided by the average lengths of caudal vertebrae 2–5), there are no significant differences between vertebral lengths in any of the four tails. As in *Compsognathus*, the dorsal neural spines are peculiar in that they are anteroposteriorly long but low, and often are fan-shaped.

The caudal centra increase in length over the first six segments, but posteriorly decrease progressively in length and all other dimensions. The first 10 tail vertebrae have neural spines, most of which slope posterodorsally. There are at least four pairs of caudal ribs in NIGP 127586, and more distal caudals have low bumps in this region that could also be interpreted as transverse processes. This could be another way to distinguish the Asian and European compsognathids, because the German specimen of *Compsognathus* apparently lacks caudal ribs and transverse processes[11]. Haemal spines are found on at least the first 47 caudals of NIGP 127586, and the anterior ones are simple spatulate structures that curve gently posteroventrally. The haemal spines are oriented more posteriorly than ventrally, and are more strongly curved.

Both specimens have 13 pairs of dorsal ribs. The ribs indicate a high but narrow body. The distal ends of the first two pairs of ribs are expanded and end in cup-like depressions that suggest the presence of a cartilaginous sternum. The gastralia are well preserved with two gastralia on each side of a segment. The median gastralia cross to form the interconnected "zig-zag" pattern characteristic of all theropods[14] and primitive birds like *Archaeopteryx* and *Confuciusornis*.

The front limb is relatively short and stout. Both NIGP 127586 and NIGP 127587 have articulated hands, something that is lacking in the two European specimens. What has been interpreted by some as the first metacarpal[11] in *Compsognathus* is the first phalanx of

表 1. 中华龙鸟和美颌龙个体大小和比例的相互比较

种类	标本编号	头骨	肱骨	桡骨	股骨	胫骨	头骨/股骨	前肢/后肢
美颌龙未定种	BSP ASI	70	39	24.7	71	87.6	0.99	0.40
美颌龙未定种	MNHN	110	67	42	110		1.00	
原始中华龙鸟	NIGP 127586	62.5	20.3	12.4	53.2	61	1.17	0.29
原始中华龙鸟	NIGP 127587	97.2	35.5	21	86.4	97	1.13	0.31

长度测量的单位是毫米，美颌龙资料来源于参考文献 11。

中华龙鸟的两件标本均有 10 节颈椎和 13 节背椎。后部的颈椎是双凹型椎体。不过我们不能确定荐椎的数量。尾巴特别长。在较小的标本（NIGP 127586）中，尾长差不多是吻肛距的两倍，有 59 节尾椎暴露，估计尾巴中段至少缺失 5 节以上的椎体（但在 GMV 2123 中存在）。在较大的标本中只有前 23 节尾椎椎体保存，但是这一段比颈椎、背椎和荐椎的总长要长。而欧洲的两个标本都没有保存完整的尾巴，但二者的尾巴明显要比身体长。当脊椎的长度进行标准化计算（除以第 2～5 节尾椎的平均长度）后，这四件标本的尾巴长度没有明显的差异。和美颌龙一样，中华龙鸟背神经棘很奇特，其前后向长但高度略低，且多呈现为扇形。

前六节尾椎的椎体长度递增，而后面尾椎的长度和其他尺寸都逐渐递减。前十节尾椎上有神经棘，多数向后背侧倾斜。在 NIGP 127586 中至少有 4 对尾肋，且很多末端尾椎在肋骨着生的部位有较低的突起，可以看作是横突。这可以作为区分亚洲和欧洲的美颌龙类的另一个方法，因为德国的美颌龙标本明显缺少尾肋和尾椎横突[11]。NIGP 127586 至少前 47 节尾椎上都具有脉弧，前面的脉弧呈轻微向后腹侧弯曲的简单的片状结构。脉弧向后的幅度比向腹部大，且弯曲得更加强烈。

两件标本均有 13 对背肋。肋骨的形态揭示了该个体的身躯高而窄。前面两对肋骨的最末端膨大，且末端具有杯状的凹陷，揭示了软骨质胸骨的存在。腹膜肋保存较好，在每一节两侧都各有两个腹膜肋。中间的腹膜肋像所有兽脚类恐龙[14]和原始鸟类（如始祖鸟和孔子鸟）中一样交叉连通呈"之"字形排列。

前肢相对短而粗壮，NIGP 127586 和 NIGP 127587 两件标本均保存着有关节的手部，这在欧洲的两标本中是缺乏的。美颌龙中被其他人鉴定为是第一掌骨[11]的

digit I, as was originally proposed by von Huene[15]. The first metacarpal is short (4.2 mm long in NIGP 127586, and double that length in 127587), and is probably the element identified as a carpal in the French specimen[12]. As is typical of all theropods, the collateral ligament pits of the first phalanx are much closer to the extensor surface of the bone than they are to the flexor surface. Both phalanx I–1 and the ungual that it supports are relatively large, each being as long as the radius, and thicker than the shafts and the distal ends of either the radius or the ulna. This unusual character seems to have been partially developed in at least phalanx I–1 of *Compsognathus*. Relative to the length of the radius, both these elements are longer in *Sinosauropteryx* than in any other known theropod except for *Mononykus*[16]. As indicated by the proposed phylogenetic placement[16] of *Mononykus*, there are too many anatomical differences between compsognathids and *Mononykus* to suggest a close relationship, and the similarities probably represent convergence.

The long (39 mm in NIGP 127586, 67.5 mm in NIGP 127587), low (22.2 mm high at both pubic and ischial peduncles in NIGP 127587) ilium is shallowly convex on the dorsal side in lateral aspect. The pubis, which is 82.8 mm long in NIGP 127587, is oriented anteroventrally, but is closer to vertical than it is in most non-avian theropods. The distal end expands into a pubic boot as in most tetanuran theropods. In the larger specimen, this expansion is 17.7 mm. As in *Archaeopteryx*[17], *Compsognathus*[11] and dromaeosaurs[18], the boot expands posteriorly from the shaft of the pubis, and the anterior expansion is moderate. The lack of the significant anterior expansion of the pubic boot may be correlated with the inclination of the shaft. The ischium is only two-thirds the length of the pubis in NIGP 127587. It tapers distally into a narrow shaft (3.2 mm in diameter), and like *Compsognathus*, there is a slight expansion at the end (6 mm in NIGP 127587). The prominent obturator process is also found in *Compsognathus*.

The shaft of the femur is gently curved. Both tibia and fibula are elongate. The astragalus and calcaneum are present in both specimens, although not clearly seen in either. There are five metatarsals, but as in other theropods and early birds, the first is reduced to a distal articular condyle, and the fifth is reduced to a proximal splint (Fig. 2b). Metatarsals II, III and IV are closely appressed and elongate, but are not co-ossified. The second and fourth metatarsals do not contact each other. Pedal phalanges are conservative in number (2–3–4–5–0) and morphology.

Inclusions within the Body Cavity

Like the German *Compsognathus*, the larger Chinese specimen has stomach contents preserved within the rib cage. This consists of a semi-articulated skeleton of a lizard, complete with skull (Fig. 4a, b). Numerous lizard skeletons have been recovered from these beds, but have yet to be described. Low in the abdomen of NIGP 127587, anterior to and slightly above the pubic boot, lies a pair of small eggs (37 × 26 mm) (Fig. 4a, c), one in front of the other. Additional eggs may lie underneath. Gastralia lie over the exposed surfaces of the eggs, and the left femur protrudes from beneath them, so there can be no doubt that

30

骨骼实际上是第一指的第一节，与最早由许纳[15]鉴别的一致。第一掌骨较短（NIGP 127586 中长 4.2 毫米，NIGP 127587 中长度为前者的两倍），它可能就是法国标本[12]中鉴定为腕骨的成分。像典型的兽脚类恐龙一样，第一指骨的侧韧带凹更靠近伸肌面而非屈肌面。第一指的第一节和其支撑的爪相对巨大，都像桡骨一样长，而且比桡骨或尺骨的骨干和末端都要粗。在美颌龙中，至少第一指的第一节有类似的特征。中华龙鸟第一指第一节及爪的长度与桡骨的长度比要比除单爪龙[16]外其他所有已知兽脚类恐龙都大。之前的系统发育结果表明[16]，美颌龙类和单爪龙之间存在许多解剖学的差异，不可能是相近的类群，二者的相似性可能为趋同所致。

髂骨长而低（NIGP 127586 中长 39 毫米，NIGP 127587 中长 67.5 毫米；NIGP 127587 中髂骨在耻骨和坐骨柄处高度均为 22.2 毫米），侧视上背部略微突起。NIGP 127587 的耻骨长 82.8 毫米，指向前腹侧，但比很多非鸟兽脚类恐龙的更接近垂直。像大多数坚尾龙类兽脚类恐龙一样，耻骨末端膨大呈靴状。在较大的标本中，这个膨大有 17.7 毫米。类似始祖鸟[17]、美颌龙[11]和驰龙类[18]，靴状突从耻骨柄向后膨大，前部适度膨大。耻骨突缺少明显的前突可能与耻骨柄的倾斜有关。NIGP 127587 的坐骨只是耻骨长度的三分之二。坐骨向远端变尖形成一个狭窄的骨干（直径 3.2 毫米），且跟美颌龙一样，在末端存在微小的膨大（NIGP 127587 中为 6 毫米）。在美颌龙中也发现了明显的闭孔突。

股骨柄略微弯曲。胫骨和腓骨相对较长。两件标本都保存了距骨和跟骨，尽管保存得都不清晰。具有 5 个跖骨，但和其他兽脚类恐龙和早期鸟类一样，第一跖骨退化为一个末端关节髁，第五跖骨退化成近端的一小块（图 2b）。第Ⅱ、Ⅲ、Ⅳ跖骨细长并紧密排列，但没有联合骨化。第Ⅱ、Ⅳ跖骨不互相接触。脚趾骨在数量（2–3–4–5–0）和形态学上相对保守。

体腔内含物

同德国的美颌龙一样，中国发现的中华龙鸟标本中较大的一块的腹腔中有胃容物保存。胃容物包括一具部分关节保存的蜥蜴骨骼，头骨完整（图 4a 和 4b）。发现中华龙鸟化石的地层已经发现过众多的蜥蜴骨骼，但还没有进行过描述研究。在 NIGP 127587 的腹部下方，耻骨靴状突的前面略上方，有一对小型的蛋化石（37 毫米 × 26 毫米）（图 4a 和 4c），一个在前一个在后。其余的蛋可能分布在下面。腹膜肋排

they were within the body cavity. It is possible that the eggs were eaten by the dinosaur. However, given their position in the abdomen behind and below the stomach contents, and the fact that they are in the wrong part of the body cavity for the egg shell to be intact, it is more likely that these were unlaid eggs of the compsognathid. Eggs have also been reported in the holotype of *Compsognathus*[19], but they are more numerous and are only 10 mm in diameter. As they were also found outside the body cavity, their identification as *Compsognathus* eggs has not been widely accepted. The presence of fewer but larger eggs in *Sinosauropteryx* casts additional doubt on this identification.

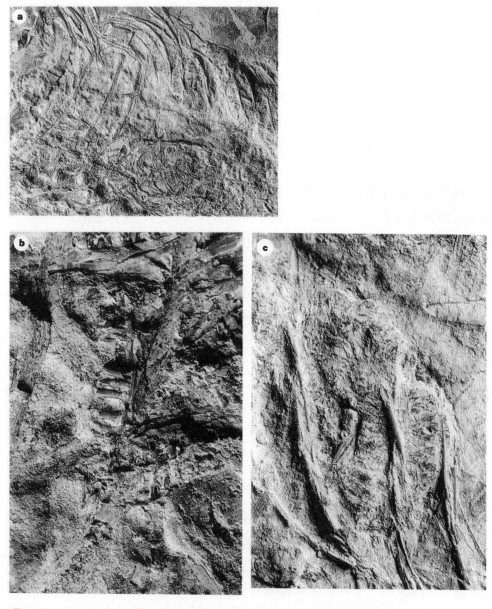

Fig. 4. Body of NIGP 127587. **a**, Stomach contents are preserved within the rib cage, and include a small lizard and a pair of eggs. **b**, A close-up of the lizard skull. **c**, A close-up of a pair of the eggs.

列在蛋化石表面，左股骨从它们下面伸出，因此它们毫无疑问处于体腔之内。蛋有可能是被恐龙吃下去的。但是，考虑到蛋的位置在腹部胃容物后侧和下面，而且如果它们是食物的话，在这个位置蛋壳不可能是完整的，因此这更像是美颌龙类还未产出的蛋。蛋化石也在美颌龙的正模标本中有过报道[19]，但是蛋的数量很多，直径却仅有10毫米。由于这些蛋都在体腔外面发现，把它们鉴定为美颌龙的蛋并没有被广泛接受。中华龙鸟腹腔内少而更大型的蛋对之前美颌龙蛋化石的鉴定提出了新的质疑。

图 4. 标本 NIGP 127587 整体。a，保存在腹腔中的胃容物，包括一只小的蜥蜴和一对蛋化石。b，蜥蜴头骨的特写。c，成对蛋化石的特写。

33

Although more than two eggs may have been present in the larger specimen of *Sinosauropteryx*, it does not seem as though many could have been held within the abdomen. It may well be that these dinosaurs laid fewer eggs than most (some species are known to have produced in excess of 40)[20]. However, it is more likely that their presence demonstrates paired ovulation, as has been suggested for *Oviraptor*[21], *Troodon*[22] and other theropods. *Sinosauropteryx* therefore probably laid eggs in pairs, with a delay for ovulation between each pair.

Integumentary Structures

One of the most remarkable features of both Chinese specimens is the preservation of integumentary structures. In the larger specimen, these structures can be seen along the dorsal surface of the neck and back, and along the upper and lower margins of the tail, but in the smaller specimen the integumentary structures are clearer (Fig. 5). They cover the top of the back half of the skull, the neck, the back, the hips and the tail. They also extend along the entire ventral margin of the tail. Small patches can be seen on the side of the skull (behind the quadrate and over the articular), behind the right humerus, and in front of the right ulna. With the exception of a small patch outside the left ribs of NIGP 127587 and several areas on the left side of the tail (lateral to the vertebrae), integumentary structures cannot be seen along the sides of the body. The structures were probably present in the living animals, as indicated by the density of the covering dorsal to the body, and by the few random patches of integumentary structures that can be seen elsewhere.

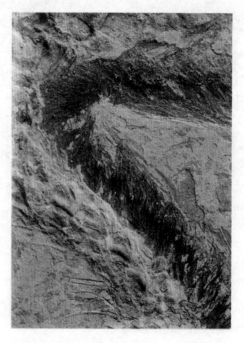

Fig. 5. Integumentary structures in the neck and dorsal sides of NIGP 127586.

尽管在大的中华龙鸟标本中可能存在多于两个蛋化石，但看上去腹腔内蛋化石的数量不会很多。很可能这些恐龙要比多数恐龙产蛋少（一些我们了解的种类产蛋会超过 40 枚）[20]。不过，这些蛋化石的出现再次证实了双排卵现象，如同在窃蛋龙 [21]、伤齿龙 [22] 和其他兽脚类恐龙中发现的那样。因此，中华龙鸟很可能每次产一对卵，每次排卵之间需要延迟一段时间。

皮肤衍生物结构

两件中国化石标本最显著的特点之一就是保存了皮肤衍生物结构。在较大的标本中，这些结构沿着颈部和背部的背面以及尾巴的上下缘分布，而在较小的标本中皮肤衍生物的结构更清晰（图 5）。它们覆盖头骨背侧的后半部、颈部、背部、臀部和尾巴，同时沿着整个尾巴的腹侧扩展。头骨（方骨后和关节骨上）、右肱骨后侧和右尺骨前侧都能够看出小片的皮肤衍生物。而在 NIGP 127587 上，除了左肋外的一小片和尾巴左侧的几个区域（椎体侧面），身体侧面的其他位置并不能观察到皮肤衍生物。这种结构可能在现生动物中也存在，类似身体背部的浓密毛发和身体其他部位可见的少量皮肤衍生物。

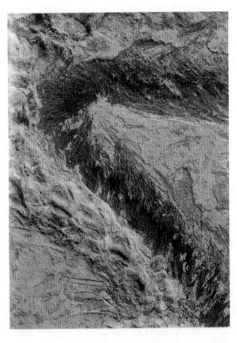

图 5. NIGP 127586 的颈部和背部的皮肤衍生物结构

Fig. 6. Integumentary structures over caudal 27 of NIGP 127587.

In the two theropods, the distances separating the integumentary structures from the underlying bones are directly proportional to the amount of skin and muscle that would have been present. As in modern animals, the integument closely adheres to the tops of the skull and hips, and becomes progressively closer to the caudal vertebrae towards the tip of the tail. In the posterior part of the neck, over the shoulders, and at the base of the tail, the integumentary structures are more distant from the underlying skeletal elements, and in life would have been separated by greater thicknesses of muscle and other soft tissues.

The orientation and frequently sinuous lines of the integumentary structures suggest they were soft and pliable, and semi-independent of each other. They frequently cross each other, and are tangled in some areas. There is an apparent tendency for the integumentary structures to clump along the tail of the smaller specimen, but this is an artefact of the splitting plane between NIGP 127586 (Fig. 2a) and GMV 2123. As both individuals were lying in the water of a lake when they were buried, it is clear that we are not looking at the normal orientation of the integumentary structures in the fossils. Under magnification, the margins of the larger structures are darker along the edges, but lighter medially, which indicates that they might have been hollow. Overall, the integumentary structures are rather coarse for such a small animal, and the thickest strands are much thicker than the hairs of the vast majority of small mammals[23]. In NIGP 127586, integumentary structures are first seen on the dorsal surface of the skull in front of the orbit. The skull is semidisarticulated, and sediment still covers the snout region, so it is possible that the integumentary structures extended more anteriorly. The most rostral integumentary structures are 5.5 mm long, and

36

图 6. NIGP 127587 的第 27 节尾椎上方的皮肤衍生物结构

在这两个兽脚类恐龙中，皮肤衍生物与下覆骨骼的间距与活着时存在的皮肤和肌肉量呈正比。像现生动物一样，皮肤衍生物紧密黏附在头骨和臀部顶部，并逐渐靠近尾巴末端的尾椎。在颈部后侧、肩部之上和尾巴的基部，皮肤衍生物距离下面的骨骼更远，可能是因为这些区域在动物活着时被非常厚的肌肉和其他软组织与骨骼分隔开。

皮肤衍生物的方向和错综复杂的排列表明它们柔软而易弯曲，相互间较为分离。它们彼此频繁交叉，在某些区域又比较紊乱。在较小的标本中，皮肤衍生物明显的趋向是沿着尾部丛生，但这是 NIGP 127586（图 2a）和 GMV 2123 之间分割面上的假象。因为两个个体被掩埋时都是在湖水中死亡，因此很显然我们在化石中看到的并非皮肤衍生物的正常方向。当放大一定倍数后，大的结构边缘是暗色的，但是中间是亮的，这显示它们中间可能有空洞。总的来说，这种小型动物皮肤衍生物结构比较粗糙，毛发最浓密部分其浓密程度超过大多数小型哺乳动物的毛发[23]。在 NIGP 127586 上，皮肤衍生物结构首先是在眼眶之前的头骨背部表面看到。头骨是半脱离的，沉积物依然覆盖在吻部，因此皮肤衍生物结构可能会更向前扩展。最靠近吻端的皮肤衍生物有 5.5 毫米长，在头骨上方又延伸了约 4 毫米。它们在肩胛骨远端的上

extend about 4 mm above the skull. They quickly lengthen to at least 21 mm above the distal ends of the scapulae. This axial length seems to stay constant along most of the back, but decreases sharply to 16 mm dorsal to the ilium. The longest integumentary structures seem to have been above the base of the tail, although it is impossible to measure any single structure. More distally along the tail, integumentary structures decrease more rapidly on the lower side of the tail than on the upper. By caudal 47, the ventral structures are 4.2 mm long, about half the length of the dorsal structures in that region.

The size distribution of the integumentary structures of NIGP 127587 follow the same general pattern as in the smaller specimen. Although the integument tends to look thinner on this specimen, it is simply because the integumentary structures are lying closer to the body. Individual measurements are consistently larger than those of NIGP 127586. The integumentary structures are 13 mm long above the skull, 23.5 mm above the fourth cervical, at least 35 mm over the scapulae, at least 40 mm over caudal 27 (Fig. 6), and at least 35 mm below caudal 25. Integumentary structures on the left side of the body are largely covered by ribs, gastralia, stomach contents and matrix, so it is only possible to say that each is more than 5 mm long. Those associated with the right ulna are 14 mm long.

Integumentary structures have also been reported in the theropod *Pelecanimimus*[24] from the Lower Cretaceous of Spain. These consist of subparallel fibres arranged perpendicular to the bones, with a less conspicuous secondary system parallel to them. As described, they seem to be similar to the integumentary structures of *Sinosauropteryx*.

Skin impressions have been found on most main types of dinosaurs, including sauropods, ankylosaurs, ornithopods, stegosaurs, ceratopsians, and several genera of large theropods. In all of these animals, there is no evidence of integumentary structures, and the skin usually has a "pebbly" surface texture. Integumentary structures have been claimed for both specimens of *Compsognathus*[11,25] though the interpretations have been questioned in both cases[11]. In the German specimen, there was supposedly a patch of skin over the abdominal region. The French specimen included some strange markings in the region of the forearm, that were originally identified as a swimming appendage formed either of dermal bone or of thick skin[12], but it is clearly not well enough preserved to be positively identified. The identification of these structures as integumentary is questionable[11], and there is nothing on the Chinese specimens to support the presence of such structures in compsognathids. Evidence of feathers in *Compsognathus* was sought[11] without success, but this lack of evidence on the German specimen of *Compsognathus* does not eliminate the possibility that they might have existed.

Discussion

The integumentary structures of *Sinosauropteryx* are extremely interesting regardless of whether they are referred to as feathers, protofeathers, or some other structure. Unfortunately, they are piled so thick that we have been unable to isolate a single one for

方迅速延长至至少 21 毫米。这个长度似乎在背部大部分区域是恒定的，但是在髂骨背侧迅速递减到 16 毫米。最长的皮肤衍生物可能在尾巴基部的上方，尽管不可能去测量所有单个结构。顺着尾巴向远端延伸，皮肤衍生物结构在尾巴下方要比在上方递减更快。到第 47 节尾椎，腹侧的结构是 4.2 毫米长，约是背部相同区域长度的一半。

NIGP 127587 的皮肤衍生物的大小分布与小的标本遵循相同的普遍模式。这件标本中的皮肤衍生物趋向是逐渐变稀，这仅仅是因为皮肤衍生物更贴近身体。每一个部位皮肤衍生物的测量尺寸都要比标本 NIGP 127586 的大。头骨上的皮肤衍生物结构长 13 毫米，第四颈椎上的长 23.5 毫米，肩胛骨上的至少 35 毫米长，在第 27 节尾椎上的至少 40 毫米长（图 6），在第 25 尾椎下的至少 35 毫米长。身体左侧的皮肤衍生物基本被肋骨、腹膜肋、胃容物和围岩所覆盖，因此，只能推测皮肤衍生物可能长于 5 毫米。附着在右侧尺骨上的皮肤衍生物是 14 毫米长。

从西班牙下白垩统发现的兽脚类恐龙似鹈鹕龙标本 [24] 同样报道过皮肤衍生物。它们是由与骨骼垂直的近似平行排列的纤维组成的，具有不显著的次级结构与之平行排列。正如描述所说，它们似乎与中华龙鸟的皮肤衍生物结构相似。

皮肤印痕在很多主要类型的恐龙中曾被发现，包括蜥脚类、甲龙类、鸟脚类、剑龙类、角龙类和一些大型兽脚类恐龙属。但是在所有的这些恐龙中都没有皮肤衍生物存在的证据，它们的皮肤表面通常具有卵纹状的纹理。虽然关于两件标本的皮肤衍生物的解释都存在质疑 [11]，但皮肤衍生物确实在两个美颌龙标本中都存在 [11,25]。在德国的标本中，推测腹部区域上方有一块皮肤。在法国的标本中，前肢区域包括一些奇怪的痕迹，最初被鉴定为游泳的附肢，可能是膜质骨或者厚的皮肤 [12]，但由于保存得不够完好而无法确切地下结论。这些结构被鉴定为皮肤衍生物是存在问题的 [11]，中国的标本没有为美颌龙存在这样的结构提供支持依据。美颌龙存在羽毛的证据并没有寻找 [11] 成功，但是仅凭德国美颌龙标本缺少羽毛证据这一事实并不能排除美颌龙存在皮肤衍生物的可能性。

讨　论

不论中华龙鸟的皮肤衍生物是羽毛、原始羽毛还是一些其他的结构，它们都是特别引人瞩目的。不巧的是，它们堆积得太厚，使我们不能单独将其分离开来

examination. Comparison with birds from the same locality shows that the same problem exists with identifying individual feathers (other than the flight feathers) and components of feathers in avian specimens. The morphological characteristics that we describe suggest that the integumentary structures seem to resemble most closely the plumules of modern birds, having relatively short quills and long, filamentous barbs. The absence of barbules and hooklets is uncommon in modern birds, but has been noted in Cretaceous specimens[26].

It has been proposed that the feathers of another recently discovered animal from the same locality in Liaoning are structurally intermediate between the integumentary structures of *Sinosauropteryx* and the feathers of *Archaeopteryx*[27]. The clearly preserved feathers of *Protarchaeopteryx robusta* are symmetrical, which indicates that the animal was not capable of flight. This is confirmed by the relatively short length of the forelimb. Both *Sinosauropteryx* and *Protarchaeopteryx* had been identified as birds because of the presence of feathers[27], but much more work needs to be done to prove that the integumentary structures of *Sinosauropteryx* have any structural relationship to feathers, and phylogenetic analysis of the skeleton clearly places compsognathids far from the ancestry of birds. Despite arguments to the contrary[28], cladistic analysis favours the notion that the bird lineage originated within theropod dinosaurs[29,30]. If this phylogenetic framework is accepted, the integumentary structures of *Sinosauropteryx* could shed light on some of the many hypotheses concerning feather origins. Three main functions have been suggested for the initial development of feathers—display, aerodynamics and insulation.

The integumentary structures of *Sinosauropteryx* have no apparent aerodynamic characteristics, but might be representative of what covered the ancestral stock of birds. It is highly unlikely that something as complex as a bird feather could evolve in one step, and many animals glide and fly with much simpler structures. Even birds secondarily simplify feathers when airborne flight ceases to be their main method of locomotion, and produce structures that are intermediate between reptilian scales and feathers[31]. The multi-branched integumentary structures of the Chinese compsognathids are relatively simple, but are suitable for modification into the more complex structures required for flight.

Feathers may have appeared first as display structures[31], but the density, distribution, and relatively short lengths of the integumentary structures of *Sinosauropteryx* suggest that they were not used for display. It is conceivable that both specimens are female, and that the males had more elaborate integumentary structures for display. It is also possible that the integumentary structures were coloured to serve a display function. Therefore, the existing *Sinosauropteryx* specimens do not support the hypothesis that feathers evolved primarily for display, but do not disprove it either.

The dense, pliable integumentary structures of the Chinese compsognathids would not have been appropriate as heat shields to screen and shade the body from the Sun's rays[32]. Although they may have been effective in protecting the body from solar radiation in warm weather, they would also have been effective in preventing an ectothermic

进行检测。和同一产地的鸟类化石相比发现，在鸟类标本中鉴别单独的羽毛（除了翼羽）和羽毛的组成时也存在同样的问题。根据我们描述的形态学特征，该皮肤衍生物结构似乎最接近现代鸟类的绒羽，具有相对较短的羽轴和长的纤维状的羽枝。现代鸟类很少缺失羽小枝和羽小钩，但这种现象在白垩纪的鸟类[26]中是较为常见的。

最近在辽宁同一产地发现的另外一种动物的羽毛被认为是介于中华龙鸟的皮肤衍生物和始祖鸟羽毛之间的中间结构[27]。粗壮原始祖鸟清晰保存的羽毛是对称的，反映出这种动物是不具飞行能力的，这一点同时也根据它的前肢长度相对较短得以证明。中华龙鸟和原始祖鸟都曾经因为羽毛的存在被鉴定为鸟类[27]，但是还有很多工作需要去做来证明中华龙鸟的皮肤衍生物同鸟类羽毛之间存在结构上的联系，而且基于骨骼形态的系统发育分析结果显示美颌龙距离鸟类的祖先很远。尽管存在反对的论断[28]，但系统发育分析仍主张鸟类起源于兽脚类恐龙[29,30]。如果这个系统发生的结果能够被接受的话，那么中华龙鸟的皮肤衍生物结构能阐明关于羽毛起源的很多假说。羽毛的早期起源被认为主要与三个方面的功能有关：展示、空气动力学和保温。

中华龙鸟的皮肤衍生物结构没有明显的空气动力学特征，可能代表鸟类体表覆盖物的原始状态。像鸟类羽毛这种复杂的结构，其演化进程基本上是不可能一步发展到位的，很多动物是利用更为简单的结构进行滑翔和飞行。甚至当一些鸟类不再有飞行的习性时，它们的羽毛会发生次生简化，形成一种介于爬行动物的鳞甲和羽毛中间的结构[31]。中国美颌龙类多分枝的皮肤衍生物结构是相对简单的，但是适合修饰成更复杂的鸟类飞行结构。

羽毛最初出现时可能是作为展示的结构[31]，但中华龙鸟皮肤衍生物的密度、分布和相对短的长度显示它们不是用来展示的。可以想象这两件标本都属于雌性个体，雄性应该具有更为精细的皮肤衍生物用来展示。皮肤衍生物也可能是色彩斑斓的，以适合展示。因此，现存的中华龙鸟标本不支持羽毛原始演化是为了展示的假说，但也不能反驳这种假说。

中国美颌龙类浓密且柔软的皮肤衍生物不适合作为保暖层以及遮蔽太阳辐射[32]。但这些皮肤衍生物在温暖的天气保护身体避免太阳辐射可能是有效的，同时也可以有效预防外温的兽脚类恐龙在太阳直射中吸收外部热量而使体温迅速升高。

theropod from rapidly warming up by basking in the sunshine. If small theropods were endothermic, they would have needed insulation to maintain high body temperatures[33-35]. The presence of dense integumentary structures may suggest that *Sinosauropteryx* was endothermic, and that heat retention was the primary function for the evolution of integumentary structures[36-38]. Recently published histological studies suggest at least some early birds were not truly endothermic[39], although they may have been physiologically intermediate between poikilothermic ectotherms and homeothermic endotherms[40].

The Chinese compsognathids have integumentary structures consisting of vertical fibres running from the base of the head along the back and around the tail extending forwards almost to the legs. There are no structures showing the fundamental morphological features of modern bird feathers, but they could be previously unidentified protofeathers which are not as complex as either down feathers or even the hair-like feathers of secondarily flightless birds. Their simplicity would not have made them ineffective for insulation when wet any more than it negates the insulatory capabilities of mammalian hair. We cannot determine whether or not the integumentary structures were arranged in pterylae, but they were long enough to cover apteria, if they existed, and could therefore still have been effective in thermoregulation. Continuous distribution is not essential to be effective in this function[28], especially if the apteria are part of a mechanism for dispersing excess heat. Finally, the aerodynamic capabilities of bird feathers are not comprised by the previous evolution of less complex protofeathers that had some other function, such as insulation.

In addition to the integumentary structures, there is dark pigmentation over the eyes of both specimens. A second region of dark pigmentation in the abdominal region of the smaller specimen might represent some soft tissues of viscera.

Multidisciplinary and multinational research is just beginning on these unique small theropods. Techniques developed to study fossil feathers[38,41] will be useful research tools as work progresses. In the meantime, the integumentary structures of *Sinosauropteryx* suggest that feathers evolved from simpler, branched structures that evolved in non-avian theropod dinosaurs, possibly for insulation.

(**391**, 147-152; 1998)

Pei-ji Chen[*], Zhi-ming Dong[†] & Shuo-nan Zhen[‡]

[*] Nanjing Institute of Geology and Palaeontology, Academia Sinica, 39 East Beijing Road, Nanjing 210008, People's Republic of China

[†] Institute of Vertebrate Paleontology and Paleoanthropology, Academia Sinica, PO Box 643, Beijing 100044, People's Republic of China

[‡] Beijing Natural History Museum, 126 Tien Qiao Street, Beijing 100050, People's Republic of China

Received 14 January; accepted 18 September 1997.

References:

1. Chen, P. J. Distribution and migration of Jehol fauna with reference to nonmarine Jurassic–Cretaceous boundary in China. *Acta Palaeontol. Sin.* **27**, 659-683 (1988).

如果小型兽脚类恐龙是内温的，它们需要隔绝外界来维持较高的体温 [33-35]。稠密的皮肤衍生物表明中华龙鸟可能是内温动物，而保暖性是皮肤衍生物结构演化的最初功能 [36-38]。最近发表的组织学研究提出，至少一些早期的鸟类不是真正的内温动物 [39]，不过它们可能在生理学上介于变温外温动物与恒温内温动物之间 [40]。

中国美颌龙类具有的皮肤衍生物结构由垂直的纤维组成，从头部基部开始沿着背部延伸，包围尾部，同时沿尾部向前延伸，几乎到达腿部。它们没有现代鸟类羽毛的基本形态结构，但是可以被看作是一种未被确认的原始羽毛，其复杂程度低于绒羽甚至次级失去飞行能力的鸟类毛发状的羽毛。它们简单的构造在潮湿情况下能有效隔热，比哺乳动物毛发潮湿情况下的保温能力要强。我们不能确定皮肤衍生物结构是否按羽区排列，但如果它们存在，它们长到足够覆盖裸区，就能有效地进行温度调节。皮肤衍生物连续分布对于这个功能的有效性并不是必不可少的 [28]，尤其是如果裸区是用来分散多余热量的机制的一部分。最后，鸟类羽毛的空气动力学功能并不为这些处于演化早期的、结构不很复杂的原始羽毛所拥有，它们有另外的功能，如保暖。

除皮肤衍生物结构外，在两件标本眼部的上方都存在深色色素沉着。此外，在较小标本的腹部也保存有深色色素沉着，可能意味着存在一些内脏的软组织。

对于这些独特的小型兽脚类恐龙进行的多学科和多国合作研究只是刚刚开始。羽毛化石研究的技术发展 [38,41] 为加快研究进展提供了有益的研究工具。同时，中华龙鸟的皮肤衍生物结构揭示了羽毛的演化是从简单、多分枝结构开始的，这种结构在非鸟兽脚类恐龙中出现可能是用于保温的。

（张玉光 翻译；汪筱林 审稿）

2. Hou, L.-H., Zhang, J.-Y., Martin, L. D. & Feduccia, A. A beaked bird from the Jurassic of China. *Nature* **377,** 616-618 (1995).

3. Hou, L.-H., Martin, L. D., Zhang, J.-Y. & Feduccia, A. Early adaptive radiation of birds: evidence from fossils from northeastern China. *Science* **274,** 1164-1167 (1996).

4. Ji, Q. & Ji, S. A. On discovery of the earliest bird fossil in China and the origin of birds. *Chinese Geol.* **233,** 30-33 (1996).

5. Chen, P. J. *et al.* Studies on the Late Mesozoic continental formations of western Liaoning. *Bull. Nanjing Inst. Geol. Palaeontol.* **1,** 22-25 (1980).

6. Chen, P. J. Nonmarine Jurassic strata of China. *Bull. Mus. N. Arizona* **60,** 395-412 (1996).

7. Dong, Z. M. Early Cretaceous dinosaur faunas in China: an introduction. *Can. J. Earth Sci.* **30,** 2096-2100 (1993).

8. Li, W. B. & Liu, Z. S. The Cretaceous palynofloras and their bearing on stratigraphic correlation in China. *Cretaceous Res.* **15,** 333-365 (1994).

9. Smith, P. E. *et al.* Dates and rates in ancient lakes: ^{40}Ar-^{39}Ar evidence for an Early Cretaceous age for the Jehol Group, northeast China. *Can. J. Earth Sci.* **32,** 1426-1431 (1995).

10. Wang, D. F. & Diao, N. C. Geochronology of Jura-Cretaceous volcanics in west Liaoning, China. *Scientific papers on geology for international exchange* **5,** 1-12 (Geological Publishing House, Beijing, 1984).

11. Ostrom, J. H. The osteology of *Compsognathus longipes* Wagner. *Zitteliana* **4,** 73-118 (1978).

12. Bidar, A., Demay, L. & Thomel, G. *Compsognathus corallestris* nouvelle espèce de dinosaurien théropode du Portlandiend de Canjuers (sud-est de la France). *Ann. Mus. d'Hist. Nat. Nice* **1,** 3-34 (1972).

13. Currie, P. J. & Zhao, X. J. A new large theropod (Dinosauria, Theropoda) from the Jurassic of Xinjiang, People's Republic of China. *Can. J. Earth Sci.* **30,** 2037-2081 (1993).

14. Claesseus, L. Dinosaur gastralia and their function in respiration. *J. Vert. Palaeontol.* **16,** 28A (1996).

15. von Huene, F. The carnivorous Saurischia in the Jura and Cretaceous formations principally in Europe. *Revista Museo Plata* **29,** 35-167 (1926).

16. Perle, A., Chiappe, L. M., Barsbold, R., Clark, J. M. & Norell, M. Skeletal morphology of *Mononykus olecranus* (Theropoda: Avialae) from the Late Cretaceous of Mongolia. *Am. Mus. Novit.* **3105,** 1-29 (1994).

17. Wellnhofer, P. Das siebte Exemplar von *Archaeopteryx* aus den Solnhofener Schichten. *Archaeopteryx* **11,** 1-48 (1993).

18. Barsbold, R. Carnivorous dinosaurs from the Cretaceous of Mongolia. *Sovmestnaya Sovetsko-Mongol'skaya Paleontol. Ekspiditsiya, Trudy* **19,** 5-119 (1983).

19. Griffiths, P. The question of *Compsognathus* eggs. *Rev. Paleobiol.* Spec. issue **7,** 85-94 (1993).

20. Carpenter, K., Hirsch, K. F. & Horner, J. R. *Dinosaur Eggs and Babies* (Cambridge Univ. Press, 1994).

21. Dong, Z. M. & Currie, P. J. On the discovery of an oviraptorid skeleton on a nest of eggs at Bayan Mandahu, Inner Mongolia, People's Republic of China. *Can. J. Earth Sci.* **33,** 631-636 (1996).

22. Varricchio, D. J., Jackson, F., Borkowski, J. J. & Horner, J. R. Nest and egg clutches of the dinosaur *Troodon formosus* and the evolution of avian reproductive traits. *Nature* **385,** 247-250 (1997).

23. Meng, J. & Wyss, A. R. Multituberculate and other mammal hair recovered from Palaeogene excreta. *Nature* **385,** 712-714 (1997).

24. Pérez-Moreno, B. P. *et al.* A unique multitoothed ornithomimosaur dinosaur from the Lower Cretaceous of Spain. *Nature* **370,** 363-367 (1994).

25. von Huene, F. Der Vermuthliche Hautpanzer des *Compsognathus longipes* Wagner. *Neues Jb. F. Min.* **1,** 157-160 (1901).

26. Grimaldi, D. & Case, G. R. A feather in amber from the Upper Cretaceous of New Jersey. *Am. Mus. Novit.* **3126,** 1-6 (1995).

27. Ji, Q. & Ji, S. A. *Protarchaeopteryx*, a new genus of Archaeopterygidae in China. *Chinese Geol.* **238,** 38-41 (1997).

28. Feduccia, A. *The Origin and Evolution of Birds* (Yale Univ. Press, New Haven, 1996).

29. Gauthier, J. in *The Origin of Birds and the Evolution of Flight* (ed. Padian, K.) 1-55 (California Acad. Sci., San Francisco, 1986).

30. Fastovsky, D. E. & Weishampel, D. B. *The Evolution and Extinction of the Dinosaurs* (Cambridge Univ. Press, 1996).

31. McGowan, C. Feather structure in flightless birds and its bearing on the question of the origin of feathers. *J. Zool. (Lond.)* **218,** 537-547 (1989).

32. Paul, G. S. *Predatory Dinosaurs of the World* (Simon and Schuster, New York, 1988).

33. Ewart, J. C. The nestling feathers of the mallard, with observations on the composition, origin, and history of feathers. *Proc. Zool. Soc. Lond.* 609-642 (1921).

34. Van Tyne, J. & Berger, A. J. *Fundamentals of Ornithology* (Wiley, New York, 1976).

35. Young, J. Z. *The Life of Vertebrates* (Oxford Univ. Press, **1950**).

36. Chinsamy, A., Chiappe, L. M. & Dodson, P. Growth rings in Mesozoic birds. *Nature* **368,** 196-197 (1994).

37. Chiappe, L. M. The first 85 million years of avian evolution. *Nature* **378,** 349-355 (1995).

38. Brush, A. H. in *Avian Biology* vol. 9 (eds Farner, D. S., King, J. R. & Parkes, K. C.) 121-162 (Academic, London, 1993).

39. Regal, P. J. The evolutionary origin of feathers. *Quart. Rev. Biol.* **50,** 35-66 (1975).

40. Ostrom, J. H. Reply to "Dinosaurs as reptiles". *Evolution* **28,** 491-493 (1974).

41. Davis, P. G. & Briggs, D. E. G. Fossilization of feathers. *Geology* **23,** 783-786 (1995).

Acknowledgements. This study was supported by NSFC. We thank L.-s. Chen and P. J. Currie (Royal Tyrrell Museum of Palaeontology) for helping to prepare the fossil materials and manuscript; M.-m. Zhang, X.-n. Mu, G. Sun, J. H. Ostram, A. Brush, L. Martin, P. Wellnhofer, N. J. Mateer, E. B. Koppelhus, D. B. Brinkman, D. A. Eberth, J. A. Ruben, L. Chiappe, S. Czerkas, R. O'Brien, D. Rimlinger, M. Vickaryous and D. Unwin for assistance and comments; and L. Mazzatenta and M. Skrepnick for help producing the photographs and drawings.

Correspondence and requests for materials should be addressed to P-j.C. (e-mail: lpsnigp@nanjing.jspta.chinamail.sprint.com).

Softening of Nanocrystalline Metals at Very Small Grain Sizes

J. Schiøtz *et al.*

Editor's Note

One of the most useful handles that metallurgists have on the properties of metals is the manipulation of the size and shape of the crystalline grains that typically form a metal's mosaic-like microstructure. It was long known that a metal's hardness increases as its grain size gets smaller—a phenomenon known as the Hall–Petch effect. This seems to stem from the way that defects called dislocations, imperfections in the crystalline packing of metal atoms that can induce ductility by moving through the crystal lattice, get stopped at grain boundaries. The Hall–Petch effect implies that nanocrystalline metals, where the grains are of nanometre scale, should be particularly hard. That expectation is generally borne out, and has led to an interest in nanocrystalline metals as very hard and strong engineering materials. But in this paper, Jakob Schiøtz and colleagues at the Technical University of Denmark suggest that the hardening process will have a fundamental limit: the Hall–Petch effect will reverse when the grains get extremely small, making the metals softer. The work relies on an ability, then relatively recent, to conduct computer simulations for huge numbers of atoms—here about 100,000, providing a block of material that contains several nanoscale grains. The softening is a result of the fact that layers of atoms at grain boundaries slide rather easily; this effect begins to dominate dislocation motions as the grains become so small that a significant fraction of the atoms are situated at their surfaces.

Nanocrystalline solids, in which the grain size is in the nanometre range, often have technologically interesting properties such as increased hardness and ductility. Nanocrystalline metals can be produced in several ways, among the most common of which are high-pressure compaction of nanometre-sized clusters and high-energy ball-milling[1-4]. The result is a polycrystalline metal with the grains randomly orientated. The hardness and yield stress of the material typically increase with decreasing grain size, a phenomenon known as the Hall–Petch effect[5,6]. Here we present computer simulations of the deformation of nanocrystalline copper, which show a softening with grain size (a reverse Hall–Petch effect[3,7]) for the smallest sizes. Most of the plastic deformation is due to a large number of small "sliding" events of atomic planes at the grain boundaries, with only a minor part being caused by dislocation activity in the grains; the softening that we see at small grain sizes is therefore due to the larger fraction of atoms at grain boundaries. This softening will ultimately impose a limit on how strong nanocrystalline metals may become.

纳米晶金属在极小晶粒尺寸时的软化现象

希厄茨等

编者按

冶金学家对金属性能最有用的一个控制手段就是改变晶体晶粒的大小和形状，通常情况下，晶粒形成了金属马赛克状的微观结构。长期以来众所周知，金属的硬度随着晶粒尺寸的减小而增加，这个现象就是霍尔–佩奇效应。这似乎来源于金属原子晶体堆积中称作位错的缺陷（沿晶格运动可产生延性）在晶界处停止运动。霍尔–佩奇效应暗示纳米晶金属——具有纳米尺度的晶粒——应该特别硬。这种预测普遍得到了证实，并使得纳米晶金属作为非常硬且强的工程材料引起了人们的兴趣。但是在本文中丹麦技术大学的雅各布·希厄茨与同事指出，其硬化过程有一个基本的限制：当晶粒极小时，金属将变软，霍尔–佩奇效应将颠倒过来。此工作依赖于当时对包含纳米尺度晶粒材料单元实施大规模原子（本文中约 100,000 个）计算机模拟的能力。软化的结果基于晶界处原子层易于滑移的事实；这一结果开始主导位错运动；这是因为随晶粒变得极小，相当大部分的原子将处于晶粒表面。

纳米晶固体——其晶粒尺寸在纳米数量级——从技术角度常常具有值得关注的性能，例如增加的硬度和延性。纳米晶金属可以通过多种方式制备，其中最常用的是纳米尺寸原子簇的高压成型与高能球磨法 [1-4]。其所得产物为具有随机取向晶粒的多晶金属。材料的硬度和屈服应力普遍随着晶粒尺寸的减小而增加，这就是我们所谓的霍尔–佩奇效应 [5,6]。本文我们要介绍的是有关纳米晶铜形变的计算机模拟，它显示晶粒达到极小尺寸时材料表现出软化（一种反霍尔–佩奇效应 [3,7]）。其塑性形变主要归因于晶界处大量原子平面的微小"滑移"，仅有一小部分是由于晶粒中的位错运动而产生的；因此我们在小晶粒尺寸时看到的软化现象归因于晶界原子所占比例更高。这种软化现象最终决定了纳米晶金属的极值强度。

T O simulate the behaviour of nanocrystalline metals with the computer, we construct nanocrystalline "samples" with structures similar to those observed experimentally: essentially equiaxed dislocation-free grains separated by narrow straight grain boundaries[1]. Each sample contains 8–64 grains in a 10.6-nm cube of material, resulting in grain sizes from 3.3 to 6.6 nm. The grains are produced by a Voronoi construction[8]: a set of grain centres are chosen at random, and the part of space closer to a given centre than to any other centre is filled with atoms in an f.c.c. (face-centred cubic) lattice with a randomly selected crystallographic orientation. A typical "sample" is shown in Fig. 1a. To mimic the system's being deep within the bulk of a large sample, the system is replicated infinitely in all three spatial directions (periodic boundary conditions). The forces between the atoms are calculated with the effective-medium theory[9,10], which suitably describes many-atom interactions in metals. The metal chosen for these simulations is copper; very similar results were obtained with palladium. Before deforming the system we "anneal" it by running a 50-ps molecular dynamics simulation at 300 K, allowing unfavourable configurations in the grain boundaries to relax. Doubling the duration of annealing does not have any significant effect, nor does an increase in the temperature to 600 K.

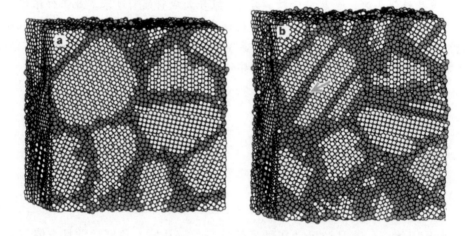

Fig. 1. A simulated nanocrystalline copper sample before (**a**) and after (**b**) 10% deformation. The system contains 16 grains and ~100,000 atoms, giving an average grain diameter of 5.2 nm. Atoms in the grain boundaries are coloured blue, atoms at stacking faults are coloured red. We clearly see stacking faults left behind by partial dislocations that have run through the grains during the deformation processes. Such stacking faults would be removed if a second partial dislocation followed the path of the first, but this is not observed in the present simulations. In the left side of the system, a partial dislocation on its way through a grain is seen (green arrow in **b**).

The main part of the simulation is a slow uniaxial deformation while minimizing the energy with respect to all atomic coordinates. The deformation is applied by expanding the simulation cell in one direction, while the size is allowed to relax in the two perpendicular directions.

The initial and final configurations of such a simulation with a total strain of 10% are shown in Fig. 1. We see how the grain boundaries have become thicker, indicating that significant activity has taken place there. In the grains a few stacking faults have appeared.

为了用计算机模拟纳米晶金属的力学行为，我们构建了类似于实验中所观测到的纳米晶"样品"：基本上等轴的无位错的晶粒被窄而直的晶界所分隔[1]。每个样品是一个包含 8～64 个晶粒的立方体，边长为 10.6 nm，因此其晶粒尺寸为 3.3 nm 到 6.6 nm。晶粒是用沃罗诺伊过程[8] 制备的：随机选定一组晶粒中心，在距离一个给定中心比任何其他中心都更近的空间区域中，以具有随机选择的晶体取向的 fcc（面心立方）晶格形式填入原子。一个典型的"样品"如图 1a 所示。为了模拟系统深入块体样品内部的情况，将该系统在全部三个空间方向上无限地复制（周期性边界条件）。原子间作用力用有效介质理论[9,10] 来计算，该理论恰当地描述了金属中多原子的相互作用。本研究模拟的金属是铜；在钯中得到了极为类似的结果。在系统形变之前，我们通过在 300 K 时运行 50 ps 的分子动力学模拟来使其"退火"，使晶界上的不利构型得以松弛。将退火的持续时间加倍没有产生任何显著影响，将温度提高到 600 K 也没有影响。

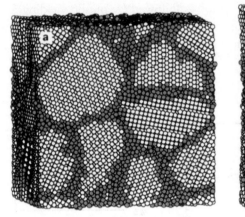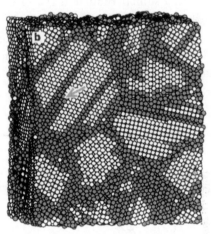

图 1. 在 10% 形变之前（a）和之后（b）的模拟的纳米晶铜样品。体系中包含 16 个晶粒，大约 100,000 个原子，平均晶粒直径为 5.2 nm。晶界上的原子标记为蓝色，堆垛层错处的原子则标记为红色。我们清楚地看到形变过程中穿过晶粒的不全位错留下的堆垛层错。如果有第二个不全位错沿第一个的路径行进，这些堆垛层错将会消失，但在当前模拟过程中并没有观测到这一点。在体系的左边，可以看到一个正在穿过晶粒的不全位错（b 中的绿色箭头）。

模拟的主要部分是在使能量相对于所有原子坐标最小化的前提下进行一次缓慢的单轴形变。形变是通过在一个方向上拉伸模拟晶胞来进行的，同时允许其尺寸在两个垂直方向上弛豫。

图 1 中显示了总应变为 10% 的模拟的初始和最终构型。我们看到晶界是怎样逐渐变厚的，它表明晶界处发生了重要的活动。晶粒中出现了少量的堆垛层错。它们

They are the signature of dislocation activity within the grains.

To facilitate the analysis of the simulations, we identify which atoms are located at grain boundaries and which are inside the grains, by determining the local crystalline order[11,12]. Atoms in local f.c.c. order are considered to be "inside" the grains; atoms in local h.c.p. (hexagonal close-packed) order are classified as stacking faults. All other atoms are considered as belonging to the grain boundaries. Unlike conventional materials, where the volume occupied by the grain boundaries is very small, a significant fraction (30–50%) of the atoms are in the grain boundaries, in agreement with theoretical estimates[2].

As the deformation takes place, we calculate the average stress in the sample as a function of the amount of deformation. For each grain size we simulated the deformation of seven different initial configurations. Fig. 2a shows the obtained average deformation curves. We see a linear elastic region with a Young's modulus around 90–105 GPa (increasing with increasing grain size), compared with 124 GPa in macrocrystalline Cu (ref. 13). This is caused by the large fraction of atoms in the grain boundaries having a lower Young's modulus[14,15]. A similar reduction is seen in simulations where the nanocrystalline metal is grown from a molten phase[16]. The elastic region is followed by plastic yielding at around 1 GPa, and finally the plastic deformation saturates at a maximal flow stress around 3 GPa. The theoretical shear stress of a perfect single crystal is approximately 6 GPa for the potential used.

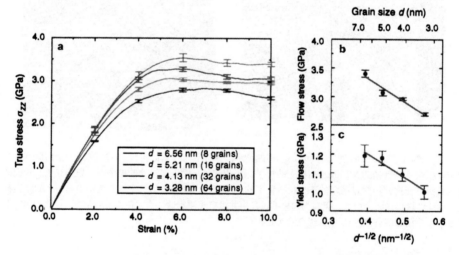

Fig. 2. The effect of grain size on deformation. **a**, The average stress in the direction of the stretch (σ_{zz}) versus strain for each grain size. Each curve is the average over seven simulations. The curves show the response of the material to mechanical deformation. In the linear part of the curve (low strains) the deformation is mainly elastic: if the tensile load is removed, the material will return to the original configuration. As the deformation is increased, irreversible plastic deformation becomes important. For large deformations plastic processes relieve the stress, and the curves level off. We see a clear grain-size dependence, which is summarized to the right. **b** and **c**, The maximal flow stress and the yield stress as a function of grain size. The yield stress decreases with decreasing grain size, resulting in a reverse Hall–Petch effect. (The maximal flow stress is the stress at the flat part of the stress–strain curves; the yield stress is defined as the stress where the strain departs 0.2% from linearity.)

是晶粒内位错活动的标志。

为了改善对模拟过程的分析，我们通过确定局部晶体有序度辨别出哪些原子位于晶界处，哪些处于晶粒内部[11,12]。我们认为具有局部 fcc 晶序的原子是位于晶粒"内部"的，具有局部 hcp(六方密堆积) 晶序的原子则归为堆垛层错。其他所有原子被视为属于晶界。不同于传统材料——其晶界所占据体积很小，所模拟的材料中处于晶界上的原子占有很大比例(30% ~ 50%)，这与理论估计是一致的[2]。

随着形变发生，我们计算了样品中的平均应力随形变量的变化。对于每一晶粒尺寸，我们模拟了七种具有不同初始构型的形变。图 2a 所示为所得到的平均形变曲线。我们看到一个杨氏模量约为 90 ~ 105 GPa(随着晶粒尺寸的增加而增大)的线性弹性区域。与之相比，粗晶铜的杨氏模量为 124 GPa(参考文献 13)。这是由于晶界处的大量原子具有较低的杨氏模量[14,15]。在对从熔融相中生长出来的纳米晶金属的模拟中也看到了类似的模量降低[16]。弹性区域后，塑性屈服出现在 1 GPa 左右，最终塑性形变在流变应力达到最大值——约为 3 GPa 时达到饱和。在所用势能的条件下，完美单晶的理论剪切应力大约为 6 GPa。

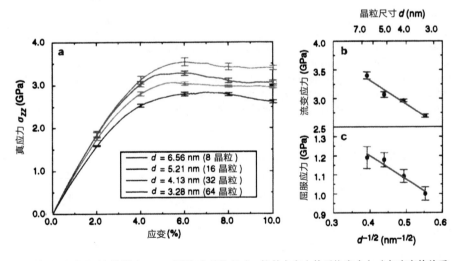

图 2. 晶粒尺寸对于形变的影响。**a**，对于每个晶粒尺寸，拉伸方向上的平均应力(σ_{zz})与应变的关系。每条曲线都是七次模拟过程的平均。曲线表示出材料对于机械形变的反应。曲线的线性部分(低应变)主要是弹性形变：如果将拉伸载荷卸载，材料会恢复其初始构型。随着形变增大，不可逆的塑性形变变得更为重要。对于大的形变，塑性过程减缓了应力，因而曲线变得平稳。我们看到一个明显的晶粒尺寸效应，并将其总结于右侧。**b** 和 **c**，极值流变应力和屈服应力随晶粒尺寸的变化函数。屈服应力随着晶粒尺寸的减小而减小，导致反霍尔-佩奇效应。(极值流变应力是位于应力–应变曲线中平坦部分的应力；屈服应力则定义为应变偏离线性 0.2% 时的应力。)

The main deformation mode is illustrated in Fig. 3b, where the relative motions of the atoms is shown. We see that most of the deformation occurs in the grain boundaries in the form of a large number of small sliding events, where only a few atoms (or sometimes a few tens of atoms) move with respect to each other. Occasionally a partial dislocation is nucleated at a grain boundary and moves through a grain. Such events are responsible for a minor part of the total deformation, but in the absence of diffusion they are required to allow for deformations of the grains, as they slide past each other. No dislocation motion is seen in Fig. 3, as none occurred at that time of the simulation. As the grain size is reduced a larger fraction of the atoms belongs to the grain boundaries, and grain-boundary sliding becomes easier. This leads to a softening of the material as the grain size is reduced (Fig. 2).

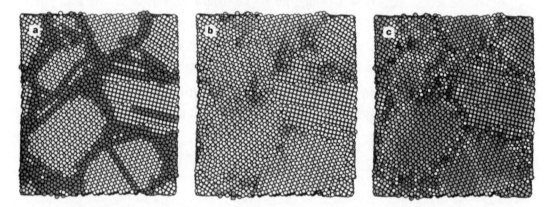

Fig. 3. Snapshot of grain structure, displacements and stresses at 8% deformation. **a**, The position of the grain boundaries (blue) and stacking faults (red) at this point in the simulation. **b**, The relative motion of the atoms in the z direction (up, in the plane of the paper) during the preceding 0.4% deformation. The green atoms move up, red atoms move down. We see many small, independent slip events in the grain boundaries; this is the main deformation mode. **c**, The stress field (the σ_{33} component) in the grains. Shades of red indicate tensile stress, shades of blue compressive stress (dark colours correspond to high stresses). The stress in the grain boundaries is seen to vary considerably on the atomic scale, and the average stress is ~10–20% lower than in the grains.

The observed deformation mode is in some ways similar to the manner in which grain boundaries carry most of the deformation in superplastic deformation[17,18]. This is consistent with recent simulations of flow speed in nanocrystalline metals[25]. However, in superplasticity the grain-boundary sliding is thermally activated, whereas here it occurs at zero temperature driven by the high stress.

In conventional metals an increase in hardness and yield strength with decreasing grain size is observed. This is called the Hall–Petch effect, and is generally considered to be caused by the grain boundaries, impeding the generation and/or motion of dislocations as the grains get smaller; this behaviour extends far into the nanocrystalline regime[3,19]. The grain sizes in the present simulations correspond to the smallest grain sizes that can be obtained experimentally. In that regime the Hall–Petch effect is often seen to cease or even to reverse, but the results depend strongly on the sample history and on the method used to vary the grain size. Many mechanisms have been proposed for this reverse Hall–Petch effect:

图 3b 描绘出主要的形变模式，其中显示了原子的相对运动。我们看到大多数形变发生在晶界处，其形式为大量的只有几个原子(或有时候是几十个原子)相对彼此运动的微小滑移事件。偶尔会有不全位错在晶界成核，并且穿过一个晶粒。这些情况只是总形变中一小部分形变的原因，但是在没有扩散存在时，需要有这种微小相互滑移才能使晶粒得以发生形变。图 3 中未看到位错运动，因为在模拟时没有发生。随着晶粒尺寸减小，晶界处的原子分数越来越大，从而使晶界的滑动变得更加容易。这导致了晶粒尺寸减小时材料的软化(图 2)。

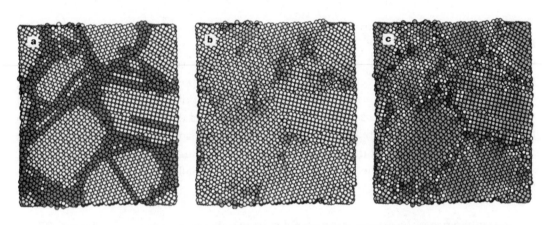

图 3. 形变 8% 时，晶粒结构、位移和应力的快照。**a**，模拟过程中该点处的晶界(蓝色)和堆垛层错(红色)的位置。**b**，在形变的前 0.4% 过程中，原子在 z 方向上(向上，纸平面中)的相对运动。绿色原子向上运动，红色原子向下运动。我们看到晶界中有很多微小独立滑动，这是主要的形变模式。**c**，晶粒中的应力场(σ_{33} 分力)。红色阴影表示拉伸应力，蓝色阴影表示压缩应力(深色对应于高应力)。可以看到，晶界处的应力在原子尺度上有相当的变化，且平均应力比晶粒内部的约低 10% ~ 20%。

在某种意义上，上述已观察到的形变模式类似于超塑性形变中晶界承担大部分形变的方式 [17,18]。这与最近对于纳米晶金属中流变速度的模拟结果 [25] 是一致的。但是在超塑性中，晶界滑动是热激活的，而在本文中，它发生在零度，受高应力所驱动。

在普通金属中，硬度和屈服强度随着晶粒尺寸减小而增加，这就是所谓的霍尔-佩奇效应。一般认为，强度增加是由于随着晶粒逐渐减小，晶界阻碍了位错的产生和(或)运动；这种行为可以推广到纳米晶体材料体系 [3,19]。当前模拟过程中的晶粒尺寸相当于用实验方法所能获得的最小晶粒尺寸。在这个范围内，经常可以看到霍尔-佩奇效应失效甚至反转，但是其结果强烈地依赖于样品的制备历史和改变晶粒尺寸的方法。对于这种反霍尔-佩奇效应已经提出了很多种机制：小晶粒尺寸

increased porosity at small grain sizes[3], suppression of dislocation pile-ups[20], dislocation motion through multiple grains[21], sliding in the grain boundaries[22], and enhanced diffusional creep in the grain boundaries[7]. Direct measurements of the creep rates seem to rule out the last mechanism[19,20], but otherwise no consensus has been reached. The present simulations indicate that the behaviour characteristic of a reverse Hall–Petch effect is possible even in the absence of porosity, and that it may be caused by sliding in the grain boundaries even in the absence of thermally activated processes. We cannot, however, see a cross-over from this "reverse" behaviour to the normal Hall–Petch regime at larger grain sizes in our simulations, because they become too computationally expensive at these larger sizes. For the same reason, we cannot provide a direct comparison with the behaviour of the bulk metal.

A direct quantitative comparison between the simulations and the experimental results should be done with caution. The main difference between the simulated strain–stress curves and experimental curves is the level of the yield stress, which is approximately twice what is observed in experiments on low-porosity samples (400 MPa)[23]. This value is, however, obtained for a grain size of around 40 nm; extrapolation to 7-nm grains gives a yield strength around 800 MPa (ref. 23), assuming that the Hall–Petch behaviour persists to these grain sizes. Experimentally produced nanocrystalline samples typically contain voids and surface defects, reducing the strength of the material. Surface defects alone have been shown to be able to reduce the strength of nanocrystalline palladium by at least a factor of five[19,24].

Another difference is the absence of thermally activated processes in the simulations. These processes give rise to a strain-rate dependence of the mechanical properties, leading to higher yield stresses at higher strain rates where there is less time available for the activated processes to occur. Thermally activated processes are not included in the simulations because the energy minimization procedure quickly removes all thermal energy. No timescale or strain rate can be directly defined in the simulations, but in a sense the procedure corresponds to a slow strain at very low temperatures: thermally activated processes are excluded but the energy created by the work is carried away fast. The above-mentioned creep measurements[19,20] indicate that diffusion does not play a major role during deformation.

During the later part of the simulated deformation larger average stresses build up within the grains than in the grain boundaries (10–20%), and larger stresses build up in the larger grains (see Fig. 3c). This results in a larger variation in the maximal flow stress than in the yield stress (Fig. 2b): when the grain size is increased the maximal flow stress increases, both because the stresses in the grains increase and because the number of atoms within the grains becomes a larger fraction of the total number of atoms.

(**391**, 561-563; 1998)

时大量增加的孔洞 [3]、位错堆积的抑制作用 [20]、穿过多晶粒的位错运动 [21]、晶界中的滑移 [22]，以及晶界处增加的扩散蠕变 [7] 等。对于蠕变速率的直接测量似乎排除了最后一种机制 [19,20]，但除此之外再未能达成任何共识。目前的模拟过程指出，反霍尔–佩奇效应所特有的行为特征甚至可能在缺乏孔洞时出现，并且它甚至可能是在没有热激活的条件下由晶界中的滑移造成的。然而，在我们的模拟过程中，并没有看到从这一"反常"行为向较大晶粒尺寸时的正常霍尔–佩奇情况的过渡，因为对于计算来说，对这些较大尺寸进行模拟的代价太大了。出于同样的理由，我们未能提供与块体金属行为的直接比较。

应该谨慎地直接定量比较模拟与实验的结果。模拟应变–应力曲线与实验所得曲线之间的主要差别是屈服应力水平，模拟值大约是低孔隙率样品实验所测数据（400 MPa）[23] 的两倍。不过，400 MPa 这个数值是从晶粒尺寸约 40 nm 的样品得到的；外推到晶粒 7 nm 时，其屈服强度约为 800 MPa（参考文献 23）（假定霍尔–佩奇效应对于这些晶粒尺寸仍然适用）。实验中制备的纳米晶样品中通常包含孔隙和表面缺陷，它们减弱了材料的强度。已经知道仅是表面缺陷就能使纳米晶钯的强度减小为原来的五分之一或更低 [19,24]。

另一个差别是在模拟中没有热激活过程。这种热激活造成力学性能对应变速率的依赖性，导致在较高应变速率时有较高的屈服应力，因为可供激活过程发生的时间变少了。模拟中不包括热激活过程是因为能量最小化过程很快地移除了所有热能。模拟中无法直接定义任何时间尺度或应变速率，但是在某种意义上，该程序对应于极低温度时的缓慢应变：热激活过程被排除，但是变形产生的能量被快速带走。前面提到过的蠕变测量实验 [19,20] 表明扩散在形变过程中并不起主要作用。

在模拟形变的后期，晶粒内部积累了比晶界处更大的平均应力（10% ~ 20%），而且晶粒越大积累的应力越大（参见图 3c）。这使得最大流变应力比屈服应力有更大的波动（图 2b）：当晶粒尺寸增加时，最大流变应力增大，这既是由于晶粒中的应力增大，也是由于晶粒内部原子的数量在总原子数量中所占比例增加。

（王耀杨 翻译；卢磊 审稿）

55

Jakob Schiøtz, Francesco D. Di Tolla* & Karsten W. Jacobsen
Center for Atomic-scale Materials Physics and Department of Physics, Technical University of Denmark, DK2800 Lyngby, Denmark
* Present address: SISSA, Via Beirut 2-4, I-34014 Grignano (TS), Italy.

Received 23 June; accepted 29 October 1997.

References:

1. Siegel, R. W. in *Encyclopedia of Applied Physics* Vol. 11, 173-199 (VCH, New York, 1994).

2. Siegel, R. W. What do we really know about the atomic-scale structure of nanophase materials? *J. Phys. Chem. Solids* **55,** 1097-1106 (1994).

3. Siegel, R. W. & Fougere, G. E. in *Nanophase Materials: Synthesis—Properties—Applications* (eds Hadjipanayis, G. C. & Siegel, R. W.) 233-261 (NATO-ASI Ser. E, Vol. 260, Kluwer, Dordrecht, 1994).

4. Gleiter, H. in *Mechanical Properties and Deformation Behavior of Materials Having Ultra-Fine Microstructures* (ed. Nastasi, M.) 3-35 (Kluwer, Dordrecht, 1993).

5. Hall, E. O. The deformation and ageing of mild steel: III Discussion of results. *Proc. Phys. Soc. Lond. B* **64,** 747-753 (1951).

6. Petch, N. J. The cleavage of polycrystals. *J. Iron Steel Inst.* **174,** 25-28 (1953).

7. Chokshi, A. H., Rosen, A., Karch, J. & Gleiter, H. On the validity of the Hall–Petch relationship in nanocrystalline materials. *Scripta Metall.* **23,** 1679-1684 (1989).

8. Voronoi, G. *Z. Reine Angew. Math.* **134,** 199 (1908).

9. Jacobsen, K. W., Nørskov, J. K. & Puska, M. J. Interatomic interactions in the effective-medium theory. *Phys. Rev. B* **35,** 7423-7442 (1987).

10. Jacobsen, K. W., Stoltze, P. & Nørskov, J. K. A semi-empirical effective medium theory for metals and alloys. *Surf. Sci.* **366,** 394-402 (1996).

11. Jónsson, H. & Andersen, H. C. Icosahedral ordering in the Lennard–Jones liquid and glass. *Phys. Rev. Lett.* **60,** 2295-2298 (1988).

12. Clarke, A. S. & Jónsson, H. Structural changes accompanying densification of random-sphere packings. *Phys. Rev. E* **47,** 3975-3984 (1993).

13. Gschneidner, K. A. Physical properties and interrelationships of metallic and semimetallic elements. *Solid State Phys.* **16,** 275-426 (1964).

14. Shen, T. D., Koch, C. C., Tsui, T. Y. & Pharr, G. M. On the elastic moduli of nanocrystalline Fe, Cu, Ni, and Cu-Ni alloys prepared by mechanical milling/alloying. *J. Mater. Res.* **10,** 2892-2896 (1995).

15. Kluge, M. D., Wolf, D., Lutsko, J. F. & Phillpot, S. R. Formalism for the calculation of local elastic constants at grain boundaries by means of atomistic simulation. *J. Appl. Phys.* **67,** 2370-2379 (1990).

16. Phillpot, S. R., Wolf, D. & Gleiter, H. Molecular-dynamics study of the synthesis and characterization of a fully dense, three-dimensional nanocrystalline material. *J. Appl. Phys.* **78,** 847-860 (1995).

17. Chokshi, A. H., Mukherjee, A. K. & Langdon, T. G. Superplasticity in advanced materials. *Mater. Sci. Eng. R* **10,** 237-274 (1993).

18. Ridley, N. (ed.) *Superplasticity: 60 years after Pearson* (Institute of Metals, London, 1995).

19. Nieman, G. W., Weertman, J. R. & Siegel, R. W. Mechanical behavior of nanocrystalline Cu and Pd. *J. Mater. Res.* **6,** 1012-1027 (1991).

20. Nieh, T. G. & Wadsworth, J. Hall–Petch relation in nanocrystalline solids. *Scripta Met. Mater.* **25,** 955- 958 (1991).

21. Lian, J., Baudelet, B. & Nazarov, A. A. Model for the prediction of the mechanical behaviour of nanocrystalline materials. *Mater. Sci. Eng. A* **172,** 23-29 (1993).

22. Langdon, T. G. The significance of grain boundaries in the flow of polycrystalline materials. *Mater. Sci. Forum* **189-190,** 31-42 (1995).

23. Suryanarayanan, R. *et al.* Mechanical properties of nanocrystalline copper produced by solution-phase synthesis. *J. Mater. Res.* **11,** 439-448 (1996).

24. Weertman, J. R. Hall–Petch strengthening in nanocrystalline metals. *Mater. Sci. Eng. A* **166,** 161-167 (1993).

25. Van Swygenhoven, H. & Caro, A. Plastic behavior of nanophase Ni: A molecular dynamics computer simulation. *Appl. Phys. Lett.* **71,** 1652-1654 (1997).

Acknowledgements. We thank J. K. Nørskov, T. Leffers, O. B. Pedersen, A. E. Carlsson and J. P. Sethna for discussions. The Center for Atomic-scale Materials Physics is sponsored by the Danish National Research Foundation.

Correspondence and requests for materials should be addressed to J.S. (e-mail: schiotz@fysik.dtu.dk).

Extraordinary Optical Transmission through Sub-wavelength Hole Arrays

T. W. Ebbesen *et al.*

Editor's Note

Light rays are generally considered to be blocked by apertures smaller than the light's wavelength. This limits the resolution of optical microscopes. But here Thomas Ebbesen of the NEC Corporation in Princeton and colleagues report the surprising ability of light to "squeeze through" an array of sub-wavelength-sized holes in a thin metal sheet. The effect stems from the interaction of light with mobile electrons on the surface of the metal, which can be excited into wavelike motions called plasmons. A plasmon wave can pass through the hole and then re-radiate light on the far side. This paper stimulated interest in the topic now called plasmonics, which takes advantage of light-plasmon applications to achieve a range of unusual effects, including "invisibility shields".

The desire to use and control photons in a manner analogous to the control of electrons in solids has inspired great interest in such topics as the localization of light, microcavity quantum electrodynamics and near-field optics[1-6]. A fundamental constraint in manipulating light is the extremely low transmittivity of apertures smaller than the wavelength of the incident photon. While exploring the optical properties of submicrometre cylindrical cavities in metallic films, we have found that arrays of such holes display highly unusual zero-order transmission spectra (where the incident and detected light are collinear) at wavelengths larger than the array period, beyond which no diffraction occurs. In particular, sharp peaks in transmission are observed at wavelengths as large as ten times the diameter of the cylinders. At these maxima the transmission efficiency can exceed unity (when normalized to the area of the holes), which is orders of magnitude greater than predicted by standard aperture theory. Our experiments provide evidence that these unusual optical properties are due to the coupling of light with plasmons—electronic excitations—on the surface of the periodically patterned metal film. Measurements of transmission as a function of the incident light angle result in a photonic band diagram. These findings may find application in novel photonic devices.

A variety of two-dimensional arrays of cylindrical cavities in metallic films were prepared and analysed for this study. Typically, a silver film of thickness $t = 0.2$ μm was first deposited by evaporation on a quartz substrate. Arrays of cylindrical holes were fabricated through the film by sputtering using a Micrion focused-ion-beam (FIB) System 9500 (50 keV Ga ions, 5 nm nominal spot diameter). The individual hole diameter d was varied between 150 nm and 1 μm and the spacing between the holes (that is, the periodicity) a_0, was between 0.6 and 1.8 μm. The zero-order transmission spectra, where the incident

通过亚波长孔洞阵列的超常光透射现象

埃贝森等

编者按

一般认为，当孔径小于光束波长时，光线将被阻碍，从而使光学显微镜的分辨率受到限制。但在本文中，普林斯顿日本电气公司研究所的托马斯·埃贝森及其同事报道了光具有"挤过"金属薄膜中亚波长尺寸孔洞阵列的惊人能力。该效应源于光与金属表面可迁移电子间的相互作用，使其激发成为与等离子体激元类似的波的运动。等离子体激元波可以通过孔洞，进而在远端发生光的再辐射。本文诱发了对目前名为等离子体光子学领域的兴趣，该领域旨在利用光−等离子体激元来实现包括"隐形盾"在内的一系列不寻常的效应。

通过与控制固体中电子类似的方式来应用和控制光子的愿望，激起了人们对于光定域化、微腔量子电动力学和近场光学等主题的强烈兴趣[1-6]。调制光时遇到的基本限制就是当孔隙小于入射光子波长时透射率将非常低。但是，在探索金属薄膜中亚微米级圆柱形谐振腔的光学性质时，我们发现，当波长大于阵列周期时，这些孔洞阵列呈现出极不寻常的零级透射光谱（入射光与所探测的光二者共线），并且没有发生衍射。更为特别的是，在波长为圆柱直径的10倍处观测到尖锐的透射峰。而且在极大值处透射率可以超过1（按孔面积归一化后），这比标准孔隙理论的预期值高出几个数量级。我们的实验为下列说法提供了证据：这些不寻常的光学性质源自有周期性排列孔洞的金属薄膜表面上光与等离子体激元的耦合——电子激发作用。以透射作为光线入射角函数的观测获得了光子带图。这些发现可能会在新型光子器件中获得应用。

在这项研究中，我们制备了具有多种二维阵列的圆柱形谐振腔的金属薄膜，并对实验结果进行了分析。典型的一例是，首先通过蒸镀在石英基质上沉积一层 $t = 0.2$ μm 厚的银膜。继而用 Micrion 聚焦离子束系统 9500(50 keV 镓离子，5 nm 标称点直径)进行溅镀，并在薄膜上加工出圆柱形孔洞阵列。单个孔直径 d 在 150 nm 到 1 μm 之间，孔洞之间的距离 a_0(亦即周期)则在 0.6 μm 和 1.8 μm 之间。零级透射

and detected light are collinear, were recorded with a Cary 5 ultraviolet–near infrared spectrophotometer with an incoherent light source, but the arrays were also studied on an optical bench for transmission, diffraction and reflection properties using coherent sources.

Figure 1 shows a typical zero-order transmission spectrum for a square array of 150 nm holes with a period a_0 of 0.9 μm in a 200 nm thick Ag film. The spectrum shows a number of distinct features. At wavelength $\lambda = 326$ nm the narrow bulk silver plasmon peak is observed which disappears as the film becomes thicker. The most remarkable part is the set of peaks which become gradually stronger at longer wavelengths, increasingly so even beyond the minimum at the periodicity a_0. There is an additional minimum at $\lambda = a_0 \sqrt{\varepsilon}$ corresponding to the metal–quartz interface (where ε is the dielectric constant of the substrate). For $\lambda > a_0 \sqrt{\varepsilon}$, there is no diffraction from the array nor from the individual holes. As expected, the first-order diffraction spots can be seen to be grazing the surface (that is, the diffraction angle approaches 90°) as the wavelength approaches the period from below (this might, in fact, enhance the coupling to be discussed in the next paragraphs). The maximum transmitted intensity occurs at 1,370 nm, nearly ten times the diameter of an individual hole in the array. Even more surprising is that the absolute transmission efficiency, calculated by dividing the fraction of light transmitted by the fraction of surface area occupied by the holes, is ≥ 2 at the maxima. In other words, more than twice as much light is transmitted as impinges directly on the holes. Furthermore, the transmittivity of the array scales linearly with the surface area of the holes. This is all the more remarkable considering that the transmission efficiency of a single sub-wavelength aperture is predicted by Bethe[7] to scale as $(r/\lambda)^4$ where r is the hole radius; accordingly for a hole of 150 nm diameter one expects a transmission efficiency on the order of 10^{-3}. In addition, the intensity (I) of the zero-order transmission from a grating is expected to decrease monotonically at larger wavelengths ($I \propto \lambda^{-1}$) (ref. 8). Therefore our results must imply that the array itself is an active element, not just a passive geometrical object in the path of the incident beam.

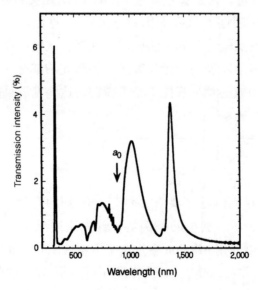

Fig. 1. Zero-order transmission spectrum of an Ag array ($a_0 = 0.9$ μm, $d = 150$ nm, $t = 200$ nm).

光谱——其中入射光和所探测的光共线——是用一台具有非相干光源的 Cary 5 型紫外–近红外分光光度计记录的，也用相干光源对光具座上的阵列就透射、衍射和反射性质进行了研究。

图 1 显示了在 200 nm 厚的银膜中，周期 a_0 为 0.9 μm 的 150 nm 孔洞方形阵列典型零级透射光谱。光谱呈现出很多不同的特征。在波长 λ = 326 nm 处可以看到本底银狭窄的等离子体峰，当薄膜厚度增加时它会消失。最值得关注的部分是在波长较长处逐渐变强的一组峰，甚至增加到超过位于周期 a_0 处的最小值。在 $λ = a_0\sqrt{ε}$ 处有一个额外的最小值，它对应于金属–石英界面（其中 ε 是基质的介电常数）。在 $λ > a_0\sqrt{ε}$ 处，阵列和单个孔洞都没有产生衍射。如同预期的那样，当波长由低处趋近周期值时，可以看到一级衍射点随之掠向表面（即衍射角接近 90°）的现象（实际上，它可能增强我们在下面段落中将要讨论到的耦合作用）。最大透射强度出现在 1,370 nm 处，这一波长大约是阵列中单独一个孔洞的直径的 10 倍。更令人吃惊的是绝对透射率——由光的透射部分除以孔洞占据的表面积即可算出——在最大处 ≥ 2。换句话说，透射出的光强比直接撞击孔洞的光强要高出一倍以上。此外，阵列的透射率随着孔洞表面积呈线性变化。如果考虑到贝特[7] 关于单个亚波长孔道透射率与 $(r/λ)^4$ 成比例的推断——其中 r 是孔洞半径，这一点就更加值得注意；据此推断，对于直径 150 nm 的孔洞来说，可以预期具有 10^{-3} 数量级的透射率。另外，预期光栅产生的零级透射强度（I）会在较大波长（$I \propto λ^{-1}$）处单调地下降（参考文献 8）。因此我们的结果必然意味着阵列本身就是一个有源元件，而不只是入射光束路径上的一个无源几何体。

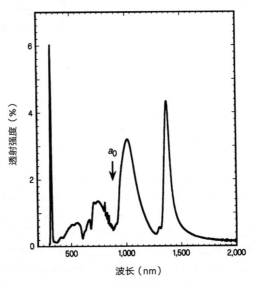

图 1. 一个银阵列（a_0 = 0.9 μm，d = 150 nm，t = 200 nm）的零级透射光谱

To understand the origin of this phenomenon, we tested the dependence on all the possible variables such as hole diameter, periodicity, thickness and type of metal. It is beyond the scope of this Letter to describe the details of all these experiments. Instead we summarize our observations and illustrate some of the key factors that determine the shape of these spectra. To begin with, the periodicity of the array determines the position of the peaks. The positions of the maxima scale exactly with the periodicity, as shown in Fig. 2a, independent of metal (Ag, Cr, Au), hole diameter and film thickness. The width of the peaks appears to be strongly dependent on the aspect ratio (t/d or depth divided by diameter) of the cylindrical holes (Fig. 2a). For $t/d = 0.2$, the peaks are very broad and just discernible and when the ratio reaches ~1, the maximum sharpness is obtained. Further narrowing might depend on the quality of the individual holes. The thickness dependence of the spectra is displayed in Fig. 2b for 0.2 and 0.5 μm Ag films. While the intensity of the bulk plasmon peak decreases rapidly in this range, that of the longer-wavelength peaks decreases approximately linearly with thickness. The spectra change significantly with the type of lattice, for example whether the array is a square or a triangular lattice.

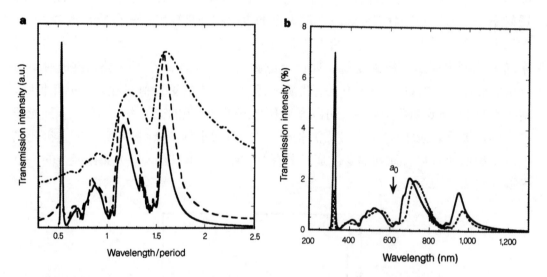

Fig. 2. Effects of parameters on zero-order transmission spectra. **a**, Spectra for various square arrays as a function of λ/a_0. Solid line: Ag, $a_0 = 0.6$ μm, $d = 150$ nm, $t = 200$ nm; dashed line: Au, $a_0 = 1.0$ μm, $d = 350$ nm, $t = 300$ nm; dashed-dotted line: Cr, $a_0 = 1.0$ μm, $d = 500$ nm, $t = 100$ nm. **b**, Spectra for two identical Ag arrays with different thicknesses. Solid line: $t = 200$ nm; dashed line: $t = 500$ nm (this spectrum has been multiplied by 1.75 for comparison). For both arrays: $a_0 = 0.6$ μm; $d = 150$ nm.

Two important clues relating this phenomenon to surface plasmons (SPs) come from the following observations. One is the absence of enhanced transmission in hole arrays fabricated in Ge films which points to the importance of the metallic film. The other clue is the angular dependence of the spectra in metallic samples. The zero-order transmission spectra change in a marked way even for very small angles, as illustrated in Fig. 3 where the spectra were recorded every 2°. The peaks change in intensity and split into new peaks which move in opposite directions. This is exactly the behaviour observed when light couples with SPs in reflection gratings[9-14]. SPs are oscillations of surface charges at the metal

为理解这种现象的来由，我们检测了它对于诸如孔洞直径、周期、金属膜厚度和类型等所有可能变量的依赖性。由于本快报篇幅有限，无法描述所有实验的细节，因此我们总结了观察结果并说明了决定这些光谱的形状的一些关键因素。首先，阵列的周期决定了峰的位置。如图 2a 所示，最大值的位置严格地与周期成比例，而不依赖于所用金属（银、铬、金）、孔洞直径和薄膜厚度。峰的宽度显然强烈地依赖于圆柱形谐振腔的纵横比（t/d，或厚度除以直径）（图 2a）。当 $t/d = 0.2$ 时，峰很宽而且刚刚可以分辨出来，而当该比值达到约 1 时具有最高清晰度。进一步的变窄可能和单个孔洞的性质有关。图 2b 中显示出光谱对于薄膜厚度的依赖性，其中银膜厚度分别为 0.2 μm 和 0.5 μm。本底等离子体激元峰的强度在这个范围内快速下降，而较长波长处的峰强度随着厚度变化大致呈线性下降。光谱随着栅格类型——例如阵列究竟是方形还是三角形栅格——发生显著变化。

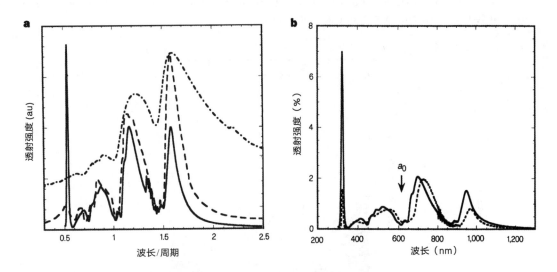

图 2. 参数对于零级透射光谱的影响。**a**，作为 λ/a_0 的函数的各种方形阵列的光谱。实线：银，$a_0 = 0.6$ μm，$d = 150$ nm，$t = 200$ nm；虚线：金，$a_0 = 1.0$ μm，$d = 350$ nm，$t = 300$ nm；点划线：铬，$a_0 = 1.0$ μm，$d = 500$ nm，$t = 100$ nm。**b**，只有厚度不同的两个相同银阵列的光谱。实线：$t = 200$ nm；虚线：$t = 500$ nm（为便于比较，此光谱已乘以倍数 1.75）。对两个阵列同时有：$a_0 = 0.6$ μm；$d = 150$ nm。

将这种现象与表面等离子体激元（SP）联系起来的两条重要线索来自下面的观测结果。其一是，由锗薄膜加工制成的孔洞阵列中不存在透射增强现象，表明了薄膜金属的重要性。另一条线索是在金属样品中光谱的角度依赖性。如图 3 所示，即使角度变化很小时，零级透射光谱也会发生明显的变化。图 3 是角度间隔为 2°时记录下来的光谱。谱峰强度发生变化，并且分裂成向相反方向移动的新谱峰。这正是光与反射光栅中表面等离子体激元耦合时观测到的行为 [9-14]。表面等离子体激元是金属

interface and are excited when their momentum matches the momentum of the incident photon and the grating as follows:

$$k_{sp} = k_x \pm nG_x \pm mG_y$$

where k_{sp} is the surface plasmon wavevector, $k_x = (2\pi/\lambda)\sin\theta$ is the component of the incident photon's wavevector in the plane of the grating and $G_x = G_y = 2\pi/a_0$ are the grating momentum wavevectors for a square array. Therefore if the angle of incidence θ is varied, the incident radiation excites different SP modes and by recording the peak energies as a function of k_x we obtain the dispersion relation shown in Fig. 4. This figure reveals the band structure of SP in the two-dimensional array and it clearly demonstrates the presence of gaps with energies around 30 to 50 meV; these are due to the lifting of the degeneracy by SPs interacting with the lattice. An extrapolation of the dispersion curves yields an intercept with the k-axis at a value which is within the experimental error ($\sim 10\%$) of that expected from the periodicity ($2\pi/a_0$). The results of Figs 3 and 4 are also sensitive to polarization as expected, but a clear assignment has not yet been possible because of inherent structure of the grains in the metal film.

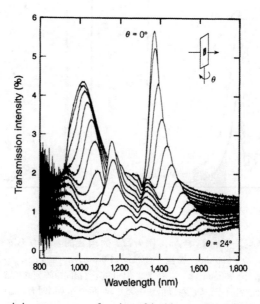

Fig. 3. Zero-order transmission spectra as a function of incident angle of the light. Spectra were taken every 2° up to 24° for a square Ag array ($a_0 = 0.9\ \mu m$, $d = 150$ nm, $t = 200$ nm). The individual spectra are offset vertically by 0.1% from one another for clarity.

界面上表面电荷的振荡，当它们的动量与入射光子以及光栅的动量满足如下表达式时将受到激发：

$$k_{sp} = k_x \pm nG_x \pm mG_y$$

其中 k_{sp} 是表面等离子体激元波矢，$k_x = (2\pi/\lambda)\sin\theta$ 是入射光子波矢在光栅平面内的分量，而 $G_x = G_y = 2\pi/a_0$ 是方形阵列的光栅动量波矢。因此，如果入射角 θ 变化，入射辐射将激发不同的表面等离子体激元模式；通过记录作为 k_x 的函数的峰值能量，我们得到了图 4 中所显示的色散关系。这幅图揭示了二维阵列中表面等离子体激元的带结构，而且它清楚地阐明了具有约 30 meV 到 50 meV 能量的带隙的存在；它们来源于表面等离子体激元与栅格相互作用导致的简并提升。色散曲线的外推得到在 k 轴上的截距，其数值在依据周期 $(2\pi/a_0)$ 预期的实验误差（约 10%）之内。如同预期，图 3 和图 4 中的结果也对极化很敏感，但是由于金属薄膜中晶粒内在结构间的差异，目前还不能做出明确的归属。

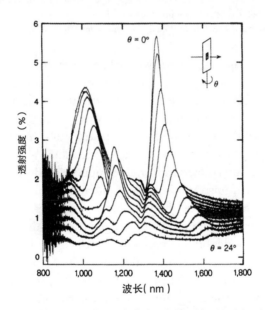

图 3. 作为光线入射角函数的零级透射光谱。对一个方形银阵列每隔 2°测一次谱图，一直测到 24°（$a_0 = 0.9\ \mu m$，$d = 150\ nm$，$t = 200\ nm$）。为清晰起见，各个谱之间在垂直方向上错开 0.1%。

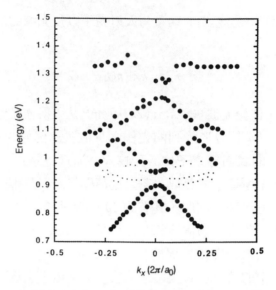

Fig. 4. Dispersion curves (solid circles) along the [10] direction of the array. These curves are extracted from the energy of the transmission peaks of Fig. 3. The momentum k_x is in the plane of the array and is given by $k_x = (2\pi/\lambda) \sin\theta$ where θ is the incident angle of the beam (the units are normalized to $2\pi/a_0$). The curves with the smaller dots correspond to peaks whose amplitudes are much weaker and may or may not be related to the band structure as they do not shift significantly with momentum.

In our experiments the coupling of light with SPs is observed in transmission rather than reflection, in contrast with previous work on SPs on reflection gratings. In those studies the coupling of light to SPs is observed as a redistribution of intensity between different diffracted orders. Even in transmission studies of wire gratings by Lochbihler[15], the effect of SPs is observed through dips in zero-order transmitted light. There is an extensive literature on the infrared properties of wire grids which show a broad transmission centred at $\lambda \approx 1.2a_0$; this has been interpreted in terms of induction effects, in analogy with electric circuits[16,17]. Our results demonstrate the strong enhancement of transmitted light due to coupling of the light with the SP of the two-dimensional array of sub-wavelength holes. Furthermore our results indicate a number of unique features that cannot be explained with existing theories. The SP modes on the metal–air interface are distinctly different from those at the metal–quartz interface. However, the spectra are identical regardless of whether the sample is illuminated from the metal or quartz side. At present we do not understand the detailed mechanism of the coupling between the SP on the front and back surfaces which results in larger than unity transmission efficiency of the holes. The thickness dependence (see Fig. 2b) suggests that the holes play an important part in mediating this coupling and that nonradiative SP modes are transferred to radiative modes by strong scattering in the holes.

In photonic bandgap arrays[2,4], the material is passive and translucent at all wavelengths except at the energies within the gap. In the present arrays, the material plays an active role (through the plasmons) and it is opaque at all wavelengths except those for which coupling occurs. The combination, or integration, of these two types of phenomena might lead to

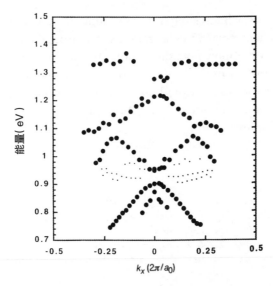

图 4. 沿阵列 [10] 方向的色散曲线（实心圆点）。这些曲线是利用图 3 中透射峰的能量得到的。动量 k_x 位于阵列平面内，由 $k_x = (2\pi/\lambda)\sin\theta$ 给出，其中 θ 是光束的入射角（单位归一化为 $2\pi/a_0$）。由较小的点组成的曲线所对应的峰，其幅度要弱很多，由于随动量变化发生的偏移并不明显，因此与带结构的关联不能确定。

在我们的实验中，光与表面等离子体激元的耦合是在透射中而不是在反射中观测到的，这与此前对反射光栅上的表面等离子体激元的研究相反。在那些研究中，光与表面等离子体激元的耦合是作为强度在不同衍射级之间的重新分配观测到的。即使是在洛赫比勒对线光栅的透射研究中 [15]，表面等离子体激元的影响也是通过零级透射光中的凹陷观测到的。关于线栅的红外性质，已经有大量文献表明它有一个以 $\lambda \approx 1.2a_0$ 为中心的宽透射；可以用类似于电路的感应效应对此进行解释 [16,17]。我们的结果表明，由于光与亚波长孔洞二维阵列的表面等离子体激元耦合，透射光将明显增强。此外，我们的结果还指出了很多用现有理论无法解释的独特性状。位于金属–空气界面上的表面等离子体激元模式明显不同于位于金属–石英界面上的。但是，无论是从金属一侧还是石英一侧对样品进行光照，所得光谱都是一样的。目前，我们还不了解导致孔洞透射效率大于 1 的前后表面上等离子体激元之间耦合的详细机制。厚度依赖性（参见图 2b）意味着孔洞对于调制这类耦合起着重要的作用，也意味着通过孔洞中的强烈散射，非辐射表面等离子体激元模式将转化为辐射模式。

在光子带隙阵列中 [2,4]，材料在除去能隙中的能量之外的所有波长处都是无源、半透明的。在当前的阵列中，材料起到有源的作用（通过等离子体激元），并且在除去发生耦合的那些波长之外的所有波长处都是不透明的。这两种现象的结合或整合

optical features that are very interesting from both fundamental and technological points of view. The demonstration of efficient light transmission through holes much smaller than the wavelength and beyond the inter-hole diffraction limit might, for example, inspire designs for novel nearfield scanning optical microscopes[6], or sub-wavelength photolithography. Theoretical analysis of the results would also be useful for gaining better insight into this extraordinary transmission phenomenon. Perhaps only then can we expect to grasp the full implications of these findings.

(**391**, 667-669; 1998)

T. W. Ebbesen[*†], H. J. Lezec[‡], H. F. Ghaemi[*], T. Thio[*] & P. A. Wolff[*§]

[*] NEC Research Institute, 4 Independence Way, Princeton, New Jersey 08540, USA

[†] ISIS, Louis Pasteur University, 67000 Strasbourg, France

[‡] Micrion Europe GmbH, Kirchenstraße 2, 85622 Feldkirchen, Germany

[§] Department of Physics, Massachusetts Institute of Technology, Cambridge, Massachusetts 02139, USA

Received 15 July; accepted 24 November 1997.

References:

1. John, S. Localization of light. *Phys. Today* 32 (May 1991).

2. Yablonovitch, E. & Leung, K. M. Hope for photonic bandgaps. *Nature* **351**, 278 (1991).

3. Dalichaouch, R., Armstrong, J. P., Schultz, S., Platzman, P. M. & McCall, S. L. Microwave localization by two-dimensional random scattering. *Nature* **354**, 53-55 (1991).

4. J. D. Joannopoulos, Meade R. D. & Winn, J. N. *Photonic Crystals* (Princeton Univ. Press, Princeton, 1995).

5. Haroche, S. & Kleppner, D. Cavity quantum electrodynamics. *Phys. Today* **24** (January 1989).

6. Betzig, E. & Trautman, J. K. Near-field optics: Microscopy, spectroscopy, and surface modification beyond the diffraction limit. *Science* **257**, 189-194 (1992).

7. Bethe, H. A. Theory of diffraction by small holes. *Phys. Rev.* **66**, 163-182 (1944).

8. Born, M. & Wolf, E. *Principles of Optics* (Pergamon, Oxford, 1980).

9. Ritchie, R. H., Arakawa, E. T., Cowan, J. J. & Hamm, R. N. Surface-plasmon resonance effect in grating diffraction. *Phys. Rev. Lett.* **21**, 1530-1533 (1968).

10. Raether, H. *Surface Plasmons* (Springer, Berlin, 1988).

11. Chen, Y. J., Koteles, E. S., Seymour, R. J., Sonek, G. J. & Ballantyne, J. M. Surface plasmons on gratings: coupling in the minigap regions. *Solid State Commun.* **46**, 95-99 (1983).

12. Kitson, S. C., Barnes, W. L. & Sambles, J. R. Full photonic band gap for surface modes in the visible. *Phys. Rev. Lett.* **77**, 2670-2673 (1996).

13. Watts, R. A., Harris, J. B., Hibbins, A. P., Preist, T. W. & Sambles, J. R. Optical excitations of surface plasmon polaritons on 90 and 60 bi-gratings. *J. Mod. Opt.* **43**, 1351-1360 (1996).

14. Derrick, G. H., McPhedran, R. C., Maystre, D. & Neviere, M. Crossed gratings: a theory and its applications. *Appl. Phys.* **18**, 39-52 (1979).

15. Lochbihler, H. Surface polaritons on gold-wire gratings. *Phys. Rev. B* **50**, 4795-4801 (1994).

16. Ulrich, R. Far-infrared properties of metallic mesh and its complimentary structure. *Infrared Phys.* **7**, 37-55 (1967).

17. Larsen, T. A survey of the theory of wire grids. *I.R.E. Trans. Microwave Theory Techniques* **10**, 191-201 (1962).

Acknowledgements. We thank S. Kishida, G. Bugmann and J. Giordmaine for their encouragement, and R. Linke, R. McDonald, M. Treacy, J. Chadi and C. Tsai for discussions. We also thank G. Lewen, G. Seidler, A. Krishnan, A. Schertel, A. Dziesiaty and H. Zimmermann for assistance.

Correspondence should be addressed to T.W.E.

可能产生无论从基础原理还是技术性角度来讲都极为有趣的光学特性。光能有效地透过比波长小很多的孔洞并突破孔洞间衍射的限制，对这一点的证明可以对诸如新型近场扫描光学显微镜[6]或者亚波长光刻法的设计等有所启发。对这些结果的理论分析可能会有助于对这种超强透射现象有更深入的理解。也许只有到那时我们才有望掌握这些发现的全部意义。

（王耀杨 翻译；宋心琦 审稿）

Potent and Specific Genetic Interference by Double-stranded RNA in *Caenorhabditis elegans*

A. Fire *et al.*

Editor's Note

Here American biologists Andrew Fire and Craig Mello describe a fundamental form of gene regulation called RNA interference, where snippets of double-stranded RNA instruct the cell to destroy genetically identical messenger RNA molecules, and so effectively silence the corresponding gene. The process occurs naturally in plants and animals, where it helps regulate gene expression and is a vital part of the immune response to viruses and other foreign genetic material. It is widely used in the laboratory to study gene function, and holds promise as a therapeutic tool for thwarting viruses and controlling the expression of aberrant, disease-causing genes. This demonstration of RNA interference in the nematode worm earned Fire and Mello a Nobel Prize eight years later.

Experimental introduction of RNA into cells can be used in certain biological systems to interfere with the function of an endogenous gene[1,2]. Such effects have been proposed to result from a simple antisense mechanism that depends on hybridization between the injected RNA and endogenous messenger RNA transcripts. RNA interference has been used in the nematode *Caenorhabditis elegans* to manipulate gene expression[3,4]. Here we investigate the requirements for structure and delivery of the interfering RNA. To our surprise, we found that double-stranded RNA was substantially more effective at producing interference than was either strand individually. After injection into adult animals, purified single strands had at most a modest effect, whereas double-stranded mixtures caused potent and specific interference. The effects of this interference were evident in both the injected animals and their progeny. Only a few molecules of injected double-stranded RNA were required per affected cell, arguing against stochiometric interference with endogenous mRNA and suggesting that there could be a catalytic or amplification component in the interference process.

DESPITE the usefulness of RNA interference in *C. elegans*, two features of the process have been difficult to explain. First, sense and antisense RNA preparations are each sufficient to cause interference[3,4]. Second, interference effects can persist well into the next generation, even though many endogenous RNA transcripts are rapidly degraded in the early embryo[5]. These results indicate a fundamental difference in behaviour between native RNAs (for example, mRNAs) and the molecules responsible for interference. We sought to test the possibility that this contrast reflects an underlying difference in RNA

秀丽隐杆线虫体内双链 RNA 强烈而特异性的基因干扰作用

法厄等

编者按

在本文中，美国生物学家安德鲁·法厄和克雷格·梅洛描述了基因调控的一种基本形式——称为 RNA 干扰，即小片段的双链 RNA 导致细胞内基因（序列）相同的信使 RNA(mRNA) 分子被破坏，从而有效地沉默相应靶基因的表达。这个过程在植物和动物体内自然发生，它有助于调控基因表达，并且也是机体对病毒和其他外来遗传物质进行免疫应答的重要组成部分。RNA 干扰在实验室中被广泛应用于基因功能的研究，并且有望成为抗病毒和控制异常基因与致病基因表达的一种治疗工具。本文中对线虫体内 RNA 干扰机制的阐述令法厄和梅洛于 8 年后获得诺贝尔奖。

在特定生物系统中，用实验方法将 RNA 导入细胞可以干扰内源基因的功能 [1,2]。研究人员认为这种干扰效应源于一种简单的反义机制，该机制依赖于导入的 RNA 与内源信使 RNA 转录本之间的杂交。RNA 干扰已经被用于操控秀丽隐杆线虫基因的表达 [3,4]。在此我们研究了干扰 RNA 结构和释放的必要条件。我们惊奇地发现，双链 RNA 产生的干扰作用比其中的任意一条单链都要强。将纯化的单链导入成年动物体内后最多只产生中等强度的干扰作用，而双链混合物的导入则能产生强烈而特异性的干扰。这种干扰作用在被注射的动物及其后代中都很明显。与内源 mRNA 产生干扰的剂量不同，仅需要几个双链 RNA 分子就可以对细胞产生干扰，这表明，在该干扰过程中可能存在具有催化或放大作用的组分。

虽然 RNA 干扰在秀丽隐杆线虫中非常有效，但是该过程的两个特征却难以得到解释。首先，单独制备的正义 RNA 链或反义 RNA 链都足以产生干扰作用 [3,4]。其次，尽管许多内源 RNA 转录本都在胚胎早期迅速地降解了，但是干扰作用却能够持续到下一代 [5]。这些结果表明产生干扰效应的 RNA 分子和天然分子（如 mRNA）在性质上具有本质差别。我们认为可能是 RNA 的根本结构差异导致了它们具有不同的

structure. RNA populations to be injected are generally prepared using bacteriophage RNA polymerases[6]. These polymerases, although highly specific, produce some random or ectopic transcripts. DNA transgene arrays also produce a fraction of aberrant RNA products[3]. From these facts, we surmised that the interfering RNA populations might include some molecules with double-stranded character. To test whether double-stranded character might contribute to interference, we further purified single-stranded RNAs and compared interference activities of individual strands with the activity of a deliberately prepared double-stranded hybrid.

The *unc-22* gene was chosen for initial comparisons of activity. *unc-22* encodes an abundant but nonessential myofilament protein[7-9]. Several thousand copies of *unc-22* mRNA are present in each striated muscle cell[3]. Semiquantitative correlations between *unc-22* activity and phenotype of the organism have been described[8]: decreases in *unc-22* activity produce an increasingly severe twitching phenotype, whereas complete loss of function results in the additional appearance of muscle structural defects and impaired motility.

Purified antisense and sense RNAs covering a 742-nucleotide segment of *unc-22* had only marginal interference activity, requiring a very high dose of injected RNA to produce any observable effect (Table 1). In contrast, a sense-antisense mixture produced highly effective interference with endogenous gene activity. The mixture was at least two orders of magnitude more effective than either single strand alone in producing genetic interference. The lowest dose of the sense-antisense mixture that was tested, ~60,000 molecules of each strand per adult, led to twitching phenotypes in an average of 100 progeny. Expression of *unc-22* begins in embryos containing ~500 cells. At this point, the original injected material would be diluted to at most a few molecules per cell.

Table 1. Effects of sense, antisense and mixed RNAs on progeny of injected animals

Gene	segment	Size (kilobases)	Injected RNA	F1 phenotype
unc-22				*unc-22*-null mutants: strong twitchers[7,8]
unc22A*	Exon 21–22	742	Sense	Wild type
			Antisense	Wild type
			Sense + antisense	Strong twitchers (100%)
unc22B	Exon 27	1,033	Sense	Wild type
			Antisense	Wild type
			Sense + antisense	Strong twitchers (100%)
unc22C	Exon 21–22†	785	Sense + antisense	Strong twitchers (100%)
fem-1				*fem-1*-null mutants: female (no sperm)[13]
fem1A	Exon 10‡	531	Sense	Hermaphrodite (98%)
			Antisense	Hermaphrodite (> 98%)
			Sense + antisense	Female (72%)
fem1B	Intron 8	556	Sense + antisense	Hermaphrodite (> 98%)
unc-54				*unc-54*-null mutants: paralysed[7,11]
unc54A	Exon 6	576	Sense	Wild type (100%)
			Antisense	Wild type (100%)
			Sense + antisense	Paralysed (100%)§

功能，并希望证实这个观点。通常情况下，用于导入的 RNA 分子都是利用噬菌体的 RNA 聚合酶来获得的 [6]。这些聚合酶尽管具有很高的特异性，但也会产生随机突变或者异位突变。DNA 转基因阵列同样也会产生一定比例的异常 RNA 分子 [3]。基于这些事实，我们总结出了干扰 RNA 分子中可能包含一些有双链特征的分子。为了验证双链分子是否可以导致干扰效应，我们进一步纯化了单链 RNA 分子，并将各个单链的干扰活性与特意准备的杂交双链分子进行了比较。

我们选择 unc-22 基因来进行初步的活性比较。unc-22 编码一种大量却非必需的肌丝蛋白 [7-9]。每个横纹肌细胞中都有数千拷贝的 unc-22 mRNA[3]。unc-22 的活性与机体表型的半定量关系已有报道 [8]：unc-22 活性降低会加剧肌肉抽搐这一表型，而 unc-22 完全丧失功能则进一步导致肌肉结构缺陷且运动能力受损。

包含 unc-22 上 742 个核苷酸的正义链和反义链 RNA 纯化片段只有微小的干扰活性，需要注射很高剂量的 RNA 才能产生可见效应（表 1）。相反，正义链–反义链混合物对内源基因活性产生高效的干扰作用。混合物产生的遗传性基因干扰比任何一条单链所产生的至少高两个数量级。实验证明，正义链–反义链混合物的最低剂量（即每个成年动物中每条单链为 60,000 个分子左右）平均可在 100 个后代中导致抽搐表型。unc-22 的表达始于包含约 500 个细胞的胚胎。此时，初始的注射物被稀释为每个细胞最多只含有几个分子。

表 1. 动物注射 RNA 分子正义链、反义链及其混合物后对后代的影响

基因	片段	大小 (kb)	注射的 RNA 链	子一代表型
unc-22				unc-22 无效突变体：剧烈抽搐 [7,8]
unc22A*	外显子 21~22	742	正义链	野生型
			反义链	野生型
			正义链 + 反义链	剧烈抽搐 (100%)
unc22B	外显子 27	1,033	正义链	野生型
			反义链	野生型
			正义链 + 反义链	剧烈抽搐 (100%)
unc22C	外显子 21~22†	785	正义链 + 反义链	剧烈抽搐 (100%)
fem-1				fem-1 无效突变体：雌体（无精子）[13]
fem1A	外显子 10‡	531	正义链	雌雄同体 (98%)
			反义链	雌雄同体 (>98%)
			正义链 + 反义链	雌体 (72%)
fem1B	内含子 8	556	正义链 + 反义链	雌雄同体 (>98%)
unc-54				unc-54 无效突变体：瘫痪 [7,11]
unc54A	外显子 6	576	正义链	野生型 (100%)
			反义链	野生型 (100%)
			正义链 + 反义链	瘫痪 (100%)§

Continued

Gene	segment	Size (kilobases)	Injected RNA	F1 phenotype
unc54B	Exon 6	651	Sense	Wild type (100%)
			Antisense	Wild type (100%)
			Sense + antisense	Paralysed (100%)§
unc54C	Exon 1–5	1,015	Sense + antisense	Arrested embryos and larvae (100%)
unc54D	Promoter	567	Sense + antisense	Wild type (100%)
unc54E	Intron 1	369	Sense + antisense	Wild type (100%)
unc54F	Intron 3	386	Sense + antisense	Wild type (100%)
hlh-1				*hlh-1*-null mutants: lumpy-dumpy larvae[16]
hlh1A	Exons1–6	1,033	Sense	Wild type (< 2% lpy-dpy)
			Antisense	Wild type (< 2% lpy-dpy)
			Sense + antisense	Lpy-dpy larvae (> 90%)‖
hlh1B	Exons1–2	438	Sense + antisense	Lpy-dpy larvae (> 80%)‖
hlh1C	Exons4–6	299	Sense + antisense	Lpy-dpy larvae (> 80%)‖
hlh1D	Intron 1	697	Sense + antisense	Wild type (< 2% lpy-dpy)
myo-3-driven GFP transgenes¶				Makes nuclear GFP in body muscle
myo-3::NLS:: gfp:: lacZ		730	Sense	Nuclear GFP–LacZ pattern of parent strain
gfpG	Exons 2–5		Antisense	Nuclear GFP–LacZ pattern of parent strain
			Sense + antisense	Nuclear GFP–LacZ absent in 98% of cells
lacZL	Exon 12–14	830	Sense + antisense	Nuclear GFP–LacZ absent in > 95% of cells
myo-3::MtLS:: gfp				Makes mitochondrial GFP in body muscle
gfpG	Exons 2–5	730	Sense	Mitochondrial-GFP pattern of parent strain
			Antisense	Mitochondrial-GFP pattern of parent strain
			Sense + antisense	Mitochondrial-GFP absent in 98% of cells
LacZL	Exon 12–14	830	Sense + antisense	Mitochondrial-GFP pattern of parent strain

Each RNA was injected into 6–10 adult hermaphrodites (0.5×10^6–1×10^6 molecules into each gonad arm). After 4–6 h (to clear prefertilized eggs from the uterus), injected animals were transferred and eggs collected for 20–22 h. Progeny phenotypes were scored upon hatching and subsequently at 12–24-h intervals.

* to obtain a semiquantitative assessment of the relationship between RNA dose and phenotypic response, we injected each *unc22A* RNA preparation at a series of different concentrations (see figure in Supplementary information for details). At the highest dose tested (3.6×10^6 molecules per gonad), the individual sense and antisense *unc22A* preparations produced some visible twitching (1% and 11% of progeny, respectively). Comparable doses of double-stranded *unc22A* RNA produced visible twitching in all progeny, whereas a 120-fold lower dose of double-stranded *unc22A* RNA produced visible twitching in 30% of progeny.

† *unc22C* also carries the 43-nucleotide intron between exons 21 and 22.

‡ *fem1A* carries a portion (131 nucleotides) of intron 10.

§ Animals in the first affected broods (laid 4–24 h after injection) showed movement defects indistinguishable from those of *unc-54*-null mutants. A variable fraction of these animals (25%–75%) failed to lay eggs (another phenotype of *unc-54*-null mutants), whereas the remainder of the paralysed animals did lay eggs. This may indicate incomplete interference with *unc-54* activity in vulval muscles. Animals from later broods frequently show a distinct partial loss-of-function phenotype, with contractility in a subset of body-wall muscles.

‖ Phenotypes produced by RNA-mediated interference with *hlh-1* included arrested embryos and partially elongated L1 larvae (the *hlh-1*-null phenotype). These phenotypes were seen in virtually all progeny after injection of double-stranded *hlh1A* and in about half of the affected animals produced after injection of double-stranded *hlh1B* and double-stranded *hlh1C*. A set of less severe defects was seen in the remainder of the animals produced after injection of double-stranded *hlh1B* and double-stranded *hlh1C*. The less severe phenotypes are characteristic of partial loss of function of *hlh-1* (B. Harfe and A.F., unpublished observations).

¶ the host for these injections, strain PD4251, expresses both mitochondrial GFP and nuclear GFP-LacZ (see Methods). This allows simultaneous assay for interference with *gfp* (seen as loss of all fluorescence) and with *lacZ* (loss of nuclear fluorescence). The table describes scoring of animals as L1 larvae. Double-stranded *gfpG* caused a loss of GFP in all but 0–3 of the 85 body muscles in these larvae. As these animals mature to adults, GFP activity was seen in 0–5 additional body-wall muscles and in the 8 vulval muscles. Lpy-dpy, lumpy-dumpy.

基因	片段	大小(kb)	注射的 RNA 链	子一代表型
unc-54B	外显子 6	651	正义链	野生型(100%)
			反义链	野生型(100%)
			正义链 + 反义链	瘫痪(100%)§
unc54C	外显子 1~5	1,015	正义链 + 反义链	幼虫及胚胎发育停滞(100%)
unc54D	启动子	567	正义链 + 反义链	野生型(100%)
unc54E	内含子 1	369	正义链 + 反义链	野生型(100%)
unc54F	内含子 3	386	正义链 + 反义链	野生型(100%)
hlh-1				hlh-1 无效突变体：矮胖粗笨幼虫 [16]
hlh1A	外显子 1~6	1,033	正义链	野生型(<2% lpy-dpy)
			反义链	野生型(<2% lpy-dpy)
			正义链 + 反义链	Lpy-dpy 幼虫(>90%)‖
hlh1B	外显子 1~2	438	正义链 + 反义链	Lpy-dpy 幼虫(>80%)‖
hlh1C	外显子 4~6	299	正义链 + 反义链	Lpy-dpy 幼虫(>80%)‖
hlh1D	内含子 1	697	正义链 + 反义链	野生型(<2% lpy-dpy)
myo-3 驱动 GFP 转移基因 ¶				
myo-3::NLS:: gfp::lacZ				肌肉产生核 GFP
gfpG	外显子 2~5	730	正义链	亲代核 GFP-LacZ 模式
			反义链	亲代核 GFP-LacZ 模式
			正义链 + 反义链	98% 细胞缺失核 GFP-LacZ
lacZL	外显子 12~14	830	正义链 + 反义链	95% 以上细胞缺失核 GFP-LacZ
myo-3::MtLS:: gfp				肌肉产生线粒体 GFP
gfpG	外显子 2~5	730	正义链	亲代线粒体 GFP 模式
			反义链	亲代线粒体 GFP 模式
			正义链 + 反义链	98% 细胞缺失线粒体 GFP
lacZL	外显子 12~14	830	正义链 + 反义链	亲代线粒体 GFP 模式

每种 RNA 分子被注入 6~10 个雌雄同体的成虫中(每个生殖腺臂注入 0.5×10^6~1×10^6 个分子)。4~6 小时后(这段时间用于清除子宫中受精前的卵子)，移出被注射动物并于 20~22 小时后收集其卵子。孵育之后即记录后代的表型，且每隔 12~24 小时记录一次。

* 为了测定 RNA 剂量和表型反应之间的半定量关系，我们以一系列的浓度梯度注射 unc22A RNA(细节详见补充信息图)。在测试的最高剂量(每个生殖腺 3.6×10^6 个分子)时，单独制备的正义链和反义链 unc22A 产生一些可见的抽搐(分别占后代的 1% 和 11%)。相等剂量的双链 unc22A RNA 在所有后代中都产生了可见的抽搐，而 1/120 剂量的双链 unc22A RNA 在 30% 的后代中产生可见抽搐。

† unc22C 同时携带了外显子 21 和 22 之间的含 43 个核苷酸的内含子。

‡ fem1A 携带内含子 10 的一部分 (131 个核苷酸)。

§ 首先受影响的一批动物(注射后 4~24 小时产卵)表现出的行动缺陷与 unc-54 无效突变体难以区分。这些动物中的一部分(25%~75%)不能产卵(unc-54 无效突变体的另一个表型)，而剩余的瘫痪动物确实产下了卵。这可能说明在外阴肌肉中对 unc-54 活性的干扰不完全。更晚的一批动物表现出明显的部分功能缺失表型，部分体壁肌肉有收缩性。

‖ 通过 RNA 介导的对 hlh-1 的干扰，产生的表型包括胚胎发育停滞和 L1 幼虫局部不能延长(hlh-1 无效表型)。这些表型在注射双链 hlh1A 后的所有后代中及注射双链 hlh1B 和双链 hlh1C 后受影响的一半动物中都可以见到。其余注射双链 hlh1B 和双链 hlh1C 的动物中看到不太严重的缺陷。不太严重的缺陷表型源自部分缺失 hlh-1 的功能(哈弗和法厄，未发表的观察结果)。

¶ 这些注射的宿主，品系为 PD4251，既表达线粒体 GFP 也表达核 GFP-LacZ(见方法部分)。这允许同时测定对 gfp(可见所有荧光丧失)和 lacZ(核荧光丧失)的干扰作用。这个表描述了 L1 幼虫的情况。双链 gfpG 引起这些幼虫中除 0~3 个以外的 85 个全身肌群中 GFP 缺失。在这些动物长到成虫时，可以在 0~5 条其他的体壁肌肉和 8 条外阴肌肉中观测到 GFP 活性。Lpy-dpy，粗笨–矮胖。

The potent interfering activity of the sense-antisense mixture could reflect the formation of double-stranded RNA (dsRNA) or, conceivably, some other synergy between the strands. Electrophoretic analysis indicated that the injected material was predominantly double-stranded. The dsRNA was gel-purified from the annealed mixture and found to retain potent interfering activity. Although annealing before injection was compatible with interference, it was not necessary. Mixing of sense and antisense RNAs in low-salt concentrations (under conditions of minimal dsRNA formation) or rapid sequential injection of sense and antisense strands were sufficient to allow complete interference. A long interval (> 1 h) between sequential injections of sense and antisense RNA resulted in a dramatic decrease in interfering activity. This suggests that injected single strands may be degraded or otherwise rendered inaccessible in the absence of the opposite strand.

A question of specificity arises when considering known cellular responses to dsRNA. Some organisms have a dsRNA-dependent protein kinase that activates a panic-response mechanism[10]. Conceivably, our sense-antisense synergy might have reflected a nonspecific potentiation of antisense effects by such a panic mechanism. This is not the case: co-injection of dsRNA segments unrelated to *unc-22* did not potentiate the ability of single *unc-22*-RNA strands to mediate inhibition (data not shown). We also investigated whether double-stranded structure could potentiate interference activity when placed in *cis* to a single-stranded segment. No such potentiation was seen: unrelated double-stranded sequences located 5' or 3' of a single-stranded *unc-22* segment did not stimulate interference. Thus, we have only observed potentiation of interference when dsRNA sequences exist within the region of homology with the target gene.

The phenotype produced by interference using *unc-22* dsRNA was extremely specific. Progeny of injected animals exhibited behaviour that precisely mimics loss-of-function mutations in *unc-22*. We assessed target specificity of dsRNA effects using three additional genes with well characterized phenotypes (Fig. 1, Table 1). *unc-54* encodes a body-wall-muscle heavy-chain isoform of myosin that is required for full muscle contraction[7,11,12]; *fem-1* encodes an ankyrin-repeat-containing protein that is required in hermaphrodites for sperm production[13,14]; and *hlh-1* encodes a *C. elegans* homologue of myoD-family proteins that is required for proper body shape and motility[15,16]. For each of these genes, injection of related dsRNA produced progeny broods exhibiting the known null-mutant phenotype, whereas the purified single RNA strands produced no significant interference. With one exception, all of the phenotypic consequences of dsRNA injection were those expected from interference with the corresponding gene. The exception (segment *unc54C* which led to an embryonic- and larval-arrest phenotype not seen with *unc-54-null* mutants) was illustrative. This segment covers the highly conserved myosin-motor domain, and might have been expected to interfere with activity of other highly related myosin heavy-chain genes[17]. The *unc54C* segment has been unique in our overall experience to date: effects of 18 other dsRNA segments (Table 1; and our unpublished observations) have all been limited to those expected from previously characterized null mutants.

正义链–反义链混合物的强干扰活性可以反映双链 RNA(dsRNA)的形成，或者可以想象为，反映了链之间的一些协同作用。电泳分析显示注射的物质主要是双链。dsRNA 是从退火的混合物中凝胶纯化得到的，并且保留了强干扰活性。虽然在注射前退火与干扰不矛盾，但并不是必需的。在低盐浓度下混合正义链和反义链的 RNA（此条件下形成的 dsRNA 最少）或快速连续注射正义和反义链都可以产生完全干扰。连续注射正义链和反义链 RNA 之间的长时间间隔（> 1 小时）会导致干扰活性显著下降。这表明已注射的单链可能被降解或因缺少互补链而难以实施干扰效应。

当考虑到已知的细胞对 dsRNA 的反应时，就产生了一个特殊问题。一些有机体具有 dsRNA 依赖性蛋白激酶，它活化一个恐慌反应机制[10]。可以想象，我们的正义–反义协同作用可能反映了由这一恐慌机制引起的反义效应的非特异性增强。事实并非如此：共同注射与 unc-22 无关的 dsRNA 片段不能增强单链 unc-22-RNA 介导的抑制作用（数据未显示）。我们也研究了将 cis 放入一个单链片段后双链结构是否可以增强干扰能力。并未观察到这种增强作用：位于一个单链 unc-22 片段 5′ 或 3′ 的无关双链序列不能激发干扰。因此，只有 dsRNA 序列位于与目标基因同源的区域内时才能观察到对干扰的增强作用。

用 unc-22 dsRNA 干扰产生的表型非常特殊。注射动物的后代表现出的行为非常类似于 unc-22 功能缺失突变体。我们用表型明确的其他三个基因来评价 dsRNA 效应的目标特异性（图 1 和表 1）。unc-54 编码的体壁肌肉重链同工型肌球蛋白是完整肌肉收缩所必需的[7,11,12]；fem-1 编码的锚定蛋白重复包含蛋白在雌雄同体动物中是产生精子所必需的[13,14]；hlh-1 编码的 myoD 家族蛋白的秀丽隐杆线虫同源物是维持适当的身体形状和运动能力所必需的[15,16]。对于每个基因来说，注射与之相关的 dsRNA 后，后代表现出已知的无效突变体表型，而提纯的单链 RNA 不能产生显著干扰。dsRNA 注射产生的所有表型皆符合对应基因干扰所预期的表型，仅有一个例外。对此例外（unc54C 片段导致胚胎及幼虫发育阻碍，这在 unc-54 无效突变体中未曾见过）作以下说明：这个片段覆盖高度保守的肌球蛋白分子马达结构域，因此预计它会干扰其他高度相关的肌球蛋白重链基因[17]。该 unc54C 片段也是我们所有实验数据中唯一出现的例外，其他 18 个 dsRNA 片段所出现的效应均可见于先前已知的无效突变表型中（表 1 和我们未发表的结果）。

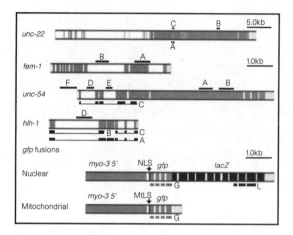

Fig. 1. Genes used to study RNA-mediated genetic interference in *C. elegans*. Intron-exon structure for genes used to test RNA-mediated inhibition are shown (grey and filled boxes, exons; open boxes, introns; patterned and striped boxes, 5' and 3' untranslated regions. *unc-22*. ref. 9, *unc*-54, ref. 12, *fem-1*, ref. 14, and *hlh-1*, ref. 15). Each segment of a gene tested for RNA interference is designated with the name of the gene followed by a single letter (for example, *unc22C*). These segments are indicated by bars and upper-case letters above and below each gene. Segments derived from genomic DNA are shown above the gene; segments derived from cDNA are shown below the gene. NLS, nuclear-localization sequence; MtLS, mitochondrial localization sequence.

The pronounced phenotypes seen following dsRNA injection indicate that interference effects are occurring in a high fraction of cells. The phenotypes seen in *unc-54* and *hlh-1* null mutants, in particular, are known to result from many defective muscle cells[11,16]. To examine interference effects of dsRNA at a cellular level, we used a transgenic line expressing two different green fluorescent protein (GFP)-derived fluorescent-reporter proteins in body muscle. Injection of dsRNA directed to *gfp* produced marked decreases in the fraction of fluorescent cells (Fig. 2). Both reporter proteins were absent from the affected cells, whereas the few cells that were fluorescent generally expressed both GFP proteins.

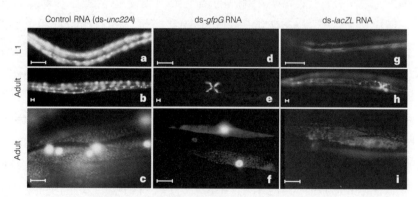

Fig. 2. Analysis of RNA-interference effects in individual cells. Fluorescence micrographs show progeny of injected animals from GFP-reporter strain PD4251. **a-c**, Progeny of animals injected with a control RNA (double-stranded (ds)-*unc22A*). **a**, Young larva, **b**, adult, **c**, adult body wall at high magnification. These GFP patterns appear identical to patterns in the parent strain, with prominent fluorescence in nuclei (nuclear-localized GFP-LacZ) and mitochondria (mitochondrially targeted GFP). **d-f**, Progeny of

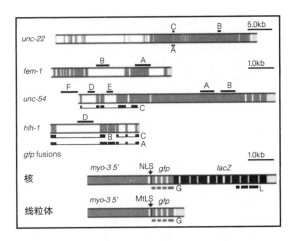

图 1. 用于研究秀丽隐杆线虫中 RNA 介导的遗传性干扰的基因。显示了用于测试 RNA 介导的抑制作用的基因内含子–外显子结构（灰色及填充的方框，外显子；空白的方框，内含子；有图案和条纹的方框，5′ 和 3′ 非翻译区。*unc-22*，参考文献 9；*unc-54*，参考文献 12；*fem-1*，参考文献 14；*hlh-1*，参考文献 15）。用于测试 RNA 干扰的基因片段用基因名加一个字母表示（如 *unc22C*）。这些片段用横条标出，并且在每个基因上部或下部用大写字母表示。从基因组 DNA 得到的片段在基因上方显示；从 cDNA 得到的片段在基因下方显示。NLS，核定位序列；MtLS，线粒体定位序列。

dsRNA 注射后所观察到的明显表型说明在大量细胞中发生了干扰效应。特别是已经知道在 *unc-54* 和 *hlh-1* 无效突变体中见到的表型是源自众多缺陷型肌细胞[11,16]。为了在细胞水平上检验 dsRNA 的干扰效果，我们使用了一个转基因株，它可以在肌肉中表达两种绿色荧光蛋白（GFP）衍生的荧光报告蛋白。注射 dsRNA 直接导致在表达绿色荧光的细胞（图 2）内 *gfp* 荧光强度的显著下降。两个报告蛋白从受影响的细胞中消失，而少量有荧光的细胞基本都表达两种 GFP 蛋白。

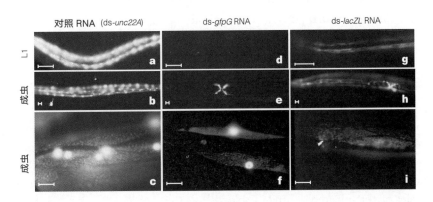

图 2. 在单个细胞中分析 RNA 干扰效应。荧光显微图显示了注射 GFP 报告品系 PD4251 的动物后代。**a~c**，注射对照 RNA（双链（ds）-*unc22A*）的动物后代。**a**，年轻幼虫，**b**，成虫，**c**，高倍放大的成虫体壁。这些子代的 GFP 表达情况与亲本一致，核（核定位 GFP-LacZ）中和线粒体（线粒体靶标 GFP）中均具有显著的荧光。**d~f**，注射 ds-*gfpG* 的动物后代。**d** 中幼虫只可见一个活性细胞，而 **e**、**f** 中的成虫的

animals injected with ds-*gfpG*. Only a single active cell is seen in the larva in **d**, whereas the entire vulval musculature expresses active GFP in the adult animal in **e**. **f**, Two rare GFP-positive cells in an adult: both cells express both nuclear-targeted GFP-LacZ and mitochondrial GFP. **g-i**, Progeny of animals injected with *ds-lacZL* RNA: mitochondrial-targeted GFP seems unaffected, while the nuclear-targeted GFP-LacZ is absent from almost all cells (for example, see larva in **g**). **h**, A typical adult, with nuclear GFP-LacZ lacking in almost all body-wall muscles but retained in vulval muscles. Scale bars represent 20 μm.

The mosaic pattern observed in the *gfp*-interference experiments was nonrandom. At low doses of dsRNA, we saw frequent interference in the embryonically derived muscle cells that are present when the animal hatches. The interference effect in these differentiated cells persisted throughout larval growth: these cells produced little or no additional GFP as the affected animals grew. The 14 postembryonically derived striated muscles are born during early larval stages and these were more resistant to interference. These cells have come through additional divisions (13–14 divisions versus 8–9 divisions for embryonic muscles[18,19]). At high concentrations of *gfp* dsRNA, we saw interference in virtually all striated body-wall muscles, with occasional lone escaping cells, including cells born during both embryonic and postembryonic development. The non-striated vulval muscles, which are born during late larval development, appeared to be resistant to interference at all tested concentrations of injected dsRNA.

We do not yet know the mechanism of RNA-mediated interference in *C. elegans*. Some observations, however, add to the debate about possible targets and mechanisms.

First, dsRNA segments corresponding to various intron and promoter sequences did not produce detectable interference (Table 1). Although consistent with interference at a post-transcriptional level, these experiments do not rule out interference at the level of the gene.

Second, we found that injection of dsRNA produces a pronounced decrease or elimination of the endogenous mRNA transcript (Fig. 3). For this experiment, we used a target transcript (*mex-3*) that is abundant in the gonad and early embryos[20], in which straightforward *in situ* hybridization can be performed[5]. No endogenous *mex-3* mRNA was observed in animals injected with a dsRNA segment derived from *mex-3*. In contrast, animals into which purified *mex-3* antisense RNA was injected retained substantial endogenous mRNA levels (Fig. 3d).

Third, dsRNA-mediated interference showed a surprising ability to cross cellular boundaries. Injection of dsRNA (for *unc-22*, *gfp* or *lacZ*) into the body cavity of the head or tail produced a specific and robust interference with gene expression in the progeny brood (Table 2). Interference was seen in the progeny of both gonad arms, ruling out the occurrence of a transient "nicking" of the gonad in these injections. dsRNA injected into the body cavity or gonad of young adults also produced gene-specific interference in somatic tissues of the injected animal (Table 2).

整个外阴肌肉都表达活性 GFP，成虫中仅有两个 GFP 活性细胞：两个细胞都表达活性核靶标 GFP-LacZ 和线粒体靶标 GFP。g~i，注射 ds-*lacZL* RNA 的动物后代：线粒体靶标 GFP 好像不受影响，而核靶标 GFP-LacZ 在几乎所有细胞中都缺失了（例如，见 g 中的幼虫）。h，一个典型的成虫，其核靶标 GFP-LacZ 几乎在所有体壁肌肉都缺失但仍存在于外阴肌肉中。比例尺代表 20 μm。

gfp 干扰实验中观察到的镶嵌状图像不是随机的。在低剂量 dsRNA 时，我们经常看到动物孵化时会出现胚胎源的肌肉细胞发生干扰。不同分化细胞中的干扰效应在整个幼虫成长过程中持续发生：这些细胞在受影响动物的生长过程中很少产生或不产生额外的 GFP。14 个源自胚胎后期的横纹肌在幼虫早期阶段产生并且更能抵抗干扰。这些细胞进行额外的分裂（13~14 次分裂，胚胎源肌肉为 8~9 次分裂 [18,19]）。在高浓度 *gfp* dsRNA 作用时，我们在所有体壁横纹肌中都看到了干扰，只有个别漏网的细胞，包括在胚胎发育和胚胎后期发育过程中出现的细胞。出现于幼虫发育晚期的外阴非横纹肌，似乎在所有检测的 dsRNA 注射浓度下都不发生干扰。

我们还不知道秀丽隐杆线虫中 RNA 介导的干扰机理。然而，一些观察结果增加到对可能靶标和机制方面的讨论中。

第一，对应于各种内含子和启动子序列的 dsRNA 片段没有产生可检测到的干扰效应（表 1）。虽然这支持了干扰发生在转录后水平的观点，但这些实验不能排除基因水平上的干扰。

第二，我们发现注射 dsRNA 导致内源 mRNA 转录本明显减少或消失（图 3）。在这个实验中，我们选用目标转录本（*mex-3*），它在生殖腺和早期胚胎中含量很丰富 [20]，在早期胚胎中可以直接进行原位杂交 [5]。在注射了来自 *mex-3* 的 dsRNA 片段的动物体内没有观察到内源性的 *mex-3* mRNA。相反，注射了纯化的 *mex-3* 反义 RNA 的动物维持基本的内源 mRNA 水平（图 3d）。

第三，dsRNA 介导的干扰显示了穿过细胞边界的惊人能力。头部或尾部体腔内注射 dsRNA（*unc-22*、*gfp* 或 *lacZ*）对孵化后代基因表达产生了特异的、稳定的干扰（表 2）。干扰主要出现在具两个性腺的后代中，且排除了在这些注射中发生短暂的"切口"的可能性。将 dsRNA 注射到年轻成虫的体腔或生殖腺后，在被注射动物的体壁组织内也产生了基因特异性干扰（表 2）。

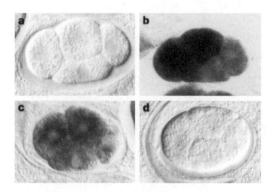

Fig. 3. Effects of *mex-3* RNA interference on levels of the endogenous mRNA. Interference contrast micrographs show *in situ* hybridization in embryos. The 1,262-nt *mex-3* cDNA clone[20] was divided into two segments, *mex-3A* and *mex-3B*, with a short (325-nt) overlap (similar results were obtained in experiments with no overlap between interfering and probe segments). *mex-3B* antisense or dsRNA was injected into the gonads of adult animals, which were fed for 24 h before fixation and *in situ* hybridization (ref. 5; B. Harfe and A.F., unpublished observations). The *mex-3B* dsRNA produced 100% embryonic arrest, whereas > 90% of embryos produced after the antisense injections hatched. Antisense probes for the *mex-3A* portion of *mex-3* were used to assay distribution of the endogenous *mex-3* mRNA (dark stain). Four-cell-stage embryos are shown; similar results were observed from the one to eight cell stage and in the germ line of injected adults. **a**, Negative control showing lack of staining in the absence of the hybridization probe. **b**, Embryo from uninjected parent (showing normal pattern of endogenous *mex-3* RNA[20]). **c**, Embryo from a parent injected with purified *mex-3B* antisense RNA. These embryos (and the parent animals) retain the *mex-3* mRNA, although levels may be somewhat less than wild type. **d**, Embryo from a parent injected with dsRNA corresponding to *mex-3B*; no *mex-3* RNA is detected. Each embryo is approximately 50 μm in length.

Table 2. Effect of site of injection on interference in injected animals and their progeny

dsRNA	Site of injection	Injected-animal phenotype	Progeny phenotype
None	Gonad or body cavity	No twitching	No twitching
None	Gonad or body cavity	Strong nuclear and mitochondrial GFP expression	Strong nuclear and mitochondrial GFP expression
unc22B	Gonad	Weak twitchers	Strong twitchers
unc22B	Body-cavity head	Weak twitchers	Strong twitchers
unc22B	Body-cavity tail	Weak twitchers	Strong twitchers
gfpG	Gonad	Lower nuclear and mitochondrial GFP expression	Rare or absent nuclear and mitochondrial GFP expression
gfpG	Body-cavity tail	Lower nuclear and mitochondrial GFP expression	Rare or absent nuclear and mitochondrial GFP expression
lacZL	Gonad	Lower nuclear GFP expression	Rare or absent nuclear-GFP expression
lacZL	Body-cavity tail	Lower nuclear GFP expression	Rare or absent nuclear-GFP expression

The GFP-reporter strain PD4251, which expresses both mitochondrial GFP and nuclear GFP-LacZ, was used for injections. The use of this strain allowed simultaneous assay for interference with *gfp* (fainter overall fluorescence), *lacZ* (loss of nuclear fluorescence) and *unc-22* (twitching). Body-cavity injections into the tail region were carried out to minimize accidental injection of the gonad; equivalent results have been observed with injections into the anterior body cavity. An equivalent set of injections was also performed into a single gonad arm. The entire progeny broods showed phenotypes identical to those described in Table 1. This included progeny of both injected and uninjected gonad arms. Injected animals were scored three days after recovery and showed somewhat less dramatic phenotypes than their progeny. This could be partly due to the persistence of products already present in the injected adult.

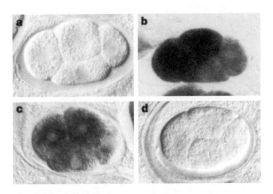

图 3. *mex-3* RNA 对内源性 mRNA 水平的干扰作用。干扰对比显微照片显示胚胎中的原位杂交结果。将 1,262 个核苷酸的 *mex-3* cDNA 克隆[20] 分为两个片段，即 *mex-3A* 和 *mex-3B*，两者有一小段 (325 个核苷酸) 重叠 (在干扰片段和探针片段之间无重叠的实验中也得到相似的结果)。将 *mex-3B* 反义链或 dsRNA 注射到成年动物生殖腺内，并在固定和进行原位杂交前饲养 24 小时 (参考文献 5；哈弗和法厄，未发表的观察结果)。*mex-3B* dsRNA 可产生 100% 的胚胎抑制，然而大于 90% 的注射了反义链的胚胎都可以孵化。*mex-3* 的部分片段 *mex-3A* 的反义探针用于分析内源性 *mex-3* mRNA (深染色) 的分布。四细胞期的胚胎被显示出来。在单细胞期到八细胞期和注射的成虫生殖系中都观察到相似结果。**a**，阴性对照显示无杂交探针时不染色。**b**，来自未注射亲本的胚胎 (显示正常内源 *mex-3* RNA 模式[20])。**c**，亲本注射了纯化的 *mex-3B* RNA 反义链产生的胚胎。这些胚胎 (和亲本动物) 保留了 *mex-3* mRNA，但水平相对野生型有一定程度降低。**d**，亲本注射了对应 *mex-3B* 的 dsRNA 产生的胚胎；未检测到 *mex-3* RNA。每个胚胎长度大约为 50 μm。

表 2. 在注射动物及其后代中注射位置对干扰的影响

dsRNA	注射位置	注射动物的表型	后代表型
无 无	生殖腺或体腔 生殖腺或体腔	不抽搐 核 GFP 及线粒体 GFP 的高表达	不抽搐 核 GFP 及线粒体 GFP 的高表达
unc22B *unc22B* *unc22B*	生殖腺 头部体腔 尾部体腔	弱抽搐 弱抽搐 弱抽搐	强烈抽搐 强烈抽搐 强烈抽搐
gfpG *gfpG*	生殖腺 尾部体腔	核 GFP 及线粒体 GFP 的低表达 核 GFP 及线粒体 GFP 的低表达	很少或无核 GFP 及线粒体 GFP 表达 很少或无核 GFP 及线粒体 GFP 表达
lacZL *lacZL*	生殖腺 尾部体腔	核 GFP 的低表达 核 GFP 的低表达	很少或无核 GFP 表达 很少或无核 GFP 表达

GFP 报告品系 PD4251 中既可以表达线粒体 GFP 又可以表达核 GFP-LacZ，因此被用于注射。使用这个品系可以同时分析 *gfp* (昏暗的整体荧光)，*lacZ* (缺失核荧光) 和 *unc-22* (抽搐) 的干扰作用。注射到体腔尾部区域以最大限度减少偶然注射到生殖腺的情况；注射到前部体腔也观察到了相同的结果。还有相同的一组只注射到单个生殖腺臂。所有孵化后代呈现的表型与表 1 所描述的相同。包含了注射和未注射生殖腺臂的后代。注射的动物恢复三天后统计的表型一定程度上不如其后代明显。这可以部分归因于被注射的成年动物体内已有产物的持久性。注射双链 *unc22B* 后，部分注射动物在标准饲养条件下产生微弱抽搐 (21 个动物中有 10 个)。左旋咪唑的处理导致这些动物发生

After injection of double-stranded *unc22B*, a fraction of the injected animals twitch weakly under standard growth conditions (10 out of 21 animals). Levamisole treatment led to twitching of 100% (21 out of 21) of these animals. Similar effects (not shown) were seen with double-stranded *unc22A*. Injections of double-stranded *gfpG* or double-stranded *lacZL* produced a dramatic decrease (but not elimination) of the corresponding GFP reporters. In some cases, isolated cells or parts of animals retained strong GFP activity. These were most frequently seen in the anterior region and around the vulva. Injections of double-stranded *gfpG* and double-stranded *lacZL* produced no twitching, whereas injections of double-stranded *unc22A* produced no change in the GFP-fluorescence pattern.

The use of dsRNA injection adds to the tools available for studying gene function in *C. elegans*. In particular, it should now be possible functionally to analyse many interesting coding regions[21] for which no specific function has been defined. Although the effects of dsRNA-mediated interference are potent and specific we have observed several limitations that should be taken into account when designing RNA-interference-based experiments. First, a sequence shared between several closely related genes may interfere with several members of the gene family. Second, it is likely that a low level of expression will resist RNA-mediated interference for some or all genes, and that a small number of cells will likewise escape these effects.

Genetic tools are available for only a few organisms. Double-stranded RNA could conceivably mediate interference more generally in other nematodes, in other invertebrates, and, potentially, in vertebrates. RNA interference might also operate in plants: several studies have suggested that inverted-repeat structures or characteristics of dsRNA viruses are involved in transgene-dependent co-suppression in plants[22,23].

There are several possible mechanisms for RNA interference in *C. elegans*. A simple antisense model is not likely: annealing between a few injected RNA molecules and excess endogenous transcripts would not be expected to yield observable phenotypes. RNA-targeted processes cannot, however, be ruled out, as they could include a catalytic component. Alternatively, direct RNA-mediated interference at the level of chromatin structure or transcription could be involved. Interactions between RNA and the genome, combined with propagation of changes along chromatin, have been proposed in mammalian X-chromosome inactivation and plant-gene co-suppression[22,24]. If RNA interference in *C. elegans* works by such a mechanism, it would be new in targeting regions of the template that are present in the final mRNA (as we observed no phenotypic interference using intron or promoter sequences). Whatever their target, the mechanisms underlying RNA interference probably exist for a biological purpose. Genetic interference by dsRNA could be used by the organism for physiological gene silencing. Likewise, the ability of dsRNA to work at a distance from the site of injection, and particularly to move into both germline and muscle cells, suggests that there is an effective RNA-transport mechanism in *C. elegans*.

Methods

RNA synthesis and microinjection. RNA was synthesized from phagemid clones by using T3

100% 的抽搐（21 个中有 21 个）。在双链 *unc22A* 实验中出现了类似结果（未显示）。注射双链 *gfpG* 或双链 *lacZL* 产生的相应 GFP 报告物明显减少（但未消失）。在某些情况下分离的细胞或部分动物体仍有很强的 GFP 活性，这在前部区域和外阴周围最为常见。注射双链 *gfpG* 和双链 *lacZL* 不产生抽搐，而注射双链 *unc22A* 后荧光模式不发生改变。

dsRNA 注射的使用为研究秀丽隐杆线虫基因功能增加了可利用的工具。特别是它可以用于针对许多研究者感兴趣但具体功能尚未确定的编码区的功能分析。虽然 dsRNA 介导的干扰效应很强并具有特异性，但我们也观察到在设计以 RNA 干扰为基础的实验时应考虑的一些局限性。第一，几个紧密相关的基因之间共享的一段序列可以干扰这个基因家族的几个成员。第二，似乎低水平表达可以使部分或全部基因对抗 RNA 介导的干扰，而且少量细胞也可借此逃避 RNA 干扰作用。

遗传工具仅对很少的生物体适用。可以想象双链 RNA 可以更普遍地在其他线虫、无脊椎动物，甚至也可能在脊椎动物中介导干扰效应。RNA 干扰在植物中也可以操作：一些研究提示植物中依赖于转基因的共抑制涉及反向重复结构或 dsRNA 病毒特征 [22,23]。

秀丽隐杆线虫的 RNA 干扰存在几种可能的机制。而简单的反义链模型也似乎不大可能：少量注射的 RNA 分子和过量的内源转录本之间发生退火，估计不能产生可观察到的表型。然而，不能排除以 RNA 为目标的过程中可能包括催化组分。也许，RNA 直接介导的干扰也包括了发生在染色质和转录水平的作用。RNA 和基因组之间的相互作用，伴随着染色质的增殖变化，被认为与哺乳动物 X 染色体失活和植物基因共抑制有关 [22,24]。如果秀丽隐杆线虫研究中的 RNA 干扰是这样的机制，在最终的 mRNA 中表现的应该是新的模板靶标区域（因为我们用内含子和启动子序列时没有观察到表型干扰）。不管它们的目标是什么，RNA 干扰机制的存在可能具有某种生物学目的。dsRNA 产生的遗传干扰可被有机体用于生理基因沉默。同样，dsRNA 能够在距注射位置一定距离处产生作用，尤其是能进入生殖腺和肌肉细胞，这表明秀丽隐杆线虫体内存在有效的 RNA 转移机制。

方　法

RNA 合成和微量注射　使用 T3 和 T7 聚合酶从噬菌粒中合成了 RNA[6]。然后通过两个

and T7 polymerase[6]. Templates were then removed with two sequential DNase treatments. When sense-, antisense-, and mixed-RNA populations were to be compared, RNAs were further purified by electrophoresis on low-gelling-temperature agarose. Gel-purified products appeared to lack many of the minor bands seen in the original "sense" and "antisense" preparations. Nonetheless, RNA species comprising < 10% of purified RNA preparations would not have been observed. Without gel purification, the "sense" and "antisense" preparations produced notable interference. This interference activity was reduced or eliminated upon gel purification. In contrast, sense-plus-antisense mixtures of gel-purified and non-gel-purified RNA preparations produced identical effects.

Sense/antisense annealing was carried out in injection buffer (ref. 27) at 37 °C for 10–30 min. Formation of predominantly double-stranded material was confirmed by testing migration on a standard (nondenaturing) agarose gel: for each RNA pair, gel mobility was shifted to that expected for dsRNA of the appropriate length. Co-incubation of the two strands in a lower-salt buffer (5 mM Tris-Cl, pH 7.5, 0.5 mM EDTA) was insufficient for visible formation of dsRNA *in vitro*. Non-annealed sense-plus-antisense RNAs for *unc22B* and *gfpG* were tested for RNA interference and found to be much more active than the individual single strands, but twofold to fourfold less active than equivalent preannealed preparations.

After preannealing of the single strands for *unc22A*, the single electrophoretic species, corresponding in size to that expected for the dsRNA, was purified using two rounds of gel electrophoresis. This material retained a high degree of interference activity.

Except where noted, injection mixes were constructed so that animals would receive an average of 0.5×10^6 to 1.0×10^6 RNA molecules. For comparisons of sense, antisense, and double-stranded RNA activity, equal masses of RNA were injected (that is, dsRNA was used at half the molar concentration of the single strands). Numbers of molecules injected per adult are approximate and based on the concentration of RNA in the injected material (estimated from ethidium bromide staining) and the volume of injected material (estimated from visible displacement at the site of injection). It is likely that this volume will vary several-fold between individual animals; this variability would not affect any of the conclusions drawn from this work.

Analysis of phenotypes. Interference with endogenous genes was generally assayed in a wild-type genetic background (N2). Features analysed included movement, feeding, hatching, body shape, sexual identity, and fertility. Interference with *gfp* (ref. 25) and *lacZ* activity was assessed using *C. elegans* strain PD4251. This strain is a stable transgenic strain containing an integrated array (ccIs4251) made up of three plasmids: pSAK4 (*myo-3* promoter driving mitochondrially targeted GFP); pSAK2 (*myo-3* promoter driving a nuclear-targeted GFP-LacZ fusion); and a *dpy-20* subclone[26] as a selectable marker. This strain produces GFP in all body muscles, with a combination of mitochondrial and nuclear localization. The two distinct compartments are easily distinguished in these cells, allowing easy distinction between cells expressing both, either, or neither of the original GFP constructs.

Gonadal injection was done as described[27]. Body-cavity injections followed a similar procedure, with needle insertion into regions of the head and tail beyond the positions of the two gonad arms. Injection into the cytoplasm of intestinal cells is also effective, and may be the least disruptive to the

连续的 DNase 处理将模板除去。当正义链、反义链和双链混合物进行比较时，通过低凝胶温度琼脂糖电泳进一步纯化 RNA。凝胶纯化产物似乎缺少在最初"正义链"和"反义链"制备物中看到的许多小条带。然而，包含 <10% 的纯化 RNA 制备物的 RNA 物质无法被观察到。不经过凝胶纯化时，"正义链"和"反义链"制备物会产生明显的干扰。这种干扰活性可被凝胶纯化降低或消除。相反，凝胶纯化的和未经凝胶纯化的正义链加反义链混合物能产生相同的作用。

37℃下在注射缓冲液（参考文献 27）中进行正义链/反义链退火 10~30 分钟。通过测试在标准（非变性）琼脂糖凝胶上的迁移来确定主要双链物质的形成：对每对 RNA，凝胶迁移到预期的 dsRNA 的适当位置。在低盐缓冲液（5 mM Tris-Cl，pH 7.5，0.5 mM EDTA）中共同孵育两条链，这不足以在体外形成可见的 dsRNA。用未经退火的 unc22B 和 gfpG 的有义链和反义链混合 RNA 测试 RNA 干扰，发现比单独用一个链的活性高很多，但比提前退火的制备物活性低 1/2 到 1/4。

unc22A 的单链提前退火后，将对应于预期的 dsRNA 大小的单一的电泳条带通过两轮凝胶电泳纯化。这种物质仍保持高度干扰活性。

除了特别指出的地方，构建的注射混合物能使动物平均接受 0.5×10^6 到 1.0×10^6 个 RNA 分子。为比较正义链、反义链和双链 RNA 活性，进行了等量的 RNA 注射（即 dsRNA 的摩尔浓度为单链 RNA 的一半）。每个成虫注射的分子数量相互接近，分子数量取决于注射物中 RNA 的浓度（通过溴化乙锭染色来确定）和注射物体积（通过注射位置的可见位移来确定）。这个体积在不同个体之间可能有几倍的变化；这一变化不影响本研究中得出的任何结论。

表型分析 内源基因的干扰一般在野生型遗传背景（N2）中分析。特征分析包括运动、进食、孵化、身体形态、性别特征和生殖能力。gfp（参考文献 25）和 lacZ 的干扰活性用秀丽隐杆线虫品系 PD4251 进行评价。这个品系是稳定的转基因品系，包含一个融合序列（ccIs4251），由三个质粒组成：pSAK4（myo-3 启动子启动以线粒体为靶标的 GFP），pSAK2（myo-3 启动子启动以核为靶标的 GFP-LacZ 融合子）和一个 dpy-20 亚克隆[26] 作为选择性标记。这个品系在全身肌肉中产生 GFP，并结合线粒体和核定位。这两个清楚的部位在这些细胞中很容易区分，在表达两种、表达其中一种和不表达初始 GFP 结构的细胞中都易于区分。

按照文献描述进行了生殖腺注射 [27]。按相似的步骤进行体腔注射，用针插入位于两个生殖腺臂上方的头部和尾部区域。注射到肠细胞的细胞质内也有效，同时可能对动物造成的

animal. After recovery and transfer to standard solid media, injected animals were transferred to fresh culture plates at 16-h intervals. This yields a series of semisynchronous cohorts in which it was straightforward to identify phenotypic differences. A characteristic temporal pattern of phenotypic severity is observed among progeny. First, there is a short "clearance" interval in which unaffected progeny are produced. These include impermeable fertilized eggs present at the time of injection. Second, after the clearance period, individuals that show the interference phenotype are produced. Third, after injected animals have produced eggs for several days, gonads can in some cases "revert" to produce incompletely affected or phenotypically normal progeny.

(**391**, 806-811; 1998)

Andrew Fire[*], SiQun Xu[*], Mary K. Montgomery[*], Steven A. Kostas[*†], Samuel E. Driver[‡] & Craig C. Mello[‡]

[*] Carnegie Institution of Washington, Department of Embryology, 115 West University Parkway, Baltimore, Maryland 21210, USA

[†] Biology Graduate Program, Johns Hopkins University, 3400 North Charles Street, Baltimore, Maryland 21218, USA

[‡] Program in Molecular Medicine, Department of Cell Biology, University of Massachusetts Cancer Center, Two Biotech Suite 213, 373 Plantation Street, Worcester, Massachusetts 01605, USA

Received 16 September; accepted 24 November 1997.

References:

1. Izant, J. & Weintraub, H. Inhibition of thymidine kinase gene expression by antisense RNA: a molecular approach to genetic analysis. *Cell* **36**, 1007-1015 (1984).

2. Nellen, W. & Lichtenstein, C. What makes an mRNA anti-sense-itive? *Trends Biochem. Sci.* **18**, 419-423 (1993).

3. Fire, A., Albertson, D., Harrison, S. & Moerman, D. Production of antisense RNA leads to effective and specific inhibition of gene expression in *C. elegans* muscle. *Development* **113**, 503-514 (1991).

4. Guo, S. & Kemphues, K. *par-1*, a gene required for establishing polarity in *C. elegans* embryos, encodes a putative Ser/Thr kinase that is asymmetrically distributed. *Cell* **81**, 611-620 (1995).

5. Seydoux, G. & Fire, A. Soma-germline asymmetry in the distributions of embryonic RNAs in *Caenorhabditis elegans*. *Development* **120**, 2823-2834 (1994).

6. Ausubel, F. *et al. Current Protocols in Molecular Biology* (Wiley, New York, 1990).

7. Brenner, S. The genetics of *Caenorhabditis elegans*. *Genetics* **77**, 71-94 (1974).

8. Moerman, D. & Baillie, D. Genetic organization in *Caenorhabditis elegans*: fine structure analysis of the *unc-22* gene. *Genetics* **91**, 95-104 (1979).

9. Benian, G., L'Hernault, S. & Morris, M. Additional sequence complexity in the muscle gene, *unc-22*, and its encoded protein, twitchin, of *Caenorhabditis elegans*. *Genetics* **134**, 1097-1104 (1993).

10. Proud, C. PKR: a new name and new roles. *Trends Biochem. Sci.* **20**, 241-246 (1995).

11. Epstein, H., Waterston, R. & Brenner, S. A mutant affecting the heavy chain of myosin in *C. elegans*. *J. Mol. Biol.* **90**, 291-300 (1974).

12. Karn, J., Brenner, S. & Barnett, L. Protein structural domains in the *C. elegans unc-54* myosin heavy chain gene are not separated by introns. *Proc. Natl Acad. Sci. USA* **80**, 4253-4257 (1983).

13. Doniach, T. & Hodgkin, J. A. A sex-determining gene, *fem-1*, required for both male and hermaphrodite development in *C. elegans*. *Dev. Biol.* **106**, 223-235 (1984).

14. Spence, A., Coulson, A. & Hodgkin, J. The product of *fem-1*, a nematode sex-determining gene, contains a motif found in cell cycle control proteins and receptors for cell-cell interactions. *Cell* **60**, 981-990 (1990).

15. Krause, M., Fire, A., Harrison, S., Priess, J. & Weintraub, H. CeMyoD accumulation defines the body wall muscle cell fate during *C. elegans* embryogenesis. *Cell* **63**, 907-919 (1990).

16. Chen, L., Krause, M., Sepanski, M. & Fire, A. The *C. elegans* MyoD homolog *HLH-1* is essential for proper muscle function and complete morphogenesis. *Development* **120**, 1631-1641(1994).

17. Dibb, N. J., Maruyama, I. N., Krause, M. & Karn, J. Sequence analysis of the complete *Caenorhabditis elegans* myosin heavy chain gene family. *J. Mol. Biol.* **205**, 603-613 (1989).

18. Sulston, J., Schierenberg, E., White, J. & Thomson, J. The embryonic cell lineage of the nematode *Caenorhabditis elegans*. *Dev. Biol.* **100**, 64-119 (1983).

19. Sulston, J. & Horvitz, H. Postembyonic cell lineages of the nematode *Caenorhabditis elegans*. *Dev. Biol.* **82**, 41-55 (1977).

20. Draper, B. W., Mello, C. C., Bowerman, B., Hardin, J. & Priess, J. R. *MEX-3* is a KH domain protein that regulates blastomere identity in early *C. elegans* embryos. *Cell* **87**, 205-216 (1996).

损害最小。恢复并转移到标准固体培养基上后，注射的动物每间隔 16 小时转移到新鲜的培养平板上。这样产生的一系列半同步同生群可以直接用于鉴定表型差异。在后代中观察到表型严重的暂时性特征样式。首先，在一个短暂的"间隙"中产生不受影响的后代。这包括在注射时已出现的不透水的受精卵。第二，在这个间隙之后产生显示干扰表型的个体。第三，注射的动物产卵几天后生殖腺在某些情况下又"恢复"产生不完全受影响或表型正常的后代。

（高如丽 李梅 翻译；彭小忠 审稿）

21. Sulston, J. *et al.* The *C. elegans* genome sequencing project: a beginning. *Nature* **356**, 37-41 (1992).

22. Matzke, M. & Matzke, A. How and why do plants inactivate homologous (*trans*) genes? *Plant Physiol.* **107**, 679-685 (1995).

23. Ratcliff, F., Harrison, B. & Baulcombe, D. A similarity between viral defense and gene silencing in plants. *Science* **276**, 1558-1560 (1997).

24. Latham, K. X chromosome imprinting and inactivation in the early mammalian embryo. *Trends Genet.* **12**, 134-138 (1996).

25. Chalfie, M., Tu, Y., Euskirchen, G., Ward, W. & Prasher, D. Green fluorescent protein as a marker for gene expression. *Science* **263**, 802-805 (1994).

26. Clark, D., Suleman, D., Beckenbach, K., Gilchrist, E. & Baillie, D. Molecular cloning and characterization of the *dpy-20* gene of *C. elegans*. *Mol. Gen. Genet.* **247**, 367-378 (1995).

27. Mello, C. & Fire, A. DNA transformation. *Methods Cell Biol.* **48**, 451-482 (1995).

Supplementary information is available on *Nature*'s World-Wide Web site (http://www.nature.com) or as paper copy from Mary Sheehan at the London editorial office of *Nature*.

Acknowledgements. We thank A. Grishok, B. Harfe, M. Hsu, B. Kelly, J. Hsieh, M. Krause, M. Park, W. Sharrock, T. Shin, M. Soto and H. Tabara for discussion. This work was supported by the NIGMS (A.F.) and the NICHD (C.M.), and by fellowship and career awards from the NICHD (M.K.M.), NIGMS (S.K.), PEW charitable trust (C.M.), American Cancer Society (C.M.), and March of Dimes (C.M.).

Correspondence and requests for materials should be addressed to A.F. (e-mail: fire@mail1.ciwemb.edu).

Total Synthesis of Brevetoxin A

K. C. Nicolaou *et al.*

Editor's Note

The synthesis of "natural products" with complex molecular structures has long been a goal of organic chemists, partly to make pharmaceutically useful new substances but also to refine the synthetic tools at the chemist's disposal. Here K. C. Nicolaou of the Scripps Research Institute in California and his collaborators scale a new peak of the art, assembling the neurotoxin brevetoxin A that is produced by the algae responsible for the poisonous blooms known as red tides. This molecule has a carbon framework containing many ring structures, and contains 22 of the stereogenic centres—molecular groups with two mirror-image forms—that pose a particular challenge to organic chemists.

Brevetoxin A is the most potent neurotoxin secreted by *Gymnodinium breve Davis*, a marine organism often associated with harmful algal blooms known as "red tides"[1-3]. The compound, whose mechanism of action involves binding to and opening of sodium channels[4-7], is sufficiently toxic to kill fish at concentrations of nanograms per ml (refs 3, 4) and, after accumulation in filter-feeding shellfish, to poison human consumers. The precise pathway by which nature constructs brevetoxin A is at present unknown[8,9], but strategies for its total synthesis have been contemplated for some time. The synthetic challenge posed by brevetoxin A reflects the high complexity of its molecular structure: 10 oxygen atoms and a chain of 44 carbon atoms are woven into a polycyclic macromolecule that includes 10 rings (containing between 5 and 9 atoms) and 22 stereogenic centres. Particularly challenging are the 7-, 8- and 9-membered rings which allow the molecule to undergo slow conformational changes and force a 90° twist at one of its rings[1-6]. Here we describe the successful incorporation of methods that were specifically developed for the construction of these rings[10,11] into an overall strategy for the total synthesis of brevetoxin A in its naturally occurring form. The convergent synthesis reported here renders this scarce neurotoxin synthetically available and, more importantly, allows the design and synthesis of analogues for further biochemical studies.

FIGURE 1 shows the synthetic strategy for the construction of brevetoxin A (**1**, Fig. 1a) in bond disconnection (Fig. 1b) and retrosynthetic (Fig. 1c) formats. This analysis led to the expectation of using D-glucose (**4**, Fig. 1c) and D-mannose (**5**, Fig. 1c) as starting materials for the construction of the requisite BCDE (**2**, Fig. 1c) and GHIJ (**3**, Fig. 1c) advanced intermediates, respectively. The union of **2** and **3** under Horner–Wittig conditions was expected to furnish, in a highly convergent manner and after ring closure, the basic ring skeleton of the target molecule, from which brevetoxin A could be

短裸甲藻毒素 A 的全合成

尼科拉乌等

编者按

合成具有复杂分子结构的天然产物一直以来都是有机化学家的目标，这不仅是为了制造出药用的新物质，而且还可以更新供化学家使用的合成工具。在本篇文章中，加利福尼亚斯克里普斯研究所的尼科拉乌和他的同事们攀登上该领域的一座新高峰，合成出神经毒素短裸甲藻毒素 A。它是由恶性繁殖而引发赤潮的藻类产生的。这个分子具有包含很多环状结构的碳骨架，含有 22 个立体异构源中心——具有 2 种镜像形式的分子团——这对有机化学家们特别具有挑战性。

短裸甲藻毒素 A 是由短裸甲藻分泌的毒性最强的神经毒素，短裸甲藻是一种海洋生物，常与被称为"赤潮"的有害的藻华有关 [1-3]。短裸甲藻毒素 A 在浓度为纳克每毫升级别时就足以毒死鱼类（参考文献 3 和 4），它的作用机制涉及结合并开启钠离子通道 [4-7]。它经滤食性贝类积累后也可使人类食用者中毒。目前还不清楚自然界生成短裸甲藻毒素 A 的精确途径 [8,9]，但是我们对其全合成策略的深入探讨已经有一段时间了。短裸甲藻毒素 A 带来的合成挑战反映了其分子结构的高度复杂性：由 10 个氧原子和一条 44 个碳原子的链组成了一个含有 10 个环（含 5 到 9 个原子）和 22 个立体异构源中心的多环大分子。最富挑战性的是 7 元、8 元和 9 元环，它们使得分子进行缓慢的构象变化并迫使其中一个环发生了 90° 的扭转 [1-6]。现在我们将阐述一种成功的方法组合，这是特别为全合成以天然形式存在的短裸甲藻毒素 A 而开发的构建环的整体策略 [10,11]。下面介绍的汇聚合成使得这种稀有神经毒素的合成得以实现，更重要的是可以为进一步的生物化学研究设计和合成类似物。

图 1 以键的切断（图 1b）和逆合成（图 1c）反应式的形式显示了短裸甲藻毒素 A（**1**，图 1a）的合成策略。这个分析提示我们可以用 D–葡萄糖（**4**，图 1c）和 D–甘露糖（**5**，图 1c）作为合成的起始原料，分别构建必需的 BCDE（**2**，图 1c）和 GHIJ（**3**，图 1c）高级中间体。经环的闭合后，以高度汇聚的方式将结构 **2** 和 **3** 在霍纳–维蒂希条件下连接有望提供目标分子的基本环状骨架，再通过去保护和官能团转化可以得到短裸甲藻毒素 A。在计划的开始阶段，我们通过分析可以明显地看到，虽

93

fashioned through deprotections and functional-group manipulations. At the outset of this project, it was also evident from this analysis that although the 5- and 6-membered rings of the target molecule could be derived via well-developed synthetic methods[12], the three larger rings were not easily accessible through conventional techniques. Thus, to construct the 7-, 8- and 9-membered rings we used two important reactions, namely the silver-promoted hydroxydithioketal cyclization reaction[10] (rings F and G) and the palladium-catalysed functionalization of lactones via coupling of their cyclic ketene acetal phosphates[11] (rings B, D and E), that were specifically developed for this purpose.

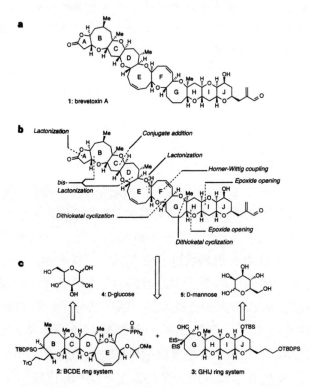

Fig. 1. Molecular structure (**a**), strategic bond disconnections (**b**), and retrosynthetic analysis (**c**) of brevetoxin A (**1**). Abbreviations for chemical groups: Me, methyl; Ph, phenyl; Tr, trityl; TBDPS, *t*-butyldiphenylsilyl; Et, ethyl; TBS, *t*-butyldimethylsilyl.

The synthesis of the BCDE ring system **2** (Figs 1c and 4) of brevetoxin A began with D-glucose (**4**, Fig. 2) and proceeded via intermediates **16** (Fig. 2) and **26** (Fig. 3). Thus, the D-glucose-derived bis(acetonide) **6**[13] (Fig. 2) was selectively cleaved with H_5IO_6 (ref. 14; for abbreviations see legends in figures) to afford the corresponding aldehyde, which reacted with MeMgBr leading to the expected secondary alcohol (84% overall yield). Swern oxidation of this alcohol, followed by reaction with allylmagnesium bromide in the presence of Ti(*i*-PrO)$_4$, led to the stereoselective formation of tertiary alcohol **7** (76.5% overall yield from **6**). Cleavage of the second acetonide moiety in **7** with EtSH-ZnCl$_2$ led to an open-chain thioketal, whose dibenzylation with NaH-BnBr and ring closure in the presence of I$_2$-aq.NaHCO$_3$ led smoothly to the 6-membered ring lactol **8** in 64.5% overall yield. Olefination of **8** with the appropriate stabilized Wittig reagent, followed by exposure to CSA

然目标分子中的 5 元和 6 元环能够用很成熟的合成方法得到 [12]，但是采用传统技术制备三个更大的环就不那么容易了。因此，为了构建 7 元、8 元和 9 元环，我们特地发明并使用了下面两种重要的反应：银促进的羟基二硫缩酮环化反应 [10]（F 和 G 环）和钯催化的经环状烯酮缩醛的磷酸酯的偶联反应对内酯的官能团化 [11]（B、D 和 E 环）。

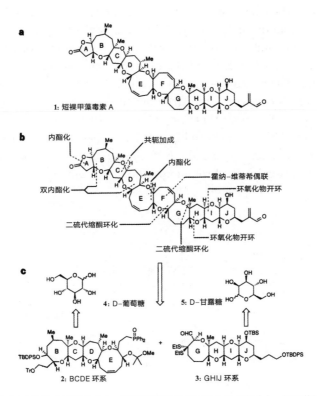

图 1. 短裸甲藻毒素 A(**1**) 的分子结构 (**a**)，键切断策略 (**b**) 和逆合成分析 (**c**)。化学基团的缩略语：Me，甲基；Ph，苯基；Tr，三苯甲基；TBDPS，叔丁基二苯基硅基；Et，乙基；TBS，叔丁基二甲基硅基。

从 D-葡萄糖 (**4**, 图 2) 开始，经由中间体 **16**(图 2) 和中间体 **26**(图 3) 合成短裸甲藻毒素 A 中的 BCDE 环系 **2**(图 1c 和图 4)。其过程是，用高碘酸选择性裂解由 D-葡萄糖衍生的双缩丙酮化合物 **6**[13](图 2) 得到相应的醛（参考文献 14；缩写请见图中示例），该醛接着与甲基溴化镁发生反应得到预期的仲醇（总收率为 84%）。此醇经过斯文氧化反应后，在四异丙氧基钛存在的条件下与烯丙基溴化镁反应，立体选择性地形成叔醇化合物 **7**(从 **6** 开始的总收率为 76.5%)。用乙硫醇-氯化锌裂解 **7** 上的第二个丙酮缩合部位得到一个开链的硫缩酮，再用氢化钠-溴化苄进行双苄基化，接着在碘-碳酸氢钠水溶液存在的条件下进行闭环，以 64.5% 的总收率顺利地得到了六元环的内半缩醛 **8**。用适当稳定化的维蒂希试剂对 **8** 进行烯化，接着用

furnished, stereoselectively, the tetrahydropyran system **9** (81% for two steps). Condensation of ketone **9** with CH$_2$=C(OMe)OTBS (ref. 15) in the presence of ZnBr$_2$ gave compound **10** as a mixture of diastereoisomers (inconsequential) in 93% yield. The terminal olefin in **10** was then converted to a methyl ketone via an oxymercuration-palladation procedure[16] (90% yield), and thence to a vinyl triflate by reaction with NaHMDS-PhNTf$_2$ furnishing intermediate **11** (93% yield). The latter compound (**11**) was coupled[17] with the zinc reagent IZnCH$_2$CH$_2$CO$_2$Me in the presence of Pd(Ph$_3$P)$_4$ catalyst to extend the "left-hand" side-chain (85% yield). The product was then sequentially deprotected (LiOH; then Li-in liquid NH$_3$) to afford the dihydroxydicarboxylic acid **12**. Yamaguchi[18] bis(lactonization) of **12** allowed the formation of bis(lactone) **13** (78.5% yield for three steps). Desilylation of **13** with HF·pyridine, followed by dehydration in the presence of Martin's sulphurane ([PhC(CF$_3$)$_2$O]$_2$SPh$_2$) (ref. 19), furnished the doubly unsaturated crystalline bis(lactone) **14** in 81% overall yield. (The melting point of **14** was 158–159 °C (ether/hexane)). An X-ray crystallographic analysis confirmed the structure of **14**. Sequential hydrogenation of the two olefinic bonds in **14**, first with Wilkinson's catalyst [(Ph$_3$P)$_3$RhCl] (**15**, 99% yield, ~4:1 ratio)[20] and then with Pd/C (**16**, 95% yield, ~19:1 ratio), established the two methyl groups in their correct stereochemical arrangement. Compound **16** was separated from its minor diastereoisomers by silica-gel flash chromatography.

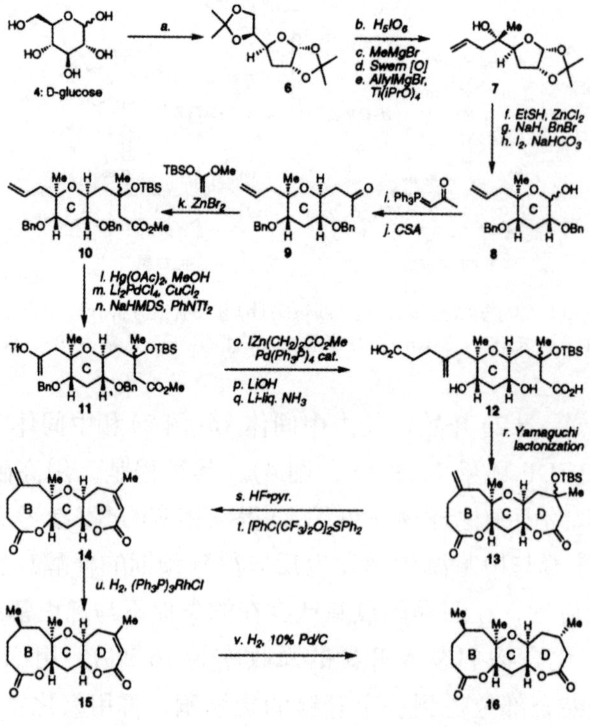

Fig. 2. Construction of the BCD bis-lactone system **16**. Reagents and conditions as follows. (a) See ref. 13; (b) H$_5$IO$_6$ (1.2 equiv.), EtOAc, 25 °C, 1 h; (c) MeMgBr (5.0 equiv.), Et$_2$O, 0 °C, 2 h, 84% for two

10-樟脑磺酸进行处理，得到立体选择性的四氢吡喃类化合物 **9**(两步收率为 81%)。在溴化锌存在的条件下用 1-甲氧基乙烯基叔丁基二甲基硅基醚(参考文献 15)与酮 **9** 进行缩合，以 93% 的收率得到非对映异构体(无关紧要)的混合物 **10**。**10** 中的末端烯烃经羟汞化-钯化过程转化为甲基酮 [16](收率为 90%)，接着与六甲基二硅基氨基钠和 N-苯基双三氟甲磺酰亚胺反应再转换成三氟甲磺酸乙烯酯，从而得到中间体 **11**(收率为 93%)。化合物 **11** 在催化剂四(三苯基膦)钯存在的条件下与锌试剂 2-甲氧羰基乙基碘化锌发生偶联反应 [17] 得到"左手"侧链延长的产物(收率为 85%)。接着对这个产物依次去保护(先是氢氧化锂，然后是锂的液氨溶液)，得到二羟基二羧酸化合物 **12**。对化合物 **12** 进行山口 [18] 双内酯化反应得到双内酯化合物 **13**(三步收率为 78.5%)。用氟化氢·吡啶对双内酯化合物 **13** 进行脱硅基化，然后在马丁硫烷([PhC(CF$_3$)$_2$O]$_2$SPh$_2$)的存在下脱水(参考文献 19)，得到双不饱和的双内酯化合物 **14** 的晶体(总收率为 81%)。**14** 的熔点为 158～159℃(乙醚/己烷)，由 X-射线晶体分析确定了其结构。依次对 **14** 的两个烯键加氢：先用威尔金森催化剂 [(Ph$_3$P)$_3$RhCl] 得到化合物 **15**(收率为 99%，比例约为 4:1) [20]，再用钯/碳得到化合物 **16**(收率为 95%，比例约为 19:1)，实现了两个甲基在立体化学上的正确排列。化合物 **16** 通过硅胶快速色谱法与其含量较少的非对映异构体分离开。

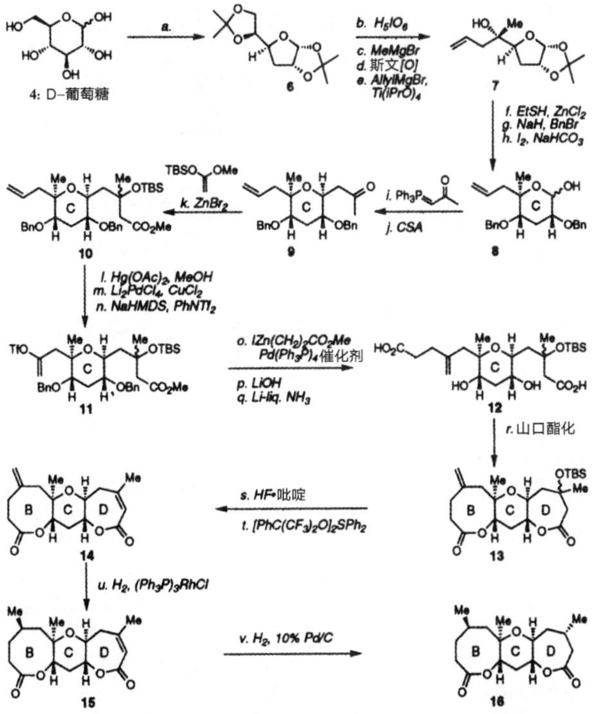

图 2. BCD 双内酯系 **16** 的构建过程。所用试剂和反应条件如下。(a)参见文献 13；(b)高碘酸(1.2 当量)，乙酸乙酯，25℃，1 h；(c)甲基溴化镁(5.0 当量)，乙醚，0℃，2 h，两步收率为 84%；(d)草酰

steps; (d) oxalyl chloride (1.8 equiv.), DMSO (2.2 equiv.), Et$_3$N (5.0 equiv.), CH$_2$Cl$_2$, $-78\,°C$, 1 h, 85%; (e) AllylMgBr (1.5 equiv.), Ti(i-PrO)$_4$ (1.5 equiv.), THF, $-78\,°C$, 2 h, 90%; (f) EtSH (20.0 equiv.), ZnCl$_2$ (5.0 equiv.), CH$_2$Cl$_2$, $0\,°C$, 1 h, 94%; (g) NaH (3.0 equiv.), BnBr (2.0 equiv.), n-Bu$_4$N$^{\oplus}$I$^{\ominus}$ (cat), THF, $0\,°C$, 5 h, 80%; (h) I$_2$ (3.3 equiv.), NaHCO$_3$ (6.5 equiv.), acetone, $25\,°C$, 1 h, 86%; (i) Ph$_3$P=CHCOMe (1.5 equiv.), toluene, $110\,°C$, 4 h; (j) CSA (0.1 equiv.), CHCl$_3$, $25\,°C$, 1 h, 81% for two steps; (k) CH$_2$=C(OMe)OTBS (1.5 equiv.), ZnBr$_2$ (0.5 equiv.), Et$_2$O, $-78\,°C$, 2 h, 93%; (l) Hg(OAc)$_2$ (1.1 equiv.), MeOH, $25\,°C$, 2 h; (m) PdCl$_2$ (0.1 equiv.), LiCl (0.2 equiv.), CuCl$_2$ (3.0 equiv.), MeOH, $70\,°C$, 2 h, 90% for two steps; (n) NaHMDS (1.2 equiv.), PhNTf$_2$ (1.2 equiv.), THF, $-78\,°C$, 1 h, 93%; (o) IZn(CH$_2$)$_2$CO$_2$Me (2.0 equiv.), Pd(Ph$_3$P)$_4$ (0.05 equiv.), DMA:benzene (1:10), $25\,°C$, 2 h, 85%; (p) LiOH (10.0 equiv.), THF:H$_2$O:MeOH (3:1:1), $65\,°C$, 3 h, 96%; (q) Li (12.0 equiv.), NH$_3$ (liq.), EtOH, $-78\,°C$, 15 min; (r) 2,4,6-trichlorobenzoyl chloride (2.2 equiv.), Et$_3$N (4.0 equiv.), THF, $0\,°C$, 30 min; then 4-DMAP (6.0 equiv.), benzene, $75\,°C$, 4 h, 82% for two steps; (s) HF·pyridine (3.0 equiv.), CH$_2$Cl$_2$, $0\,°C$, 3 h; (t) [PhC(CF$_3$)$_2$O]$_2$SPh$_2$ (1.5 equiv.), CH$_2$Cl$_2$, $0\,°C$, 15 min, 81% for two steps; (u) H$_2$, (Ph$_3$P)$_3$RhCl (0.1 equiv.), benzene, $25\,°C$, 5 h, 99%, ~4:1 ratio; (v) H$_2$, 10% Pd/C (0.1 equiv.), EtOAc, $25\,°C$, 12 h, 95%, ~19:1 ratio. CSA, 10-camphorsulphonic acid; DMA, N,N-dimethylacetamide; 4-DMAP, 4-N-dimethylaminopyridine; DMF, N,N-dimethylformamide; DMSO, dimethylsulphoxide; NMO, 4-methylmorpholine-N-oxide; TBS, t-BuMe$_2$Si; THF, tetrahydrofuran; Tf, triflate. NaHMDS, sodium hexamethyldisilazide; PhNHTf$_2$, N-phenyltrifluoromethanesulphonimide. For selected physical data for compound **14**, see Supplementary Information.

Fig. 3. Construction of the BCDE lactone **26**. Reagents and conditions as follows. (a) KHMDS (4.0 equiv.), (PhO)$_2$P(O)Cl (6.0 equiv.), HMPA (6.0 equiv.), THF, $-78\,°C$, 1 h, 80%; (b) Me$_3$SnSnMe$_3$ (8.0 equiv.), LiCl (6.0 equiv.), Pd(Ph$_3$P)$_4$ (0.1 equiv.), THF, $70\,°C$, 2 h, 81%; (c) n-BuLi (1.55 M in hexanes, 4.0 equiv.), Cu-C≡C-n-Pr (4.4 equiv.), HMPT (8.8 equiv.), THF, $-78 \to -30\,°C$, 2 h; (d) TfOCH$_2$CH$_2$OBn (7.0 equiv.), THF, $-78 \to -30\,°C$, 12 h, 65% for two steps; (e) thexylborane (4.0 equiv.), THF, $-30 \to 0\,°C$, 3 h; then H$_2$O$_2$ (20.0 equiv.), NaOH (20.0 equiv.), H$_2$O, $0\,°C$, 2 h, 80%; (f) t-BuPh$_2$SiCl (1.3 equiv.), imidazole (3.0 equiv.), DMF, $25\,°C$, 8 h, 90% (five recycles); (g) 10% Pd/C (10% wt. equiv.), H$_2$, MeOH, $25\,°C$, 2 h, 93%; (h) cyclohexanone dimethylketal (1.3 equiv.), PPTS (0.1 equiv.), CH$_2$Cl$_2$, $25\,°C$, 2 h, 85%; (i) PivCl (1.3 equiv.), 4-DMAP (1.4 equiv.), CH$_2$Cl$_2$, $25\,°C$, 0.5 h, 99%; (j) CSA (0.1 equiv.), MeOH:CH$_2$Cl$_2$ (2:1), $25\,°C$, 0.5 h, 88%; (k) TrCl·4-DMAP (4.0 equiv.), CH$_2$Cl$_2$, $40\,°C$, 12 h, 95%; (l) Ac$_2$O (1.5 equiv.), 4-DMAP (0.5 equiv.), CH$_2$Cl$_2$, $25\,°C$, 15 min, 99%; (m) TFA (1.5 equiv.), MeOH (10 equiv.), CH$_2$Cl$_2$, $25\,°C$, 30 min, 90%; (n) TPAP (0.1 equiv.), NMO (2.0 equiv.), CH$_2$Cl$_2$, $25\,°C$, 30 min, 97%; (o) $^{\ominus}$Br$^{\oplus}$PPh$_3$(CH$_2$)$_3$CO$_2$Me (2.5 equiv.), KHMDS (2.0 equiv.), THF, $-78\,°C$, 1 h, 83%; (p) LiOH (10.0 equiv.), THF:H$_2$O:MeOH

氯(1.8 当量)，二甲基亚砜(2.2 当量)，三乙胺(5.0 当量)，二氯甲烷，−78℃，1 h，收率为 85%；(e)烯丙基溴化镁(1.5 当量)，四异丙氧基钛(1.5 当量)，四氢呋喃，−78℃，2 h，收率为 90%；(f)乙硫醇(20.0 当量)，氯化锌(5.0 当量)，二氯甲烷，0℃，1 h，收率为 94%；(g)氢化钠(3.0 当量)，溴化苄(2.0 当量)，碘化四正丁基铵(催化剂)，四氢呋喃，0℃，5 h，收率为 80%；(h)碘(3.3 当量)，碳酸氢钠(6.5 当量)，丙酮，25℃，1 h，收率为 86%；(i)2−氧代亚丙基三苯基膦(1.5 当量)，甲苯，110℃，4 h；(j)10−樟脑磺酸(0.1 当量)，三氯甲烷，25℃，1 h，两步收率为 81%；(k)1−甲氧基乙烯基叔丁基二甲基硅醚(1.5 当量)，溴化锌(0.5 当量)，乙醚，−78℃，2 h，收率为 93%；(l)乙酸汞(1.1 当量)，甲醇，25℃，2 h；(m)氯化钯(0.1 当量)，氯化锂(0.2 当量)，氯化铜(3.0 当量)，甲醇，70℃，2 h，两步收率为 90%；(n)六甲基二硅基氨基钠(1.2 当量)，N−苯基双三氟甲磺酰亚胺(1.2 当量)，四氢呋喃，−78℃，1 h，收率为 93%；(o)2−甲氧羰基乙基碘化锌(2.0 当量)，四(三苯基膦)钯(0.05 当量)，N,N−二甲基乙酰胺：苯(1:10)，25℃，2 h，收率为 85%；(p)氢氧化锂(10.0 当量)，四氢呋喃：水：甲醇(3:1:1)，65℃，3 h，收率为 96%；(q)锂(12.0 当量)，液氨，乙醇，−78℃，15 min；(r)2,4,6−三氯苯甲酰氯(2.2 当量)，三乙胺(4.0 当量)，四氢呋喃，0℃，30 min；接着 4−二甲氨基吡啶(6.0 当量)，苯，75℃，4 h，两步收率为 82%；(s)氟化氢·吡啶(3.0 当量)，二氯甲烷，0℃，3 h；(t)二(双三氟甲基苄氧基)二苯基硫砜(1.5 当量)，二氯甲烷，0℃，15 min，两步收率为 81%；(u)氢气，三(三苯基膦)氯化铑(0.1 当量)，苯，25℃，5 h，收率为 99%，比例约 4:1；(v)氢气，10% 钯/碳(0.1 当量)，乙酸乙酯，25℃，12 h，收率为 95%，比例约 19:1。CSA，10−樟脑磺酸；DMA，N,N−二甲基乙酰胺；4−DMAP，4−二甲氨基吡啶；DMF，N,N−二甲基甲酰胺；DMSO，二甲基亚砜；NMO，4−甲基吗啉−N−氧化物；TBS，叔丁基二甲基硅基；THF，四氢呋喃；Tf，三氟甲磺酰基。NaHMDS，六甲基二硅基氨基钠；PhNHTf₂，N−苯基双三氟甲磺酰亚胺。关于化合物 14 的重要物理数据，请参见补充信息。

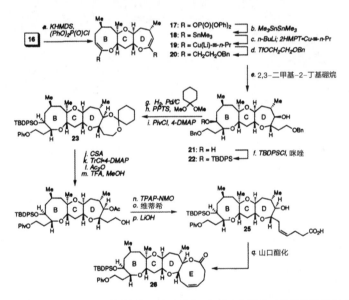

图 3. BCDE 内酯 26 的构建过程。所用试剂和反应条件如下。(a)六甲基二硅基氨基钾(4.0 当量)，二苯氧基磷酰氯(6.0 当量)，六甲基磷酰胺(6.0 当量)，四氢呋喃，−78℃，1 h，收率为 80%；(b)六甲基二锡(8.0 当量)，氯化锂(6.0 当量)，四(三苯基膦)钯(0.1 当量)，四氢呋喃，70℃，2 h，收率为 81%；(c)正丁基锂(1.55 M 的己烷溶液，4.0 当量)，正戊炔基铜(4.4 当量)，六甲基磷酰三胺(8.8 当量)，四氢呋喃，−78℃→−30℃，2 h；(d)三氟甲磺酸(2−苄氧基)乙基酯(7.0 当量)，四氢呋喃，−78℃→−30℃，12 h，两步收率为 65%；(e)2,3−二甲基−2−丁硼烷(4.0 当量)，四氢呋喃，−30℃→0℃，3 h；接着双氧水(20.0 当量)，氢氧化钠(20.0 当量)，水，0℃，2 h，收率为 80%；(f)叔丁基二苯基氯硅烷(1.3 当量)，咪唑(3.0 当量)，N,N−二甲基甲酰胺，25℃，8 h，收率为 90%(循环五次)；(g)10% 钯/碳(10% 重量当量)，氢气，甲醇，25℃，2 h，收率为 93%；(h)环己酮二甲缩酮(1.3 当量)，对甲苯磺酸吡啶盐(0.1 当量)，二氯甲烷，25℃，2 h，收率为 85%；(i)2,2−二甲基丙酰氯(1.3 当量)，4−二甲氨基吡啶(1.4 当量)，二氯甲烷，25℃，0.5 h，收率为 99%；(j)10−樟脑磺酸(0.1 当量)，甲醇：二氯甲烷(2:1)，25℃，0.5 h，收率为 88%；(k)三苯甲基氯·4−二甲氨基吡啶(4.0 当量)，二氯甲烷，40℃，12 h，收率为 95%；(l)乙酸酐(1.5 当量)，4−二甲氨基吡啶(0.5 当量)，二氯甲烷，25℃，15 min，收

(3:2:1), 0 °C, 4 h; then 4-DMAP (4.0 equiv.), benzene, 80 °C, 2 h, 89% for two steps. 4-DMAP, 4-N-dimethylaminopyridine; DMF, N,N-dimethylformamide; KHMDS, potassium hexamethyldisilazide; HMPT, hexamethylphosphorous triamide; NMO, 4-methyl-morpholine-N-oxide; Piv, pivalate; PPTS, pyridinium p-toluenesulphonate; TPAP, tetra-n-propylammonium perruthenate; TBDPS, t-BuPh$_2$Si; Tr, trityl. For selected physical data for compound **26**, see Supplementary Information.

Fig. 4. Construction of the BCDE phosphine oxide **2**. Reagents and conditions as follows. (a) KHMDS (3.0 equiv.), (PhO)$_2$P(O)Cl (5.0 equiv.), THF, −78 °C, 1 h; (b) (n-Bu)$_3$SnCH−CH$_2$ (3.0 equiv.), LiCl (5.0 equiv.), Pd(Ph$_3$P)$_4$ (0.1 equiv.), THF, 75 °C, 2 h, 81% for two steps; (c) O$_2$, TPP (0.01 equiv.), CCl$_4$, 25 °C, 0.3 h; (d) Al(Hg) (excess), H$_2$O:THF (1:29), 25 °C, 2 h, 58% for two steps; (e) t-Bu$_2$SiCl (1.5 equiv.), imidazole (10.0 equiv.), CH$_2$Cl$_2$, 25 °C, 1 h, 91%; (f) TPAP (0.1 equiv.), NMO (2.0 equiv.), 4-Å MS, CH$_2$Cl$_2$, 25 °C, 1 h, 82%; (g) [(Ph$_3$P)CuH]$_6$ (2.0 equiv.), benzene, 25 °C, 72 h, 70%; (h) DIBAL (2.5 equiv.), CH$_2$Cl$_2$, −78 °C, 30 min, 95%, ~5:1 ratio; (i) TrCl·4-DMAP (15.0 equiv.), CH$_2$Cl$_2$, 40 °C, 24 h, 75% pure trityl ether after silica-gel chromatography; (j) POCl$_3$ (0.01 equiv.), CH$_2$=CMe(OMe) (solvent), 25 °C, 6 h, 95%; (k) neutral alumina-1% H$_2$O activated by heating (excess), hexane, 25 °C, 2 h, 96%; (l) MsCl (2.0 equiv.), Et$_3$N (4.0 equiv.), CH$_2$Cl$_2$, 0 °C, 15 min, 99%; (m) Ph$_2$PLi (3.0 equiv.), HMPA (3.0 equiv.), THF, 0 °C, 30 min; then 5% aq. H$_2$O$_2$ (excess), 93%. DIBAL, diisobutylaluminium hydride; 4-DMAP, 4-N-dimethylaminopyridine; HMPA, hexamethylphosphoramide; KHMDS, potassium hexamethyldisilazide; MS, molecular sieves; NMO, 4-methylmorpholine-N-oxide; TBDPS, t-BuPh$_2$Si; TBS, t-BuMe$_2$Si; THF, tetrahydrofuran; TPAP, tetra-n-propylammonium perruthenate; TPP, tetraphenylporphyrin; Tr, trityl. Ms, CH$_3$SO$_2$ or methanesulphonyl; imid., imidazole. For selected physical data for compound **2**, see Supplementary Information.

The conversion of bis(lactone) **16** to the tetracyclic 9-membered ring lactone **26** relied on a two-directional approach similar to that previously reported from these laboratories[21] and on a newly developed method for the transformation of lactones to cyclic ethers involving cyclic ketene acetal phosphates[11]. Figure 3 summarizes the sequence for this

率为 99%；(m) 三氟乙酸 (1.5 当量)，甲醇 (10.0 当量)，二氯甲烷，25℃，30 min，收率为 90%；(n) 四正丙基过钌酸铵 (0.1 当量)，4-甲基吗啉-N-氧化物 (2.0 当量)，二氯甲烷，25℃，30 min，收率为 97%；(o) 溴化三苯基-3-甲氧羰基正丙基磷鎓盐 (2.5 当量)，六甲基二硅基氨基钾 (2.0 当量)，四氢呋喃，-78℃，1 h，收率为 83%；(p) 氢氧化锂 (10.0 当量)，四氢呋喃：水：甲醇 (3:2:1)，0℃，4 h；接着 4-二甲氨基吡啶 (4.0 当量)，苯，80℃，2 h，两步收率为 89%。4-DMAP，4-二甲氨基吡啶；DMF，N,N-二甲基甲酰胺；KHMDS，六甲基二硅基氨基钾；HMPT，六甲基磷酰三胺；NMO，4-甲基吗啉-N-氧化物；Piv，2,2-二甲基丙酰基；PPTS，对甲苯磺酸吡啶盐；TPAP，四正丙基过钌酸铵；TBDPS，叔丁基二苯基硅基；Tr，三苯甲基。关于化合物 26 的重要物理数据，请参见补充信息。

图 4. BCDE 膦氧化物 2 的构建过程。所用试剂和反应条件如下。(a) 六甲基二硅基氨基钾 (3.0 当量)，二苯氧基磷酰氯 (5.0 当量)，四氢呋喃，-78℃，1 h；(b) 三正丁基乙烯基锡 (3.0 当量)，氯化锂 (5.0 当量)，四 (三苯基膦) 钯 (0.1 当量)，四氢呋喃，75℃，2 h，两步收率为 81%；(c) 氧气，四苯基卟啉 (0.01 当量)，四氯化碳，25℃，0.3 h；(d) 铝汞齐 (过量)，水：四氢呋喃 (1:29)，25℃，2 h，两步收率为 58%；(e) 二叔丁基氯硅烷 (1.5 当量)，咪唑 (10.0 当量)，二氯甲烷，25℃，1 h，收率为 91%；(f) 四正丙基过钌酸铵 (0.1 当量)，4-甲基吗啉-N-氧化物 (2.0 当量)，4 Å 分子筛，二氯甲烷，25℃，1 h，收率为 82%；(g) 三苯基膦氢化铜六聚物 (2.0 当量)，苯，25℃，72 h，收率为 70%；(h) 二异丁基氢化铝 (2.5 当量)，二氯甲烷，-78℃，30 min，收率为 95%，比例约为 5:1；(i) 三苯基甲基氯·4-二甲氨基吡啶 (15.0 当量)，二氯甲烷，40℃，24 h，硅胶色谱法纯化后收率为 75% 纯三苯基甲基醚；(j) 三氯氧磷 (0.01 当量)，2-甲氧基丙烯 (溶剂)，25℃，6 h，收率为 95%；(k) 含有 1% 水分的中性氧化铝 (加热活化，过量)，己烷，25℃，2 h，收率为 96%；(l) 甲磺酰氯 (2.0 当量)，三乙胺 (4.0 当量)，二氯甲烷，0℃，15 min，收率为 99%；(m) 二苯基膦化锂 (3.0 当量)，六甲基磷酰胺 (3.0 当量)，四氢呋喃，0℃，30 min；接着 5% 双氧水溶液 (过量)，收率为 93%。DIBAL，二异丁基氢化铝；4-DMAP，4-二甲氨基吡啶；HMPA，六甲基磷酰胺；KHMDS，六甲基二硅基氨基钾；MS，分子筛；NMO，4-甲基吗啉-N-氧化物；TBDPS，叔丁基二苯基硅基；TBS，叔丁基二甲基硅基；THF，四氢呋喃；TPAP，四正丙基过钌酸铵；TPP，四苯基卟啉；Tr，三苯甲基。Ms，甲磺酰基；imid.，咪唑。关于化合物 2 的重要物理数据，请参见补充信息。

　　由双内酯化合物 16 到含有 9 元环内酯的四环化合物 26 的转化依赖于一个双向的反应方法 (类似于一些实验室先前报道的方法 [21]) 和一个新开发的使用环状烯酮缩醛的磷酸酯将内酯转换为环醚的方法 [11]。图 3 中总结了这个转化的顺序。将双内酯

conversion. The bis(lactone) **16** was added to a solution of KHMDS and (PhO)$_2$P(O)Cl, furnishing bis(phosphate) **17** (80% yield), which was coupled with Me$_3$SnSnMe$_3$ in the presence of Pd(Ph$_3$P)$_4$ catalyst and LiCl to afford the bis(vinylstannane) **18** (81% yield). Reaction of **18** with *n*-BuLi, followed by addition of *n*-PrC $\equiv$ CCu · 2HMPT (ref. 22) and TfOCH$_2$CH$_2$OBn, led to bis(vinylether) **20** in 65% overall yield via mixed cuprate **19**. Hydroboration of **20** with thexylborane, followed by basic hydrogen peroxide work-up (80% yield) and selective monosilylation (90% yield, five recycles) as previously described[21], led to compound **22** via **21**. Having simultaneously functionalized bis(lactone) **16** on both sides, and then successfully differentiating the two ends, we were now in a position to resume the stepwise manipulation of the molecule towards the 9-membered ring lactone **26**. Thus, hydrogenolysis of the two benzyl ethers in **22** (H$_2$, 10% Pd/C, 93% yield) and ketalization with cyclohexanone dimethylketal in the presence of PPTS (85% yield), followed by pivalate formation (PivCl, 4-DMAP, 99% yield) at the remaining primary alcohol, gave compound **23**. The ketal was then disassembled under acidic conditions (CSA, 88% yield), and the two hydroxyl groups so released were differentiated by trityl ether formation (TrCl · 4-DMAP, primary alcohol, 95% yield), acetylation (Ac$_2$O, 4-DMAP, secondary alcohol, 99% yield) and trityl ether cleavage (TFA, 90% yield) to afford hydroxy acetate **24**. The 9-membered ring lactone was then built on the "right side" of the molecule by (1) TPAP-NMO oxidation[23] of the primary alcohol to the corresponding aldehyde (97% yield); (2) olefination with the ylide derived from Br$^{\ominus}$Ph$_3$P$^{\oplus}$(CH$_2$)$_3$CO$_2$Me and KHMDS (83% yield); (3) saponification of both the methylester and the acetate groups (LiOH) to afford hydroxyacid **25**; and (4) Yamaguchi[18] lactonization, furnishing the targeted intermediate **26** (89% for two steps).

The functionalization of the 9-membered ring lactone to the desired E ring ether functionality was achieved via sequential application of the cyclic ketene acetal phosphate-palladium coupling technology[11] and a selective[24] singlet oxygen 4+2 cycloaddition reaction of a constructed conjugated diene system. Figure 4 outlines the chemistry leading from **26** to the targeted phosphine oxide **2**. Conversion of **26** to its ketene acetal phosphate by reaction with KHMDS and (PhO)$_2$P(O)Cl, followed by palladium-catalysed coupling with tri-*n*-butylvinyltin, resulted, via **27**, in the formation of conjugated diene **28** (81% overall yield), thus accomplishing the crucial transformation of a lactone to a cyclic ether[11]. The selective functionalization of the diene system in **28** in the presence of the additional double bond within the 9-membered ring, was achieved by a novel 4+2 cycloaddition reaction[24] involving singlet O$_2$, generated by light in the presence of tetraphenylporphyrin (TPP) as a sensitizer. The formed endoperoxide **29** (~1:1 ratio of diastereoisomers, inconsequential) was then converted to the corresponding diol (58% yield for two steps) by reductive cleavage of the O—O bond with Al(Hg) in THF-H$_2$O.

化合物 **16** 加入到六甲基二硅基氨基钾和二苯氧基磷酰氯的溶液中得到双磷酸酯化合物 **17**(收率为 80%)，接着在四(三苯基膦)钯催化剂和氯化锂的存在下与六甲基二锡进行偶联得到双乙烯基锡烷化合物 **18**(收率为 81%)。用正丁基锂与化合物 **18** 反应，接着加入正戊炔基铜·二(六甲基磷酰三胺)(参考文献 22)和三氟甲磺酸(2–苄氧基)乙基酯，经过一个混合的铜盐化合物 **19** 以 65% 的总收率得到双烯醚化合物 **20**。用 2,3–二甲基–2–丁基硼烷对化合物 **20** 进行硼氢化反应，接着用碱性过氧化氢处理(收率为 80%)并以先前报道的方式进行选择性单硅烷化反应[21](收率 90%，循环五次)，经过化合物 **21** 得到化合物 **22**。我们已经对双内酯化合物 **16** 的两个侧链同时进行了官能化，并成功地区分了两个末端，现在处于为了得到 9 元环内酯化合物 **26** 继续对分子进行逐步操作的过程中。于是，对化合物 **22** 中的两个苄基醚进行氢解(H_2，10% 钯/碳，收率为 93%)，然后在对甲苯磺酸吡啶盐存在的条件下用环己酮二甲缩酮进行缩酮化反应(收率为 85%)，接着在剩余的伯醇上用 2,2–二甲基丙酰氯和 4–二甲氨基吡啶进行 2,2–二甲基丙酰化(收率为 99%)，得到化合物 **23**。在酸性条件下缩酮进行分解(10–樟脑磺酸，收率 88%)，释放出的两个羟基经以下反应得到区分：用三苯甲基氯·4–二甲氨基吡啶进行伯醇的三苯甲基醚化(收率为 95%)，用乙酸酐和 4–二甲氨基吡啶进行仲醇的乙酰化(收率为 99%)，然后用三氟乙酸脱掉三苯甲基(收率为 90%)。最后得到羟基乙酸酯化合物 **24**。接下来在这个分子的"右侧"进行 9 元环内酯的构建：(1)在伯醇上用四正丙基过钌酸铵和 4–甲基吗啉–N–氧化物进行氧化[23]得到相应的醛(收率为 97%)；(2)用叶立德(由溴化三苯基–3–甲氧羰基正丙基磷鎓盐和六甲基二硅基氨基钾得到)进行烯化反应(收率为 83%)；(3)用氢氧化锂对甲酯和乙酸酯基团进行皂化反应得到羟基酸化合物 **25**；(4)通过山口[18]内酯化，得到目标中间产物 **26**(两步收率为 89%)。

对 9 元环内酯进行官能化以便获得醚化的 E 环是通过如下反应实现的：依次使用环烯酮缩醛的磷酸酯–钯偶联技术[11]和在所构建的共轭二烯上进行选择性[24]单重态氧的 4+2 环加成反应。图 4 概述了从化合物 **26** 到目标膦氧化物 **2** 的化学过程。将 **26** 转化为它的烯酮缩醛磷酸酯是通过与六甲基二硅基氨基钾和二苯氧基磷酰氯反应，经过化合物 **27**，接着在钯催化下与三正丁基乙烯基锡偶联，形成了共轭二烯 **28**(总收率为 81%)，从而完成了内酯到环醚的关键性转化[11]。在 9 元环中多余双键存在的情况下，通过一种新的单重态氧参与的 4+2 环加成反应[24]实现了对化合物 **28** 的双烯系统的选择性官能化，其中单重态氧是以四苯基卟啉作为敏化剂经光照生成的。生成的内过氧化物 **29**(非对映异构体的比例约为 1:1，无关紧要)经 O—O 键在四氢呋喃和水中被铝汞齐还原断裂，转化成相应的二醇化合物(两步的收率为 58%)。用叔丁基二甲基氯硅烷和咪唑对所得二醇中的伯羟基进行选择性硅烷化(收率为

The primary hydroxyl group of the resulting diol was selectively silylated with TBSCl-imid. (91% yield), and the secondary alcohol was oxidized with TPAP-NMO (ref. 23; 82% yield), leading to enone **30**.

The proper functionality and stereochemistry was established on ring E by sequential reduction of the double bond ([(Ph$_3$P)CuH]$_6$, 70% yield)[25] and of the carbonyl group (DIBAL). The DIBAL reduction also cleaved the pivalate group, furnishing diol **31** (95% yield, ~5:1 ratio of diastereoisomers, separated at a later stage). The resulting diol (**31**) was protected by sequential formation of a trityl ether at the primary position (TrCl · 4-DMAP, separation of diastereoisomers with silica-gel flash chromatography, 75% pure trityl ether) and of a mixed methoxyketal at the secondary position [CH$_2$ = CMe(OMe), POCl$_3$ catalyst], before selective removal of the TBS group (alumina)[26] to afford primary alcohol **32** (91% yield for two steps). Finally, mesylation of **32** (MsCl, Et$_3$N), followed by displacement with LiPPh$_2$ and oxidation with hydrogen peroxide, furnished the desired phosphine oxide **2** (BCDE ring system) in 92% overall yield.

The construction of the other requisite advanced intermediate, GHIJ ring system **3**, is shown in Fig. 5. Thus, D-mannose (**5**) was converted to tricyclic dithioketal-aldehyde **33** by a modification of our previously published[27] multi-step sequence (full details will be published elsewhere). Generation of the ylide from phosphonium salt **34**[27] (n-BuLi) and coupling with aldehyde **33** (THF, HMPA, −78 °C) resulted, after selective removal of the TBS group, in the stereoselective formation of Z-olefin **35** in 73% overall yield. Hydroxydithioketal cyclization[10] of **35**, facilitated by AgClO$_4$ in the presence of NaHCO$_3$, SiO$_2$, 4-Å MS in MeNO$_2$, resulted in the formation of the mixed 8-membered ring thioketal **36** in 92% yield. The introduction of the α-methyl group at the fusion of rings G and H was accomplished by oxidation of **36** to sulphone **37** (mCPBA, 93% yield) followed by reaction with AlMe$_3$ to afford **38** (94% yield). The well-defined conformation of the intermediate oxonium species in the latter reaction is presumed to be responsible for the observed stereochemical control. The benzylidene group in **38** was then removed by exposure to EtSH in the presence of Zn(OTf)$_2$ and NaHCO$_3$ (92% yield), and the resulting diol was differentiated by silylation (TBSCl-imid., primary silylether, 92% yield), followed by acetylation (Ac$_2$O, pyridine, secondary acetate, 95% yield), leading to compound **39**. Hydrogenation of the double bond in **39** in the presence of Pd(OH)$_2$/C resulted in concomitant debenzylation, furnishing the corresponding saturated alcohol in 90% yield. The newly liberated secondary hydroxyl group was then protected as a TBS ether (TBSOTf-2,6-lutidine, 87% yield), and the acetate group on ring G was cleaved by the action of K$_2$CO$_3$ in MeOH (93% yield). The resulting secondary alcohol was then oxidized to ketone **40** with TPAP-NMO (ref. 23) in 94% yield. Finally, exposure of **40** to EtSH in the presence of Zn(OTf)$_2$ led to the corresponding dithioketal with concomitant desilylation of the neighbouring oxygen (92% yield), and was followed by oxidation (SO$_3$ · pyridine, DMSO, Et$_3$N)[28] to afford the targeted GHIJ ring system **3**, in 87% overall yield.

91%），再用四正丙基过钌酸铵和 4–甲基吗啉–*N*–氧化物对仲醇进行氧化（参考文献 23；收率为 82%），得到烯酮化合物 **30**。

通过依次还原双键（三苯基膦氢化铜六聚物，收率 70%）[25] 和羰基（二异丁基氢化铝）在 E 环上建立起正确的官能团和立体化学。二异丁基氢化铝同样也会脱除 2,2–二甲基丙酰基，并得到二醇化合物 **31**（收率为 95%，非对映异构体的比例约为 5∶1，在后面的阶段进行分离）。依次对所得二醇 **31** 进行保护，在伯醇上进行三苯甲基醚化（三苯甲基氯·4–二甲氨基吡啶，用硅胶快速色谱法分离非对映异构体，得到 75% 纯的三苯甲基醚），在仲醇上进行混合甲氧基缩酮化（2–甲氧基丙烯，三氯氧磷作催化剂），接着用氧化铝选择性地脱除叔丁基二甲基硅基 [26]，得到伯醇化合物 **32**（两步的收率为 91%）。最后用甲磺酰氯/三乙胺对 **32** 进行甲磺酰化，接着用二苯基膦化锂置换甲磺酰基，再用过氧化氢氧化，得到预期的膦氧化物 **2**（BCDE 环系，总收率为 92%）。

图 5 列出了另一个必需的高级中间体（GHIJ 环系 **3**）的构建过程。其过程是，D–甘露糖（**5**）通过改进的我们先前报道 [27] 的多步连续反应方式转化成二硫缩酮–醛的三环化合物 **33**（合成细节将另行报道）。镂盐 **34**[27] 与正丁基锂反应生成的叶立德与醛化合物 **33** 在 −78℃、六甲基磷酰胺和四氢呋喃的条件下进行偶联，然后选择性脱去叔丁基二甲基硅基，立体选择性地得到 *Z* 构型烯烃化合物 **35**（总收率为 73%）。在碳酸氢钠、二氧化硅、4 Å 分子筛的硝基甲烷溶液中用高氯酸银对 **35** 进行羟基二硫缩酮的环化反应 [10]，结果形成了含有 8 元环的混合硫缩酮化合物 **36**（收率为 92%）。用间氯过氧苯甲酸对化合物 **36** 进行氧化得到砜化物 **37**（收率为 93%），接着再与三甲基铝反应得到化合物 **38**（收率为 94%），从而在 G 环和 H 环的连接处引入了 *α*–甲基。在后面的反应中形成的氧鎓中间体具有明确的构象，我们推测是它控制着所观察到的立体化学。接着在三氟甲磺酸锌和碳酸氢钠的存在下用乙硫醇处理使化合物 **38** 脱掉苯亚甲基得到二醇（收率为 92%），两个羟基发生不同反应，其中伯羟基用叔丁基二甲基氯硅烷和咪唑进行硅烷基化（伯硅醚收率为 92%），接着用乙酸酐和吡啶进行仲羟基的乙酰化（仲乙酸酯收率为 95%），得到化合物 **39**。在氢氧化钯/碳存在的条件下，化合物 **39** 的双键进行氢化并伴随着脱苄基化，以 90% 的收率得到相应的饱和醇，然后对新产生的仲羟基用叔丁基二甲基硅基三氟甲磺酸酯和 2,6–二甲基吡啶进行保护形成叔丁基二甲基硅醚（收率为 87%），接着在碳酸钾的甲醇溶液中脱去 G 环上的乙酸酯（收率为 93%）。再用四正丙基过钌酸铵和 4–甲基吗啉–*N*–氧化物（参考文献 23）对得到的仲醇进行氧化，得到酮化合物 **40**（收率为 94%）。最后，在三氟甲磺酸锌的条件下用乙硫醇处理化合物 **40** 得到相应的二硫缩酮化合物并伴随着邻近氧原子的脱硅烷化（收率为 92%），用三氧化硫·吡啶在三乙胺和二甲基亚砜条件下进行氧化 [28] 得到了目标的 GHIJ 环系化合物 **3**（总收率为 87%）。

Fig. 5. Construction of the GHIJ ring system **3**. Reagents and conditions as follows. (a) See ref. 27; (b) add *n*-BuLi (1.55 M in hexanes, 1.0 equiv.) to **34** (1.0 equiv.) and HMPA (10.0 equiv.) in THF; then add **33**, THF, −78 °C, 1 h; then 25 °C, 1.5 h, 88%; (c) TBAF (1.1 equiv.), THF, 25 °C, 7 h, 83%; (d) AgClO₄ (3.0 equiv.), NaHCO₃ (3.0 equiv.), SiO₂, 4-Å MS, MeNO₂, 25 °C, 3 h, 92%; (e) *m*CPBA (2.0 equiv.), CH₂Cl₂, 0 °C, 2 h, 93%; (f) AlMe₃ (3.0 equiv.), CH₂Cl₂, −78 °C, 1 h, 94%; (g) EtSH (20 equiv.), Zn(OTf)₂ (0.3 equiv.), NaHCO₃ (0.5 equiv.), CH₂Cl₂, 25 °C, 4 h, 92%; (h) TBSCl (1.1 equiv.), imidazole (1.5 equiv.), CH₂Cl₂, 25 °C, 1 h, 92%; (i) Ac₂O (1.1 equiv.), 4-DMAP (1.3 equiv.), CH₂Cl₂, 25 °C, 2.5 h, 95%; (j) H₂, 10% Pd(OH)₂/C (0.15 equiv.), AcOH, 25 °C, 5 h, 90%; (k) TBSOTf (2.5 equiv.), 2,6-lutidine (3.0 equiv.), CH₂Cl₂, 0 °C, 0.5 h, 87%; (l) K₂CO₃ (0.5 equiv.), MeOH, 25 °C, 4 h, 93%; (m) TPAP (0.05 equiv.), NMO (2.0 equiv.), CH₂Cl₂, 25 °C,1 h, 94%; (n) EtSH (20.0 equiv.), Zn(OTf)₂ (0.3 equiv.), CH₂Cl₂, 25 °C, 2.5 h, 92%; (o) SO₃ · pyridine (3.0 equiv.), DMSO (10.0 equiv.), Et₃N (10.0 equiv.), CH₂Cl₂, 0 °C, 1.5 h, 87%. 4-DMAP, 4-*N*-dimethylaminopyridine; DMF, *N*,*N*-dimethylformamide; DMSO, dimethyl sulphoxide; HMPA, hexamethyphosphoramide; *m*CPBA, *m*-chloroperbenzoic acid; NMO, 4-methylmorpholine-*N*-oxide; TBAF, tetra-*n*-butylammonium fluoride; TBDPS, *t*-BuPh₂Si; TBS, *t*-BuMe₂Si; Tf, triflate; THF, tetrahydrofuran; TPAP, tetra-*n*-propylammonium perruthenate; imid., imidazole. For selected physical data for compound **3**, see Supplementary Information.

The union of fragments **2** and **3** and the final stages of the total synthesis of brevetoxin A (**1**) are depicted in Fig. 6. Thus, generation of the anion of phosphine oxide **2** with *n*-BuLi (THF, −78 °C), followed by addition of aldehyde **3**, gave a mixture of two diastereomeric hydroxyphosphine oxides (**41a** and **41b**, ~1:1 ratio, 75% total yield). Treatment of this mixture (**41a**+**41b**) with KH in DMF, followed by acidic cleavage of the mixed ketal, resulted in the formation of the desired *Z*-olefin **42** as a single isomer and in 72% yield. The choice of the phosphine oxide (Horner–Wittig reaction)[29] as opposed to the more commonly used phosphonium salt for fragment **2** was necessitated by the inability of the latter to react, via its ylide, with the sterically hindered aldehyde **3**. Furthermore, the placement of the mixed methoxyketal on ring E of the fragment **2** was crucial for the exclusive *Z*-stereoselectivity observed in the coupling reaction. It is presumed that the two oxygens of this group act synergistically with the E-ring oxygen in complexing the lithium atom in a well-defined structure, which leads to the exclusive formation of only the *anti* adducts **41a** and **41b**. Both

图 5. GHIJ 环系化合物 3 的构建过程。所用试剂和反应条件如下。(a) 参见文献 27；(b) 将正丁基锂 (1.55 M 的己烷溶液，1.0 当量) 加入到化合物 34(1.0 当量) 和六甲基磷酰胺 (10.0 当量) 的四氢呋喃溶液中；接着加入化合物 33，四氢呋喃，−78℃，1 h；然后 25℃，1.5 h，收率为 88%；(c) 四正丁基氟化铵 (1.1 当量)，四氢呋喃，25℃，7 h，收率为 83%；(d) 高氯酸银 (3.0 当量)，碳酸氢钠 (3.0 当量)，二氧化硅，4 Å 分子筛，硝基甲烷，25℃，3 h，收率为 92%；(e) 间氯过氧苯甲酸 (2.0 当量)，二氯甲烷，0℃，2 h，收率为 93%；(f) 三甲基铝 (3.0 当量)，二氯甲烷，−78℃，1 h，收率为 94%；(g) 乙硫醇 (20.0 当量)，三氟甲磺酸锌 (0.3 当量)，碳酸氢钠 (0.5 当量)，二氯甲烷，25℃，4 h，收率为 92%；(h) 叔丁基二甲基氯硅烷 (1.1 当量)，咪唑 (1.5 当量)，二氯甲烷，25℃，1 h，收率为 92%；(i) 乙酸酐 (1.1 当量)，4−二甲氨基吡啶 (1.3 当量)，二氯甲烷，25℃，2.5 h，收率为 95%；(j) 氢气，10% 氢氧化钯/碳 (0.15 当量)，乙酸，25℃，5 h，收率为 90%；(k) 叔丁基二甲基硅基三氟甲磺酸酯 (2.5 当量)，2,6−二甲基吡啶 (3.0 当量)，二氯甲烷，0℃，0.5 h，收率为 87%；(l) 碳酸钾 (0.5 当量)，甲醇，25℃，4 h，收率为 93%；(m) 四正丙基过钌酸铵 (0.05 当量)，4−甲基吗啉−N−氧化物 (2.0 当量)，二氯甲烷，25℃，1 h，收率为 94%；(n) 乙硫醇 (20.0 当量)，三氟甲磺酸锌 (0.3 当量)，二氯甲烷，25℃，2.5 h，收率为 92%；(o) 三氧化硫·吡啶 (3.0 当量)，二甲基亚砜 (10.0 当量)，三乙胺 (10.0 当量)，二氯甲烷，0℃，1.5 h，收率为 87%。4−DMAP，4−二甲氨基吡啶；DMF，N,N−二甲基甲酰胺；DMSO，二甲基亚砜；HMPA，六甲基磷酰胺；mCPBA，间氯过氧苯甲酸；NMO，4−甲基吗啉−N−氧化物；TBAF，四正丁基氟化铵；TBDPS，叔丁基二苯基硅基；TBS，叔丁基二甲基硅基；Tf，三氟甲磺酰基；THF，四氢呋喃；TPAP，四正丙基过钌酸铵；imid.，咪唑。关于化合物 3 的重要物理数据，请参见补充信息。

　　图 6 中列出的是片断 2 和 3 的整合及短裸甲藻毒素 A(1) 全合成的最后几步反应。也就是，用正丁基锂在 −78℃，四氢呋喃作为溶剂的条件下与膦氧化物 2 反应生成阴离子，接着与醛化物 3 加成，得到两个非对映异构的羟基膦氧化物的混合物 41a 和 41b(比例约为 1∶1，总收率为 75%)。用含氢化钾的 N,N−二甲基甲酰胺溶液处理此混合物 (41a+41b)，接着酸性裂解该混合缩酮，得到了预期的 Z−型烯烃 42 的单一异构体 (收率为 72%)。之所以选择霍纳−维蒂希反应[29] 的膦氧化物，而不选择更常用的鏻盐来合成片断 2，是因为后者形成的叶立德不能与具有空间位阻的醛化物 3 发生反应。此外，片断 2 中 E 环上的混合甲氧基缩酮的位置对偶联反应中专一的 Z−型立体选择性具有决定作用。我们推测，这个基团上的两个氧原子与 E 环上的氧原子在与锂原子形成具有明确结构的复合物时发生协同作用，使得仅生成反

diastereoisomers furnish the same *Z*-olefin (whereas the *syn* diastereoisomers would form the *E*-olefin). Ring closure of hydroxydithioketal[10] **42**, induced by AgClO$_4$, furnished the 8-membered ring mixed thioketal in 80% yield. Oxidation of this mixed thioketal with *m*CPBA led to the corresponding sulphone (85% yield), which upon exposure to BF$_3$ · Et$_2$O in the presence of Et$_3$SiH (ref. 10) at −78 °C, resulted in the reductive removal of the ethyl sulphone moiety and concomitant cleavage of the trityl ether, furnishing the BCDEFGHIJ ring skeleton (compound **43**, 80% yield) of brevetoxin A. The stereochemistry of the newly

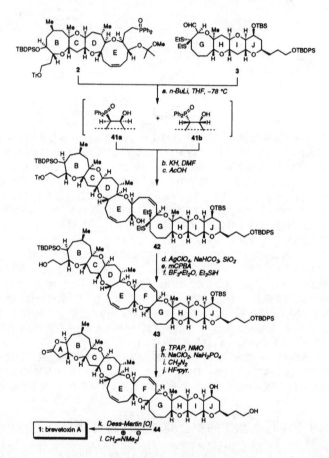

Fig. 6. Total synthesis of brevetoxin A (**1**). Reagents and conditions as follows. (a) Add *n*-BuLi (1.55 M in hexanes, 1.0 equiv.) to **2** (1.0 equiv.) in THF at −78 °C, then **3**, 20 min, 75%; (b) KH (2.0 equiv.), DMF, 25 °C, 0.5 h; (c) 20% aq. AcOH, THF, 25 °C, 6 h, 72% for two steps; (d) AgClO$_4$ (3.0 equiv.), NaHCO$_3$ (3.0 equiv.), SiO$_2$, 4-Å MS, MeNO$_2$, 25 °C, 2 h, 80%; (e) *m*CPBA (2.2 equiv.), CH$_2$Cl$_2$, 0 °C, 1 h, 85%; (f) BF$_3$ · Et$_2$O (3.0 equiv.), Et$_3$SiH, CH$_2$Cl$_2$, −78 °C, 30 min, 80%; (g) TPAP (0.1 equiv.), NMO (2.0 equiv.), CH$_2$Cl$_2$, 25 °C, 30 min; (h) NaClO$_2$ (2.0 equiv.), NaH$_2$PO$_4$ (3.0 equiv.), 2-methyl-2-butene (5.0 equiv.), *t*-BuOH:H$_2$O (5:1), 25 °C, 30 min; (i) CH$_2$N$_2$, Et$_2$O, 25 °C, 15 min, 80% for three steps; (j) HF · pyridine (10.0 equiv.), CH$_2$Cl$_2$, 0 °C, 2 h, 90%; (k) Dess–Martin periodinane (1.5 equiv.), CH$_2$Cl$_2$, 25 °C, 1 h, 80%; (l) CH$_2$=N$^{\oplus}$Me$_2$I$^{\ominus}$(10.0 equiv.), Et$_3$N (15.0 equiv.), CH$_2$Cl$_2$, 25 °C, 5 h, 90%. DMF, *N,N*-dimethylformamide; *m*CPBA, *m*-chloroperbenzoic acid; NMO, 4-methylmorpholine-*N*-oxide; TBDPS, *t*-BuPh$_2$Si; TBS, *t*-BuMe$_2$Si; THF, tetrahydrofuran; TPAP, tetra-*n*-propylammonium perruthenate; Tr, trityl. Selected physical data for synthetic brevetoxin A (**1**): retention factor R_f = 0.45 (silica, 1:1 ethyl acetate:hexane); specific rotation $[α]_D^{22}$ + 75.5 (*c* 0.2, CH$_2$Cl$_2$); infrared spectra (pure compound) $ν_{max}$ 3,425.1, 2,920.5, 2,849.2,

式加成产物 **41a** 和 **41b**。这两种非对映异构体具有同样的 *Z*−型烯键（而顺式加成的非对映异构体则会形成 *E*−型烯键）。用高氯酸银引导羟基二硫缩酮化合物 **42** 进行闭环反应 [10]，以 80% 的收率得到 8 元环的混合硫缩酮产物。然后用间氯过氧苯甲酸氧化该混合硫缩酮得到相应的砜化物（收率为 85%），接着在 −78℃、三乙基硅烷存在的条件下（参考文献 10），用三氟化硼·乙醚还原去除其乙基砜的部分，并伴随着三苯甲基醚键的断裂，生成短裸甲藻毒素 A 的 BCDEFGHIJ 环系骨架（化合物 **43**，收

图 6. 短裸甲藻毒素 A(**1**) 的全合成。所用试剂和反应条件如下。(a) 在 −78℃ 下，将正丁基锂 (1.55 M 的己烷溶液，1.0 当量) 加入到化合物 **2**(1.0 当量) 的四氢呋喃溶液中，接着加入化合物 **3**，20 min，收率为 75%；(b) 氢化钾 (2.0 当量)，*N,N*−二甲基甲酰胺，25℃，0.5 h；(c) 20% 的乙酸水溶液，四氢呋喃，25℃，6 h，两步收率 72%；(d) 高氯酸银 (3.0 当量)，碳酸氢钠 (3.0 当量)，二氧化硅，4 Å 分子筛，硝基甲烷，25℃，2 h，收率为 80%；(e) 间氯过氧苯甲酸 (2.2 当量)，二氯甲烷，0℃，1 h，收率为 85%；(f) 三氟化硼·乙醚 (3.0 当量)，三乙基硅烷，二氯甲烷，−78℃，30 min，收率为 80%；(g) 四正丙基过钌酸铵 (0.1 当量)，4−甲基吗啉−*N*−氧化物 (2.0 当量)，二氯甲烷，25℃，30 min；(h) 亚氯酸钠 (2.0 当量)，磷酸二氢钠 (3.0 当量)，2−甲基−2−丁烯 (5.0 当量)，叔丁醇∶水 (5∶1)，25℃，30 min；(i) 重氮甲烷，乙醚，25℃，15 min，三步收率为 80%；(j) 氟化氢·吡啶 (10.0 当量)，二氯甲烷，0℃，2 h，收率为 90%；(k) 戴斯−马丁氧化剂 (1.5 当量)，二氯甲烷，25℃，1 h，收率为 80%；(l) *N,N*−二甲基亚甲基碘化铵 (10.0 当量)，三乙胺 (15.0 当量)，二氯甲烷，25℃，5 h，收率为 90%。DMF，*N,N*−二甲基甲酰胺；*m*CPBA，间氯过氧苯甲酸；NMO，4−甲基吗啉−*N*−氧化物；TBDPS，叔丁基二苯基硅基；TBS，叔丁基二甲基硅基；THF，四氢呋喃；TPAP，四正丙基过钌酸铵；Tr，三苯甲基。关于合

1,788.2, 1,689.5, 1,455.1, 1,378.7, 1,314, 1,265.7, 1,208.3, 1,079.9, 897.1, 737.0 cm^{-1}; ^{1}H NMR (C$_6$D$_6$, 600 MHz, 45 °C) δ 9.29 (s, 1 H), 5.90 (s, 1 H), 5.87 (dd, J = 10.8, 5.2 Hz, 1 H), 5.78–5.67 (bm, 2 H), 5.71–5.64 (m, 1 H), 5.43 (s, 1 H), 4.55 (dd, J = 8.0, 5.4 Hz, 1 H), 3.98–3.91 (m, 3 H), 3.74 (b, 1 H), 3.59–3.54 (m, 1 H), 3.52 (ddd, J = 10.6, 7.9, 2.3 Hz, 1 H), 3.46–3.34 (m, 4 H), 3.29 (bd, J = 8.5 Hz, 1 H), 3.19–3.10 (m, 3 H), 2.94 (ddd, J = 12.0, 8.0, 4.1 Hz, 1 H), 2.89 (ddd, J = 11.9, 9.2, 4.5 Hz, 1 H), 2.80 (dd, J = 11.9, 4.6 Hz, 1 H), 2.78 (dd, J = 9.6, 2.7 Hz, 1 H), 2.72–2.40 (bm, 4 H), 2.65 (dd, J = 10.7, 10.7 Hz, 1 H), 2.40 (ddd, J = 15.2, 10.8, 4.0 Hz, 1 H), 2.36 (dd, J = 16.9, 8.2 Hz, 1 H), 2.31 (dd, J = 11.6, 4.2 Hz, 1 H), 2.25–2.21 (m, 1 H), 2.24–2.20 (m, 1 H), 2.21 (dd, J = 16.9, 10.3 Hz, 1 H), 2.18–2.10 (m, 1 H), 2.05–1.92 (m, 6 H), 1.92–1.86 (m, 1 H), 1.86–1.74 (m, 3 H), 1.72–1.52 (m, 7 H), 1.44–1.38 (m, 1 H), 1.37–1.24 (m, 1 H), 1.24 (s, 3 H), 1.20 (d, J = 7.1 Hz, 3 H), 0.98–0.95 (m, 1 H), 0.96 (s, 3 H), 0.77 (d, J = 6.8 Hz, 3 H); ^{13}C NMR (C$_6$D$_6$, 150 MHz, 45 °C) δ 193.5, 171.0, 148.9, 139.0, 134.4, 128.5, 127.4, 124.8, 92.0, 88.5, 86.7, 85.1, 84.7, 84.1, 83.8, 82.2, 82.0, 80.4, 79.9, 78.9, 77.8, 76.7, 76.2, 71.9, 71.1, 69.6, 66.3, 62.4, 52.9, 45.6, 43.3, 38.2, 37.5, 36.9, 36.2, 35.9, 33.8, 32.3, 31.4, 29.0, 27.6, 27.2, 21.4, 19.7, 16.8, 15.0 (carbons 17, 20 and 28 were observed at best as weak broad signals at 34.5, 34.5 and 30.9, respectively, as previously reported[30,31]; mass spectra: calculated for C$_{49}$H$_{70}$O$_{13}$ (M + Na$^+$) 889.4714, found 889.4747. For selected physical data for compounds **42** and **43**, see Supplementary Information.

generated stereocenter at the FG ring junction of **43** was confirmed by comparison of its NMR data to those of brevetoxin A[1,3,30,31] and of a model system[24]. The fusion of ring A onto the rest of the brevetoxin A framework was accomplished by stepwise oxidation of the primary alcohol in **43** to the corresponding aldehyde (TPAP-NMO)[23] and carboxylic acid (NaClO$_2$-NaH$_2$PO$_4$, 2-methyl-2-butene)[32], followed by methyl ester formation (CH$_2$N$_2$, 80% for three steps) and exposure to HF·pyridine. The latter reaction resulted in the removal of all three silyl groups and spontaneous ring closure to form the γ-lactone ring, furnishing the ABCDEFGHIJ ring system **44** (90% yield) of brevetoxin A. Finally, exposure of diol **44** to carefully controlled amounts of Dess–Martin[33] reagent gave, selectively, the corresponding hydroxy aldehyde (80% yield), whose reaction with CH$_2$ = N$^{\oplus}$Me$_2$I$^{\ominus}$ (ref. 34) proceeded smoothly to afford brevetoxin A (**1**) in 90% yield. Synthetic brevetoxin A was identical in all respects with naturally occurring brevetoxin A (as determined by thin-layer chromatography, high-performance liquid chromatography, ^{1}H and ^{13}C NMR, optical rotation $[\alpha]_D^{22}$, infrared spectroscopy and mass spectrometry)[1-3,30,31].

The total synthesis of brevetoxin A (**1**) became possible only after a long campaign which involved several thwarted strategies and the development of a number of new synthetic technologies. Particularly rewarding were the solutions found to the synthetic challenges posed by the 7-, 8- and 9-membered rings of the target molecule[10,11] and the convergent strategy finally developed to reach the complex and aesthetically pleasing structure of brevetoxin A. Through the described chemistry, brevetoxin A (**1**) joins brevetoxin B[35-38] in yielding to total synthesis. The work described here opens up the possibility of preparing suitable molecular probes that may contribute to the development of sensing devices[39] for this and related biotoxins of the "red tides"[40-45] and enhance our understanding of this phenomenon.

(**392**, 264-269; 1998)

成的短裸甲藻毒素 A(**1**)的重要物理数据如下：薄层保留系数 $R_f = 0.45$（硅胶，乙酸乙酯：己烷 $= 1 : 1$）；比旋光度 $[\alpha]_D^{22} +75.5$（c 0.2，二氯甲烷）；红外光谱（纯化合物）ν_{max} 3,425.1，2,920.5，2,849.2，1,788.2，1,689.5，1,455.1，1,378.7，1,314，1,265.7，1,208.3，1,079.9，897.1，737.0 cm^{-1}；^{1}H 核磁共振谱（氘代苯，600 MHz，45 ℃）δ 9.29 (s, 1 H)，5.90 (s, 1 H)，5.87 (dd, $J = 10.8, 5.2$ Hz, 1 H)，5.78 ~ 5.67 (bm, 2 H)，5.71 ~ 5.64 (m, 1 H)，5.43 (s, 1 H)，4.55 (dd, $J = 8.0, 5.4$ Hz, 1 H)，3.98 ~ 3.91 (m, 3 H)，3.74 (b, 1 H)，3.59 ~ 3.54 (m, 1 H)，3.52 (ddd, $J = 10.6, 7.9, 2.3$ Hz, 1 H)，3.46 ~ 3.34 (m, 4 H)，3.29 (bd, $J = 8.5$ Hz, 1 H)，3.19 ~ 3.10 (m, 3 H)，2.94 (ddd, $J = 12.0, 8.0, 4.1$ Hz, 1 H)，2.89 (ddd, $J = 11.9, 9.2, 4.5$ Hz, 1 H)，2.80 (dd, $J = 11.9, 4.6$ Hz, 1 H)，2.78 (dd, $J = 9.6, 2.7$ Hz, 1 H)，2.72 ~ 2.40 (bm, 4 H)，2.65 (dd, $J = 10.7, 10.7$ Hz, 1 H)，2.40 (ddd, $J = 15.2, 10.8, 4.0$ Hz, 1 H)，2.36 (dd, $J = 16.9, 8.2$ Hz, 1 H)，2.31 (dd, $J = 11.6, 4.2$ Hz, 1 H)，2.25 ~ 2.21 (m, 1 H)，2.24 ~ 2.20 (m, 1 H)，2.21 (dd, $J = 16.9, 10.3$ Hz, 1 H)，2.18 ~ 2.10 (m, 1 H)，2.05 ~ 1.92 (m, 6 H)，1.92 ~ 1.86 (m, 1 H)，1.86 ~ 1.74 (m, 3H)，1.72 ~ 1.52 (m, 7 H)，1.44 ~ 1.38 (m, 1 H)，1.37 ~ 1.24 (m, 1 H)，1.24 (s, 3 H)，1.20 (d, $J = 7.1$ Hz, 3 H)，0.98 ~ 0.95 (m, 1 H)，0.96 (s, 3 H)，0.77 (d, $J = 6.8$ Hz, 3 H)；^{13}C 核磁共振谱（氘代苯，150 MHz，45 ℃）δ 193.5，171.0，148.9，139.0，134.4，128.5，127.4，124.8，92.0，88.5，86.7，85.1，84.7，84.1，83.8，82.2，82.0，80.4，79.9，78.9，77.8，76.7，76.2，71.9，71.1，69.6，66.3，62.4，52.9，45.6，43.3，38.2，37.5，36.9，36.2，35.9，33.8，32.3，31.4，29.0，27.6，27.2，21.4，19.7，16.8，15.0（正如先前的报道 [30,31]，碳 17，20 和 28 分别在 34.5，34.5 和 30.9 处峰型最佳，为弱的宽峰）；质谱：C$_{49}$H$_{70}$O$_{13}$（M+Na$^+$）的计算值为 889.4714，测定值为 889.4747。关于化合物 **42** 和 **43** 的重要物理数据，请参见补充信息。

率为 80%）。在化合物 **43** 中 FG 环的连接处形成了新的立构中心，其立体化学是通过将它的核磁共振数据与短裸甲藻毒素 A 的 [1,3,30,31] 和一个模型系统 [24] 中的核磁共振数据进行比较而得到证实的。经过下面的过程将 A 环结合到短裸甲藻毒素 A 骨架的其余部分上：对化合物 **43** 上的伯醇进行逐步氧化，先用四正丙基过钌酸铵和 4-甲基吗啉-N-氧化物氧化成相应的醛 [23]，再用亚氯酸钠-磷酸二氢钠和 2-甲基-2-丁烯进一步氧化成羧酸 [32]，然后用重氮甲烷进行甲酯化（三步的收率为 80%），接着再用氟化氢·吡啶处理以脱去三个甲硅烷基，并同时发生闭环反应形成 γ-内酯环，得到短裸甲藻毒素 A 的 ABCDEFGHIJ 环系化合物 **44**（收率为 90%）。最后，用严格控制加入量的戴斯-马丁 [33] 试剂选择性地氧化二醇化合物 **44**，得到相应的羟基醛化合物（收率为 80%），再与 N,N-二甲基亚甲基碘化铵（参考文献 34）顺利地发生反应，以 90% 的收率得到短裸甲藻毒素 A(**1**)。合成的短裸甲藻毒素 A 与天然存在的短裸甲藻毒 A 在各方面的性质都是一致的（经薄层色谱，高效液相色谱，^{1}H 和 ^{13}C 核磁共振谱，比旋光度 $[\alpha]_D^{22}$，红外光谱和质谱验证）[1-3,30,31]。

只有经过长期努力（其间经历了数次失败的策略和若干新合成技术的发展），才实现了短裸甲藻毒素 A(**1**)的全合成。尤其有意义的是找到了应对目标分子中 7 元、8 元和 9 元环的合成挑战的方法 [10,11]，以及最终开发出的构建了短裸甲藻毒素 A 这一复杂而美观结构的汇聚式合成策略。通过前述的化学合成，短裸甲藻毒素 A(**1**)和短裸甲藻毒素 B [35-38] 都可以通过全合成得到。这里发表的研究工作为制备适当的分子探针开拓了可行性，这些探针可能有助于开发出传感设备 [39]，用于检测与"赤潮"相关的生物毒素 [40-45]，并增进我们对该现象的理解。

（刘振明 翻译；许家喜 审稿）

K. C. Nicolaou, Zhen Yang, Guo-qiang Shi, Janet L. Gunzner, Konstantinos A. Agrios & Peter Gärtner

Department of Chemistry and The Skaggs Institute for Chemical Biology, The Scripps Research Institute, 10550 North Torrey Pines Road, La Jolla, California 92037, USA and Department of Chemistry and Biochemistry, University of California, San Diego, 9500 Gilman Drive, La Jolla, California 92093, USA

Received 24 November 1997; accepted 19 January 1998.

References:

1. Shimizu, Y., Chou, H.-N. & Bando, H. Structure of brevetoxin A (GB-1 toxin), the most potent toxin in the Florida red tide organism *Gymnodinium breve* (*Ptychodiscus brevis*). *J. Am. Chem. Soc.* **108,** 514-515 (1986).

2. Shimizu, Y., Bando, H., Chou, H.-N., Van Duyne, G. & Clardy, J. C. Absolute configuration of brevetoxins. *J. Chem. Soc., Chem. Commun.* 1656-1658 (1986).

3. Pawlak, J. *et al.* Structure of brevetoxin A as constructed from NMR and MS data. *J. Am. Chem. Soc.* **109,** 1144-1150 (1987).

4. Rein, K. S., Baden, D. G. & Gawley, R. E. Conformational analysis of the sodium channel modulator, brevetoxin A, comparison with brevetoxin B conformations, and a hypothesis about the common pharmacophore of the "site 5" toxins. *J. Org. Chem.* **59,** 2101-2106 (1994).

5. Rein, K. S., Lynn, B., Baden, D. G. & Gawley, R. E. Brevetoxin B: chemical modifications, synaptosome binding, toxicity, and an unexpected conformational effect. *J. Org. Chem.* **59,** 2107-21013 (1994).

6. Yasumoto, T. & Murata, M. Marine toxins. *Chem. Rev.* **93,** 1897-1909 (1993).

7. Gawley, R. E. *et al.* The relationship of brevetoxin "length" and A-ring functionality to binding and activity in neuronal sodium channels. *Chem. Biol.* **2,** 533-541 (1995).

8. Chou, H.-N. & Shimizu, Y. Biosynthesis of brevetoxins. Evidence for the mixed origin of the backbone carbon chain and the possible involvement of dicarboxylic acids. *J. Am. Chem. Soc.* **109,** 2184-2185 (1987).

9. Lee, M. S., Qin, G.-W., Nakanishi, K. & Zagorski, M. G. Biosynthesis studies on brevetoxins, potent neurotoxins produced by the dinoflagellate *Gymnodinium breve. J. Am. Chem. Soc.* **111,** 6234-6241 (1989).

10. Nicolaou, K. C., Prasad, C. V. C., Hwang, C.-K., Duggan, M. E. & Veale, C. A. Cyclizations of hydroxydithioketals. New synthetic technology for the construction of oxocenes and related medium ring systems. *J. Am. Chem. Soc.* **111,** 5321-5330 (1989).

11. Nicolaou, K. C., Shi, G.-Q., Gunzner, J. L., Gärtner, P. & Yang, Z. Palladium-catalyzed functionalization of lactones via their cyclic ketene acetal phosphates. Efficient new synthetic technology for the construction of medium and large cyclic ethers. *J. Am. Chem. Soc.* **119,** 5467-5468 (1997).

12. Nicolaou, K. C., Prasad, C. V. C., Somers, P. K. & Hwang, C.-K. Activation of 6-*endo* over 5-*exo* hydroxy epoxide openings. Stereo- and ringselective synthesis of tetrahydrofuran and tetrahydropyran systems. *J. Am. Chem. Soc.* **111,** 5330-5334 (1989).

13. Freeman, J. P. in *Organic Syntheses (Collective Vol. VII)* 139-141 (Wiley, New York, 1990).

14. Xie, M., Beges, D. A. & Robins, M. J. Efficient "dehomologation" of di-*O*-isopropylidenehexofuranose derivatives to give *O*-isopropylidenepentofuranose by sequential treatment with periodic acid in ethyl acetate and sodium borohydride. *J. Org. Chem.* **61,** 5178-5179 (1996).

15. Mukaiyama, T., Banno, K. & Narasaka, K. New cross-aldol reactions. Reactions of silyl enol ethers with carbonyl compounds activated by titanium tetrachloride. *J. Am. Chem. Soc.* **96,** 7503-7509 (1974).

16. Rodeheaver, G. T. & Hunt, D. F. Conversion of olefins into ketones with mercuric acetate and palladium chloride. *Chem. Commun.* 818-819 (1971).

17. Tamura, Y., Ochiai, H., Nakamura, T. & Yoshida, Z. Arylation and vinylation of 2-carboethoxyethylzinc iodide and 3-carboethoxypropylzinc iodide catalyzed by palladium. *Tetrahedr. Lett.* **27,** 955-958 (1986).

18. Inanaga, J., Hirata, K., Saeki, H., Katsuki, T. & Yamaguchi, M. A rapid esterification by mixed anhydride and its application to large-ring lactonization. *Bull. Chem. Soc. Jpn* 52, 1989-1993 (1979).

19. Arhart, R. J. & Martin, J. C. Sulfuranes, V. The chemistry of sulfur (IV) compounds. Dialkoxydiarylsulfuranes. *J. Am. Chem. Soc.* **94,** 4997-5003 (1972).

20. Hoveyda, A. H., Evans, D. A. & Fu, G. C. Substrate-directable chemical reactions. *Chem. Rev.* **93,** 1307-1370 (1993).

21. Nicolaou, K. C., McGarry, D. G. & Sommers, P. K. New synthetic strategies for the construction of medium-size cyclic ethers. Stereocontrolled synthesis of the BCD ring framework of brevetoxin A. *J. Am. Chem. Soc.* **112,** 3696-3697 (1990).

22. Corey, E. J. & Beames, D. J. Mixed cuprate reagents of type R1R2CuLi which allow selective group transfer. *J. Am. Chem. Soc.* **94,** 7210-7211 (1972).

23. Ley, S. V., Norman, J., Griffith, W. P. & Marsden, S. P. Tetrapropylammonium perruthenate, Pr$_4$N$^+$RuO$_4^-$, TPAP: a catalytic oxidant for organic synthesis. *Synthesis* 639-666 (1994).

24. Nicolaou, K. C. *et al.* New synthetic technology for the construction of 9-membered ring cyclic ethers. Construction of the EFGH ring skeleton of brevetoxin A. *J. Am. Chem. Soc.* **119,** 8105-8106 (1997).

25. Mahoney, W. S., Brestensky, D. M. & Stryker, J. M. Selective hydride-mediated conjugate reduction of α,β-unsaturated carbonyl compounds using [(Ph$_3$P)CuH]$_6$. *J. Am. Chem. Soc.* **119,** 8105-8106 (1997).

26. Feixas, J., Capdevila, A. & Guerrero, A. Utilization of neutral alumina as a mild reagent for the selective cleavage of primary and secondary silyl ethers. *Tetrahedron* **50,** 8539-8550 (1994).

27. Nicolaou, K. C. *et al.* Novel strategies for the construction of complex polycyclic ether frameworks. Stereocontrolled synthesis of the FGHIJ ring system of brevetoxin A. *Angew. Chem. Int. Edn. Engl.* **30,** 299-303 (1991).

28. Parikh, J. R. & Doering, W. von E. Sulfur trioxide in the oxidation of alcohols by dimethyl sulfoxide. *J. Am. Chem. Soc.* **89,** 5505-5507 (1967).

29. Clayden, J. & Warren, S. Stereocontrol in organic synthesis using the diphenylphosphoryl group. *Angew. Chem. Int. Edn. Engl.* **35,** 241-270 (1996).

30. Zagorski, M. G., Nakanishi, K., Qin, G.-W. & Lee, M. S. Assignment of ^{13}C NMR peaks of brevetoxin A: application of two-dimensional Hartmann-Hahn spectroscopy. *J. Org. Chem.* **53,** 4156-4157 (1988).

31. Lee, M. S., Nakanishi, K. & Zagorski, M. G. Full proton assignment of BTX-A, $C_{49}H_{70}O_{13}$. *New J. Chem.* **11,** 753-756 (1987).

32. Klindgren, B. O. & Nilsson, T. Preparation of carboxylic acids from aldehydes (including hydroxylated benzaldehydes) by oxidation with chlorite. *Acta Chem. Scand.* **27,** 888-890 (1973).

33. Dess, D. B. & Martin, J. C. Readily accessible 12-I-5 oxidant for the conversion of primary and secondary alcohols to aldehydes and ketones. *J. Org. Chem.* **48,** 4155-4156 (1983).

34. Schreiber, J., Maag, H., Hashimoto, N. & Eschenmoser, A. Dimethyl(methylene)ammonium iodide. *Angew. Chem. Int. Edn. Engl.* **10,** 330-331 (1971).

35. Nicolaou, K. C. *et al.* Total synthesis of brevetoxin B. 1. First generation strategies and new approaches to oxepane systems. *J. Am. Chem. Soc.* **117,** 10227-10238 (1995).

36. Nicolaou, K. C. *et al.* Total synthesis of brevetoxin B. 2. Second generation strategies and construction of the dioxepane region [DEFG]. *J. Am. Chem. Soc.* **117,** 10239-10251 (1995).

37. Nicolaou, K. C. *et al.* Total synthesis of brevetoxin B. 3. Final strategy and completion. *J. Am. Chem. Soc.* **117,** 10252-10263 (1995).

38. Nicolaou, K. C. The total synthesis of brevetoxin B: a twelve year odyssey in organic synthesis. *Angew. Chem. Int. Edn. Engl.* **35,** 589-607 (1996).

39. Homaka, Y. Recent methods for detection of seafood toxins: recent immunological method for ciguatoxin and related polyethers. *Food Addit. Contam.* **10,** 71-82 (1993).

40. Anderson, D. M. Turning back the harmful red tide. *Nature* **388,** 513-514 (1997).

41. Barker, R. *And the Waters Turned to Blood* (Simon & Schuster, New York, 1997).

42. Anderson, D. M. Red tides. *Sci. Am.* **271,** 62-68 (1994).

43. Hall, S. & Strichartz, G. (eds) (American Chemical Society, Washington DC, 1990).

44. Okaichi, T., Anderson, D. M. & Nemoto, T. (eds) *Red Tides. Biology, Environmental Science, and Toxicology* (Elsevier, New York, 1989).

45. Anderson, D. M. & White, A. W. Marine biotoxins at the top of the food chain. *Oceanus* **35,** 55-61 (1992).

Supplementary Information is available on *Nature*'s World-Wide Web site (http://www.nature.com) and as paper copy from Mary Sheehan at the London editorial office of *Nature*.

Acknowledgements. We are indebted to those of our collaborators whose early contributions on this project made its success possible; their names will appear in the full account of this work. We thank K. Nakanishi and Y. Shimizu for samples of natural brevetoxin A, and D. H. Huang, G. Siuzdak and R. Chadha for the NMR, mass spectroscopic and X-ray crystallographic assistance, respectively. This work was supported by the National Institutes of Health USA (GM), The Skaggs Institute for Chemical Biology, Novartis, Merck, Hoffmann-La Roche, Schering Plough, DuPont Merck, the American Chemical Society (graduate fellowship sponsored by Aldrich to J.L.G.) and the Foundation for the Promotion of Scientific Investigations (postdoctoral fellowship to P.G.). Z.Y. and G.S. contributed equally to this project.

Correspondence and requests for materials should be addressed to K.C.N. at The Scripps Research Institute (e-mail: kcn@scripps.edu).

Global-scale Temperature Patterns and Climate Forcing over the Past Six Centuries

M. E. Mann *et al.*

Editor's Note

To understand current climate change, we need to know how it compares with changes in the historical past, which would have had natural rather than human causes. Here Michael Mann and his coworkers use a wide range of different climate indicators in the geological and biological records to reconstruct global climate over the past 600 years. They conclude that the average temperature in the Northern Hemisphere fluctuated only slightly before an abrupt twentieth-century increase. This "hockey-stick" graph (Figure 5b) became notorious when it was later found to have been derived from flawed calculations. But although climate sceptics have implied that it undermines the case for global warming, many other lines of evidence in fact show it to be real.

Spatially resolved global reconstructions of annual surface temperature patterns over the past six centuries are based on the multivariate calibration of widely distributed high-resolution proxy climate indicators. Time-dependent correlations of the reconstructions with time-series records representing changes in greenhouse-gas concentrations, solar irradiance, and volcanic aerosols suggest that each of these factors has contributed to the climate variability of the past 400 years, with greenhouse gases emerging as the dominant forcing during the twentieth century. Northern Hemisphere mean annual temperatures for three of the past eight years are warmer than any other year since (at least) AD 1400.

KNOWING both the spatial and temporal patterns of climate change over the past several centuries remains a key to assessing a possible anthropogenic impact on post-industrial climate[1]. In addition to the possibility of warming due to increased concentrations of greenhouse gases during the past century, there is evidence that both solar irradiance and explosive volcanism have played an important part in forcing climate variations over the past several centuries[2,3]. The unforced "natural variability" of the climate system may also be quite important on multidecadal and century timescales[4,5]. If a faithful empirical description of climate variability could be obtained for the past several centuries, a more confident estimation could be made of the roles of different external forcings and internal sources of variability on past and recent climate. Because widespread instrumental climate data are available for only about one century, we must use proxy climate indicators combined with any very long instrumental records that are available to obtain such an empirical description of large-scale climate variability during past centuries. A variety of studies have sought to use a "multiproxy" approach to understand long-term climate variations, by analysing a widely distributed set of proxy and instrumental

过去六个世纪全球尺度上的
气温分布型与气候强迫

曼等

编者按

为理解当前的气候变化，我们需要把其与历史上自然的而非人为因素导致的变化进行比较。本文中，迈克尔·曼和他的同事们运用一系列不同的地质记录和生物记录的气候指标重建了过去 600 年间的全球气候。他们认为，北半球的平均温度在二十世纪突然升高，但在此之前的变化都只是轻微的波动。后来，这种"曲棍球棍"图（图5b）因被发现是一种错误的计算结果而变得臭名昭著。然而，尽管气候怀疑论者已经暗示它削弱了全球变暖的论据，但仍有许多其他的证据表明它是真实存在的。

利用广泛分布的高分辨率气候代用指标的多变量校正结果，本文对过去六个世纪中每年的地表温度分布型在空间上进行了全球恢复重建。重建结果与代表温室气体浓度、太阳辐照度以及火山气溶胶变化的时间序列记录之间的时间相关性表明，上述各因子对过去 400 年中发生的气候变化都有影响，其中温室气体从二十世纪开始成为最主要的影响因素。在北半球，过去八年中有三年的年均气温均创造了（至少）从公元 1400 年以来的最高纪录。

认识过去几个世纪中气候变化的空间和时间模态仍然是评估后工业化时代人类对气候影响的关键所在 [1]。有证据显示，除了在过去的一个世纪中温室气体浓度的增加导致了气候变暖以外，在过去的几个世纪中，太阳辐照度与火山爆发在影响气候变化中也起着重要作用 [2,3]。气候系统的非强迫"自然变率"也可能在数十年乃至上百年的时间尺度上具有重要意义 [4,5]。倘若能够对过去几个世纪中的气候变化做出较可靠的经验描述，那么就可据此对引起过去和近期气候变化的各外部强迫及内在因素所起的作用给出一个更为可信的评估。由于采用大范围仪器进行气候观测仅开展了约一个世纪的时间，我们必须将气候代用指标与已有的较长期的仪器观测资料相结合，以获得过去几个世纪中大尺度气候变化的经验性描述。已有大量研究试图采用"多种代用资料"法来推断长期的气候变化，通过分析广泛分布的各类代用资料和仪器观测的气候指标 [1,5-8] 来深入研究全球的长期气候变化。在上述前人研究的基础

climate indicators[1,5-8] to yield insights into long-term global climate variations. Building on such past studies, we take a new statistical approach to reconstructing global patterns of annual temperature back to the beginning of the fifteenth century, based on the calibration of multiproxy data networks by the dominant patterns of temperature variability in the instrumental record.

Using these statistically verifiable yearly global temperature reconstructions, we analyse the spatiotemporal patterns of climate change over the past 500 years, and then take an empirical approach to estimating the relationship between global temperature changes, variations in volcanic aerosols, solar irradiance and greenhouse-gas concentrations during the same period.

Data

We use a multiproxy network consisting of widely distributed high-quality annual-resolution proxy climate indicators, individually collected and formerly analysed by many palaeoclimate researchers (details and references are available: see Supplementary Information). The network includes (Fig. 1a) the collection of annual-resolution dendroclimatic, ice core, ice melt, and long historical records used by Bradley and Jones[6] combined with other coral, ice core, dendroclimatic, and long instrumental records. The long instrumental records have been formed into annual mean anomalies relative to the 1902–80 reference period, and gridded onto a 5°×5° grid (yielding 11 temperature grid-point series and 12 precipitation grid-point series dating back to 1820 or earlier) similar to that shown in Fig. 1b. Certain densely sampled regional dendroclimatic data sets have been represented in the network by a smaller number of leading principal components (typically 3–11 depending on the spatial extent and size of the data set). This form of representation ensures a reasonably homogeneous spatial sampling in the multiproxy network (112 indicators back to 1820).

Potential limitations specific to each type of proxy data series must be carefully taken into account in building an appropriate network. Dating errors in a given record (for example, incorrectly assigned annual layers or rings) are particularly detrimental if mutual information is sought to describe climate patterns on a year-by-year basis. Standardization of certain biological proxy records relative to estimated growth trends, and the limits of constituent chronology segment lengths (for example, in dendroclimatic reconstructions), can restrict the maximum timescale of climate variability that is recorded[9], and only a limited subset of the indicators in the multiproxy network may thus "anchor in" the longest-term trends (for example, variations on timescales greater than 500 years). However, the dendroclimatic data used were carefully screened for conservative standardization and sizeable segment lengths. Moreover, the mutual information contained in a diverse and widely distributed set of independent climate indicators can more faithfully capture the consistent climate signal that is present, reducing the compromising effects of biases and weaknesses in the individual indicators.

116

上，我们采用了一种新的统计方法重建了十五世纪以来全球气温的年际变化分布型，其中多代用数据网的校正，是根据（近期的）器测资料反映的主要气温变化模态来进行的。

本文利用上述经统计学验证的年际全球气温重建资料，分析了过去 500 年中气候变化的时空模态，之后又利用经验方法研究了同时期内全球温度变化与火山气溶胶、太阳辐照度以及温室气体浓度变化之间的相互关系。

数　　据

我们所采用的是由广泛分布的高质量年际分辨率气候代用指标组成的多种代用资料网，而这些代用指标都是许多古气候学家各自收集和之前分析过的（详细内容及参考文献见补充信息）。该资料网包括（图 1a）布拉德利和琼斯[6]所采用的年际分辨率的年轮气候学、冰芯、冰融体以及长期历史记录，同时还结合了其他学者关于珊瑚、冰芯、年轮气候学以及长期的仪器观测记录的资料。其中长期的仪器观测记录均处理为相对于 1902～1980 年间的年平均值的距平值，并插值为 5°×5° 的网格（形成 11 个温度格点序列和 12 个降雨量格点序列，一直追溯到 1820 年甚至更早），如图 1b 所示。在数据网络中，对于某些取样较为密集的区域性年轮气候学数据库就通过主成分信息进行稀疏化处理（根据空间范围和数据库的大小一般取代表性的 3～11 个点不等）来表示。这样就保证了所形成的多种代用资料网在空间取样上的合理均匀性（其中 112 个指标可回溯至 1820 年）。

在建立合适的数据网的过程中，必须仔细考虑到每种代用指标数据序列的潜在局限性。倘若通过寻找共有信息以得出逐年的气候分布型，那么存在于给定记录中的测年误差（比如，年层或年轮的判定错误）将极具破坏性。基于生长趋势估计的某类生物代用记录的标准化处理以及（用以在）年代学中使用的指标片段的（时间）长度（例如树轮重建），限制了我们在记录中能看到的最长时间尺度的气候变化特征[9]，因此在多种代用资料网中只有非常有限的一部分代用指标能用于"锚定"长尺度（例如时间尺度大于 500 年）的气候。不过我们所用的年轮数据均已经过严格的筛选，以实现守恒标准化以及较长的分割长度。此外，各类气候指标库分布广泛且相互独立，它们所包含的共有信息更能准确地获取已有的一致性的气候信号，进而降低因单个指标的偏差和不足而带来的相互影响。

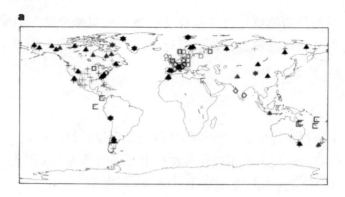

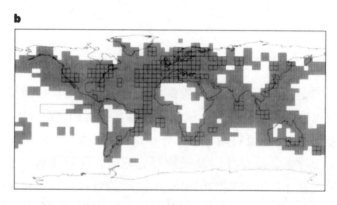

Fig. 1. Data used in this study. **a**, Distribution of annual-resolution proxy indicators used in this study. Dendroclimatic reconstructions are indicated by "tree" symbols, ice core/ice melt proxies by "star" symbols and coral records by "C" symbols. Long historical records and instrumental "grid-points" series are shown by squares (temperature) or diamonds (precipitation). Groups of "+" symbols indicate principal components of dense tree-ring sub-networks, with the number of such symbols indicating the number of retained principal components. Sites are shown dating back to at least 1820 (red), 1800 (blue-green), 1750 (green), 1600 (blue) and 1400 (black). Certain sites (for example, the Quelccaya ice core) consist of multiple proxy indicators (for example, multiple cores, and both $\delta^{18}O$ isotope and accumulation measurements). **b**, Distribution of the 1,082 nearly continuous available land air/sea surface temperature grid-point data available from 1902 onward, indicated by shading. The squares indicate the subset of 219 grid-points with nearly continuous records extending back to 1854 that are used for verification. Northern Hemisphere (NH) and global (GLB) mean temperature are estimated as areally weighted (that is, cosine latitude) averages over the Northern Hemisphere and global domains respectively.

Monthly instrumental land air and sea surface temperature[10] grid-point data (Fig. 1b) from the period 1902–95 are used to calibrate the proxy data set. Although there are notable spatial gaps, this network covers significant enough portions of the globe to form reliable estimates of Northern Hemisphere mean temperature, and certain regional indices of particular importance such as the "NINO3" eastern tropical Pacific surface temperature index often used to describe the El Niño phenomenon. The NINO3 index is constructed from the eight grid-points available within the conventional NINO3 box (5° S to 5° N, 90–150° W).

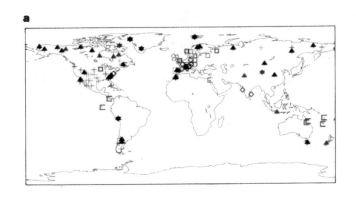

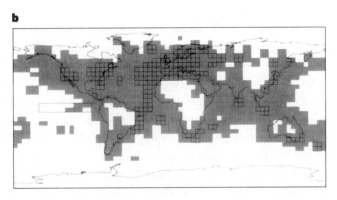

图 1. 本研究所采用的数据。**a**，本研究中年分辨率代用指标的分布。年轮气候重建以"树形"符号表示，冰芯/冰融体代用指标以"星状"符号表示，而珊瑚记录则用"C"表示。长期记录和仪器记录的"格点"序列则分别以方框（温度）和菱形（降雨量）表示。"+"群表示密集的树木年轮子网络的主成分，其数量代表所保留下来的主成分的数目。图中还给出了可分别回溯到 1820 年（红）、1800 年（蓝绿）、1750 年（绿）、1600 年（蓝）和 1400 年（黑）的站位分布。有些观测站（例如魁尔克亚冰芯）包含多个代用指标（如，多个岩芯，同时测定 $\delta^{18}O$ 同位素和加积速率两项指标）。**b**，1902 年以后的所有 1,082 个近乎连续的陆地气温/海表温度格点数据的分布，在图中以阴影表示。方框表示由 219 个格点组成的子库，是检验过程中所采用的可回溯到 1854 年的连续记录。北半球（NH）和全球（GLB）平均气温是分别根据北半球和全球范围内的面积加权（即纬度的余弦值）平均值估算而来的。

利用 1902～1995 年间仪器记录的陆地气温和海表温度 [10] 的月际格点数据（图 1b）对代用指标数据库作了校正。尽管空间上还存明显的空白，但该数据网络已对全球所有重要区域实现了覆盖，由此得出的北半球平均气温是可靠的，而某些区域性指标则具有特殊意义，比如，"NINO3"赤道东太平洋表层温度指数常用来描述厄尔尼诺现象。NINO3 指数是从传统 NINO3 区（南纬 5°到北纬 5°，西经 90°到西经 150°）中抽出的 8 个格点构成的。

Multiproxy Calibration

Although studies have shown that well chosen regional paleoclimate reconstructions can act as surprisingly representative surrogates for large-scale climate[11-13], multiproxy networks seem to provide the greatest opportunity for large-scale palaeoclimate reconstruction[6] and climate signal detection[1,5]. There is a rich tradition of multivariate statistical calibration approaches to palaeoclimate reconstruction, particularly in the field of dendroclimatology where the relative strengths and weaknesses of various approaches to multivariate calibration have been well studied[14,15]. Such approaches have been applied to regional dendroclimatic networks to reconstruct regional patterns of temperature[16,17] and atmospheric circulation[18-20] or specific climate phenomena such as the Southern Oscillation[21]. Largely because of the inhomogeneity of the information represented by different types of indicators in a true "multiproxy" network, we found conventional approaches (for example, canonical correlation analysis, CCA, of the proxy and instrumental data sets) to be relatively ineffective. Our approach to climate pattern reconstruction relates closely to statistical approaches which have recently been applied to the problem of filling-in sparse early instrumental climate fields, based on calibration of the sparse sub-networks against the more widespread patterns of variability that can be resolved in shorter data sets[22,23]. We first decompose the twentieth-century instrumental data into its dominant patterns of variability, and subsequently calibrate the individual climate proxy indicators against the time histories of these distinct patterns during their mutual interval of overlap. One can think of the instrumental patterns as "training" templates against which we calibrate or "train" the much longer proxy data (that is, the "trainee" data) during the shorter calibration period which they overlap. This calibration allows us to subsequently solve an "inverse problem" whereby best estimates of surface temperature patterns are deduced back in time before the calibration period, from the multiproxy network alone.

Implicit in our approach are at least three fundamental assumptions. (1) The indicators in our multiproxy trainee network are linearly related to one or more of the instrumental training patterns. In the relatively unlikely event that a proxy indicator represents a truly local climate phenomenon which is uncorrelated with larger-scale climate variations, or represents a highly nonlinear response to climate variations, this assumption will not be satisfied. (2) A relatively sparse but widely distributed sampling of long proxy and instrumental records may nonetheless sample most of the relatively small number of degrees of freedom in climate patterns at interannual and longer timescales. Regions not directly represented in the trainee network may nonetheless be indirectly represented through teleconnections with regions that are. The El Niño/Southern Oscillation (ENSO), for example, shows global-scale patterns of climatic influence[24], and is an example of a prominent pattern of variability which, if captured, can potentially describe variability in regions not directly sampled by the trainee data. (3) Patterns of variability captured by the multiproxy network have analogues in the patterns we resolve in the shorter instrumental data. This last assumption represents a fairly weak "stationarity" requirement—we do not require that the climate itself be stationary. In fact, we expect that some sizeable trends

多代用指标的校正

虽然研究显示仔细筛选出的区域性古气候重建出乎意料地可以代替大尺度的气候变化[11-13]，但多代用资料网似乎才是最有可能完成大尺度古气候重建[6]和气候信号探测[1,5]的方法。多元统计校正法用于古气候重建已有许多先例，特别是在年轮气候学领域，各种多元校正法的优缺点已有详细研究[14,15]。此类方法也已应用到区域性年轮气候指标网络以重建区域性温度分布型[16,17]、大气环流[18-20]以及特殊的气候现象（如南方涛动）[21]等。不过我们发现，主要由于在真正的"多代用"数据网中不同类型指标所反映的信息是不一致的，从而传统方法（例如，代用指标和仪器数据库的典型相关分析，即CCA）的效果相对较差。而我们所采用的气候分布型重建方法则与统计方法紧密相关，该统计方法近来已在解决早期仪器气候记录稀疏地方的插值问题方面得到了应用，其原理是，根据分布更为广泛的短期数据库所反映的变化模式[22,23]对稀疏的子网络加以校正。首先，我们将二十世纪的仪器观测数据分解，得出其主要变化模式。然后根据不同分布型的时间序列，对各个气候代用指标加以校正，校正在各指标共同的重叠区间上进行。我们可以将仪器记录下的分布型看作"训练"模板，我们利用它来校正或"训练"相互重叠的更短的校正期间内的长期代用指标数据（即"训练"数据）。接下来我们可以利用该校正来解决一个"反向问题"，从而仅仅根据多代用指标就可推出校正期间以前的表层温度分布型。

该方法包括至少三项基本前提假设。（1）多代用训练数据网中的指标与所采用的一个或多个仪器训练分布型呈线性相关。为数极少的情况，即代用指标代表的是真实的局部气候现象（与大尺度气候变化并不相关），或者代表的是对气候变化的强烈非线性响应时，是不能满足上述假设条件的。（2）相对稀疏但分布广泛的长期代用指标还是能获取年际或更长时间尺度上的气候分布型中自由度较小的大部分数据。训练数据网中无法直接表示的区域可以通过与其他区域的遥相关间接表示。例如，厄尔尼诺和南方涛动（ENSO）对气候分布型的影响是全球范围的[24]，也是主要的变化模态的一个实例，倘若找到了它的变化，就可以大致得出未能通过训练数据进行直接取样区域的变化情况。（3）利用多代用指标网得到的变化模态与我们根据短期器测数据从分布型得出的结果相似。最后这一项假设代表了一种比较低的"稳定"要求——我们并不是要求气候本身固定不变。事实上，我们希望通过重建找出那些较大的气候变化。不过，我们确实也假定上世纪气候的基本空间变化模态与前几个世

in the climate may be resolved by our reconstructions. We do, however, assume that the fundamental spatial patterns of variation which the climate has shown during the past century are similar to those by which it has varied during past recent centuries. Studies of instrumental surface-temperature patterns suggest that such a form of stationarity holds up at least on multidecadal timescales, during the past century[23]. The statistical cross-validation exercises we describe later provide the best evidence that these key underlying assumptions hold.

We isolate the dominant patterns of the instrumental surface-temperature data through principal component analysis[25] (PCA). PCA provides a natural smoothing of the temperature field in terms of a small number of dominant patterns of variability or "empirical eigenvectors". Each of these eigenvectors is associated with a characteristic spatial pattern or "empirical orthogonal function" (EOF) and its characteristic evolution in time or "principal component" (PC). The ranking of the eigenvectors orders the fraction of variance they describe in the (standardized) multivariate data during the calibration period. The first five of these eigenvectors describe a fraction $\beta = 0.93$ (that is, 93%) of the global-mean (GLB) temperature variations, 85% of the Northern Hemisphere-mean (NH) variations, 67% of the NINO3 index, and 76% of the non-trend-related (DETR) NH variance (see Methods for a description of the β statistic used here as a measure of resolved variance). A sizeable fraction of the total multivariate spatiotemporal variance (MULT) in the raw (instrumental) data (27%) is described by these five eigenvectors, or about 30% of the standardized variance (no. 1 = 12%, no. 2 = 6.5%, no. 3 = 5%, no. 4 = 4%, no. 5 = 3.5%). Figure 2 shows the EOFs of the first five eigenvectors. The associated PCs and their reconstructed counterparts (RPCs) are discussed in the next section. The first eigenvector, associated with the significant global warming trend of the past century, describes much of the variability in the global (GLB = 88%) and hemispheric (NH = 73%) means. Subsequent eigenvectors, in contrast, describe much of the spatial variability relative to the large-scale means (that is, much of the remaining MULT). The second eigenvector is the dominant ENSO-related component, describing 41% of the variance in the NINO3 index. This eigenvector shows a modest negative trend which, in the eastern tropical Pacific, describes a "La Niña"-like cooling trend[26], which opposes warming in the same region associated with the global warming pattern of the first eigenvector. The third eigenvector is associated largely with interannual-to-decadal scale variability in the Atlantic basin and carries the well-known temperature signature of the North Atlantic Oscillation (NAO)[27] and decadal tropical Atlantic dipole[28]. The fourth eigenvector describes a primarily multidecadal timescale variation with ENSO-scale and tropical/subtropical Atlantic features, while the fifth eigenvector is dominated by multidecadal variability in the entire Atlantic basin and neighbouring regions that has been widely noted elsewhere[29-34].

We calibrate each of the indicators in the multiproxy data network against these empirical eigenvectors at annual-mean resolution during the 1902–80 training interval. Although the seasonality of variability is potentially important—many extratropical proxy indicators, for example, reflect primarily warm-season variability[6,7]—we seek in the present study to resolve only annual-mean conditions, exploiting the seasonal climate persistence, and the fact that the mutual information from data reflecting various seasonal windows should

纪是相似的。对器测地表温度时空分布模态的研究显示，在上世纪，这样的稳定形态至少在数十年的时间尺度上是成立的 [23]。后文我们将会给出统计交叉验证结果，而它正是上述暗含的关键假设成立的最好证据。

利用主成分分析法 [25]（PCA）将器测地表温度数据显示的主要分布型从中提取出来。PCA 提供一个自然平滑的温度场，它以少数几个主变化模态或"经验特征向量"的形式表示。每个特征向量都对应着一个与之相关的特征空间分布型（或"经验正交函数"，EOF）及其时间演化特征（或"主成分"，PC）。在（标准化）多元数据分析中，特征向量的级别表示其在校正期间内贡献的方差的大小。前五个特征向量反映的依次是：全球平均（GLB）温度变化 $\beta = 0.93$（即 93%），北半球（NH）的平均温度变化 85%，NINO3 指数 67%，北半球非趋势相关（DETR）方差 76%（这里 β 用来衡量解释方差，其统计学意义见方法部分）。这五个特征向量就可以表示出原始（仪器观测）数据中总多元时空方差（MULT）（27%）的大部分，约贡献标准方差的 30%（no.1 = 12%，no.2 = 6.5%，no.3 = 5%，no.4 = 4%，no.5 = 3.5%）。图 2 所示为前五个特征向量的经验正交函数。相关的多个主成分及其对应的重建主成分（RPC）将在下节讨论。第一个特征向量，与上世纪全球显著变暖的趋势相关，体现了全球（GLB = 88%）和半球（NH = 73%）平均值的主要变化。而后面的特征向量则表示了不同地区相对于大尺度平均值的主要空间变化（即主要剩余 MULT）。第二个特征向量主要是与 ENSO 相关的分量，占 NINO3 指数方差的 41%。该特征向量表现出轻微的负相关变化趋势，在热带东太平洋地区，代表了类似于拉尼娜现象的变冷趋势 [26]，这与反映全球变暖分布型的第一个特征向量在同一地区显示出的变暖趋势相反。第三个特征向量主要反映了大西洋海盆在年际、十年尺度上的变化，带有众所周知的北大西洋涛动（NAO）的温度变化 [27] 和十年际热带大西洋偶极子 [28] 等特征。第四个特征向量反映的主要是 ENSO 和热带、副热带大西洋在数十年尺度上的变化，而第五个特征向量则反映了整个大西洋海盆及其邻近地区数十年尺度上的变化，这在其他文章中 [29-34] 已得到了广泛的关注。

利用上述经验特征向量对多代用数据网中的每个指标在 1902～1980 年间的值逐一作了年均分辨率尺度上的校正。虽然季节性尺度的变化具有潜在的重要意义（例如，许多温带代用指标主要反映了暖季变化性 [6,7]），但在本研究中我们借助季节性气候的持续性，仅关注年均变化情况，而反映了不同季节性气候的数据的共有信息

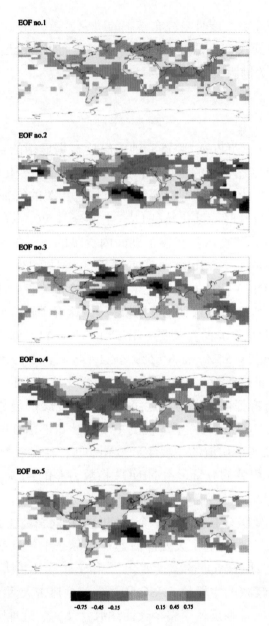

Fig. 2. Empirical orthogonal functions (EOFs) for the five leading eigenvectors of the global temperature data from 1902 to 1980. The gridpoint areal weighting factor used in the PCA procedure has been removed from the EOFs so that relative temperature anomalies can be inferred from the patterns.

provide complementary information regarding annual mean climate conditions[10]. Following this calibration, we apply an overdetermined optimization procedure to determine the best combination of eigenvectors represented by the multiproxy network back in time on a year-by-year basis, with a spatial coverage dictated only by the spatial extent of the instrumental training data. From the RPCs, spatial patterns and all relevant averages or indices can be readily determined. The details of the entire statistical approach are described in the Methods section.

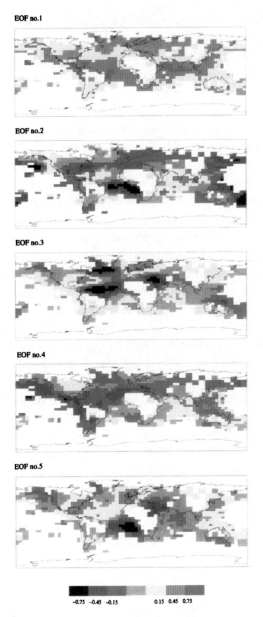

图 2. 1902～1980 年间全球温度数据的五个主要特征向量的经验正交函数（EOF）。主成分分析中所采用的格点面积加权因子已从经验正交函数中剔除，因而可从其分布型上推断出相对温度距平值。

也可为年均气候的相关情况提供补充信息[10]。根据这一校正方法，我们用一个超定优选方案来确定特征向量的最优组合，并以多代用数据网中的逐年数据表示，其空间尺度仅取决于仪器记录的训练数据的空间范围。从数个 RPC 中可以确定出空间分布型和所有相关的平均值或指标。整个统计方法的详细描述见方法部分。

The skill of the temperature reconstructions (that is, their statistical validity) back in time is established through a variety of complementary independent cross-validation or "verification" exercises (see Methods). We summarize here the main results of these experiments (details of the quantitative results of the calibration and verification procedures are available; see Supplementary Information).

(1) In the reconstructions from 1820 onwards based on the full multiproxy network of 112 indicators, 11 eigenvectors are skilfully resolved (nos 1–5, 7, 9, 11, 14–16) describing ~70–80% of the variance in NH and GLB mean series in both calibration and verification. (Verification is based here on the independent 1854–1901 data set which was withheld; see Methods.) Figure 3 shows the spatial patterns of calibration β, and verification β and the squared correlation statistic r^2, demonstrating highly significant reconstructive skill over widespread regions of the reconstructed spatial domain. 30% of the full spatiotemporal variance in the gridded data set is captured in calibration, and 22% of the variance is verified in cross-validation. Some of the degradation in the verification score relative to the calibration score may reflect the decrease in instrumental data quality in many regions before the twentieth century rather than a true decrease in resolved variance. These scores thus compare favourably to the 40% total spatiotemporal variance that is described by simply filtering the raw 1902–80 instrumental data with 11 eigenvectors used in calibration, suggesting that the multiproxy calibrations are describing a level of variance in the data reasonably close to the optimal "target" value. Although a verification NINO3 index is not available from 1854 to 1901, correlation of the reconstructed NINO3 index with the available Southern Oscillation index (SOI) data from 1865 to 1901 of $r = -0.38$ ($r^2 = 0.14$) compares reasonably with its target value given by the correlation between the actual instrumental NINO3 and SOI index from 1902 to 1980 ($r = -0.72$). Furthermore, the correspondence between the reconstructed NINO3 index warm events and historical[35] El Niño chronology back to 1820 (see Methods) is significant at the 98% level.

(2) The calibrations back to 1760, based on 93 indicators, continue to resolve at least nine eigenvectors (nos 1–5, 7, 9, 11, 15) with no degradation of calibration or verification resolved variance in NH, and only slight degradation in MULT (calibration ~27%, verification ~17%). Our reconstructions are thus largely indistinguishable in skill back to 1760.

(3) The network available back to 1700 of 74 indicators (including only two instrumental or historical indicators) skilfully resolves five eigenvectors (nos 1, 2, 5, 11, 15) and shows some significant signs of decrease in reconstructive skill. In this case, ~60–70% of NH variance is resolved in calibration and verification, ~14–18% of MULT in calibration, and 10–12% of MULT in verification. The verification r of NINO3 with the SOI is in the range of $r \approx -0.25$ to -0.35, which is statistically significant (as is the correspondence with the historical[35] chronology back to 1700) but notably inferior to the later calibrations. In short, both spatial patterns and large-scale means are skilfully resolved, but with significantly less resolved variance than in later calibrations.

向回追溯的温度重建方法的可行性（即其统计有效性）是通过各种补充的独立交叉验证或"检验"练习（见方法）确定的。在此我们总结了上述实验的主要结果（校正和检验过程中详细的定量结果见补充信息）。

（1）根据 112 个指标组成的完整多代用数据网得到的可追溯至 1820 年的重建结果，可分辨出 11 个特征向量（nos 1~5，7，9，11，14~16），在 NH 和 GLB 平均值序列的校正和检验中方差贡献率约为 70%~80%。（本文检验是根据 1854~1901年保留下来的独立数据库进行的，见方法部分。）图 3 所示为校正 β，检验 β 以及平方相关系数 r^2 的空间分布型，证明在重建的空间范围所覆盖的广大区域上该方法均具有很高的可行性。校正范围覆盖了网格数据库中总时空变化的 30%，且有 22%通过了交叉验证法的检验。检验得分相对于校正得分的降低可能反映了二十世纪以前仪器测定数据质量较低而非解释方差的真正降低。由此，上述得分与 40% 的总时空方差就顺利对应起来了，其中总时空方差以校正中所用的 11 个特征向量对1902~1980 年仪器测定的原始数据进行简单筛选后的结果来表示，说明多代用指标校正反映了数据方差水平，与理想"目标"值非常接近。虽然没有 1854~1901 年间 NINO3 指数的检验结果，但重建 NINO3 指数与 1865~1901 年间南方涛动指数（SOI）的相关系数 $r = -0.38 (r^2 = 0.14)$，与 1902~1980 年间实际仪器记录中 NINO3和 SOI 指数的相关系数（$r = -0.72$）给出的目标值是可比的。另外，重建 NINO3 指数变暖事件与历史上 [35] 自 1820 年以来发生的厄尔尼诺现象（见方法）之间相关性显著，置信水平为 98%。

（2）基于 93 项指标可回溯至 1760 年的校正结果，仍可分辨出至少 9 个特征向量（nos 1~5，7，9，11，15），而北半球的校正方差和检验方差都没有下降，并且MULT 也只有微弱的下降（校正方差约 27%，检验方差约为 17% 左右）。因此，我们所得到的自 1760 年以来的重建结果从效果上来说没有很大区别。

（3）由 74 个指标（仅包含两个仪器测定或历史指标）组成的、可回溯到 1700 年的数据网，可分辨出 5 个特征向量（nos 1，2，5，11，15），在重建效果上则表现出明显的下降迹象。此时，校正和检验过程中北半球的方差约占 60%~70%，而 MULT 的校正方差约占 14%~18%，检验方差为 10%~12%。根据 SOI 得到的 NINO3 的检验系数 r 约在 -0.25 到 -0.35 之间，具有统计学意义（回溯到历史上 [35]1700 年的也是如此），但明显低于后续的校正结果。总之，空间分布型和大尺度平均值都可很好地解译出来，但与后来的校正结果相比，解释方差较小。

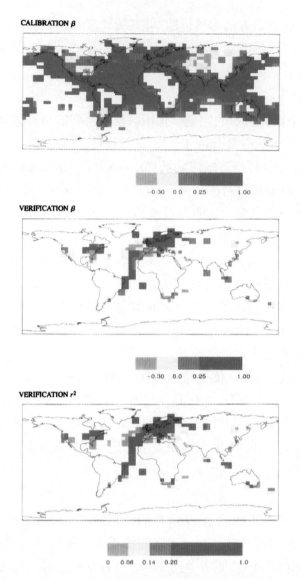

Fig. 3. Spatial patterns of reconstruction statistics. Top, calibration β (based on 1902–80 data); middle, verification (based on 1854–1901 data) β; bottom, verification r^2 (also based on 1854–1901 data). For the β statistic, values that are insignificant at the 99% level are shown in grey negative; but 99% significant values are shown in yellow, and significant positive values are shown in two shades of red. For the r^2 statistic, statistically insignificant values (or any grid-points with unphysical values of correlation $r < 0$) are indicated in grey. The colour scale indicates values significant at the 90% (yellow), 99% (light red) and 99.9% (dark red) levels (these significance levels are slightly higher for the calibration statistics which are based on a longer period of time). A description of significance level estimation is provided in the Methods section.

(4) The network of 57 indicators back to 1600 (including one historical record) skilfully resolves four eigenvectors (nos 1, 2, 11, 15). 67% of NH is resolved in calibration, and 53% in verification. 14% of MULT is resolved in calibration, and 12% of MULT in verification. A significant, but modest, level of ENSO-scale variability is resolved in the calibrations.

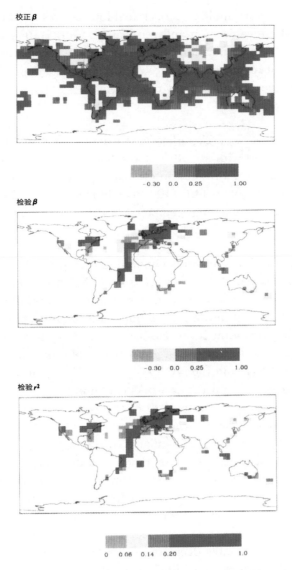

图3. 重建数据的空间分布型。上图，校正 β（以 1902～1980 年的数据为基础）；中图，检验 β（以 1854～1901 年的数据为基础）；下图，检验 r^2（同样以 1902～1980 年的数据为基础）。对 β 来说，置信水平为 99% 时，不具有显著性的值以灰色的负数表示，具有显著性的值以黄色表示，而显著相关的正值则以两种红色色调来表示。对于 r^2，不具有统计学意义的值（或 $r < 0$ 的非自然值格点）以灰色表示。彩色图例分别表示置信水平为 90%（黄色）、99%（浅红色）和 99.9%（深红色）时显著相关的值（对于在长时间尺度上得到的校正数据来说，上述显著性水平略高）。有关显著性水平估计的详细描述见方法部分。

(4) 可回溯到 1600 年的 57 个指标（包括一个历史记录）组成的数据网则可分辨出 4 个特征向量（nos 1，2，11，15）。北半球的校正方差贡献为 67%，检验方差为 53%。校正 MULT 为 14%，检验 MULT 为 12%。校正过程中发现 ENSO 变化的影响很显著，但贡献不是很大。

(5) The network of 24 proxy indicators back to 1450 resolves two eigenvectors (nos 1, 2) and ~40–50% of NH in calibration and verification. Only ~10% of MULT is resolved in calibration and ~5% in verification. There is no skilful reconstruction of ENSO-scale variability. Thus spatial reconstructions are of marginal usefulness this far back, though the largest-scale quantities are still skilfully resolved.

(6) The multiproxy network of 22 indicators available back to 1400 resolves only the first eigenvector, associated with 40–50% of resolved variance in NH in calibration and verification. There is no useful resolution of spatial patterns of variability this far back. The sparser networks available before 1400 show little evidence of skill in reconstructing even the first eigenvector, terminating useful reconstruction at the initial year AD 1400.

(7) Experiments using trainee networks containing only proxy (that is, no instrumental or historical) indicators establish the most truly independent cross-validation of the reconstruction as there is in this case neither spatial nor temporal dependence between the calibration and verification data sets. Such statistically significant verification is demonstrated at the grid-point level (calibration and verification resolved variance ~15% for the MULT statistic), at the largest scales (calibration and verification resolved variance ~60–65% for NH) and the NINO3-scale (90–95% statistical significance for all verification diagnostics). In contrast, networks containing only the 24 long historical or instrumental records available back to 1820 resolve only ~30% of NH in calibration or verification, and the modest multivariate calibration and verification resolved variance scores of MULT (~10%) are artificially inflated by the high degree of spatial correlation between the instrumental "multiproxy" predictor and instrumental predictand data. No evidence of skilful ENSO-scale reconstruction is evident in these latter reconstructions. In short, the inclusion of the proxy data in the "multiproxy" network is essential for the most skilful reconstructions. But certain sub-components of the proxy dataset (for example, the dendroclimatic indicators) appear to be especially important in resolving the large-scale temperature patterns, with notable decreases in the scores reported for the proxy data set if all dendroclimatic indicators are withheld from the multiproxy network. On the other hand, the long-term trend in NH is relatively robust to the inclusion of dendroclimatic indicators in the network, suggesting that potential tree growth trend biases are not influential in the multiproxy climate reconstructions. The network of all combined proxy and long instrumental/historical indicators provide the greatest cross-validated estimates of skilful reconstruction, and are used in obtaining the reconstructions described below.

Temperature Reconstructions

The reconstructions discussed here are derived using all indicators available, and using the optimal eigenvector subsets determined in the calibration experiments described above (11 from 1780–1980, 9 from 1760–1779, 8 from 1750–1759, 5 from 1700–1749, 4

(5) 可回溯到 1450 年的 24 个代用指标解出了 2 个特征向量（nos 1，2），其中北半球的校正方差和检验方差贡献均在 40%～50% 左右。校正过程中仅有约 10% 的 MULT 可解释，而检验过程中则仅为 5% 左右。在 ENSO 尺度的变化方面没有获得有效的重建结果。因此，回溯到这一时期所获得的空间重建结果的用处不大，不过所得出的最大尺度还是有效的。

(6) 可回溯到 1400 年的 22 个指标组成的多代用数据网仅能解出第一个特征向量，其中，北半球的校正和检验方差贡献为 40%～50%。关于这一时期的空间变化分布型没有得出更多有效信息。1400 年以前的数据网则更稀疏，即便对第一个特征向量也未能在其中找到有效证据，因此可以认为有效的重建结果止于最初的公元 1400 年。

(7) 利用仅包含代用指标（即没有仪器记录和历史记录）的训练数据网进行实验建立的交叉验证结果的独立性最好，因为在这种情况下，校正和检验数据库之间无论是时间上还是空间上都没有相关性。此类显著性检验在格点水平上（MULT 统计结果的校正和检验解释方差约为 15%），最大尺度上（北半球的校正和检验解释方差贡献率约为 60%～65%），以及 NINO3 尺度（所有方差检验在置信水平为 90%～95% 时都具有统计学意义）都得到了验证。相反，可回溯至 1820 年的仅包含 24 个长期历史记录和仪器记录的数据网中，北半球通过校正和检验的置信水平仅为 30% 左右，而 MULT 在多元校正和检验下的方差得分则更低（约 10%），而且其中还存在人为增大的因素，因为仪器"多代用"预报因子和仪器预报值之间在空间上具有高度的相关性。在后一种重建结果中，没有找到 ENSO 尺度变动的有效证据。总而言之，要获得最佳重建结果，"多代用"网中代用指标数据所包含的内容是最重要的。但某些代用指标组成的子库（例如树木年轮指标）在解决大尺度温度分布型时显得尤为重要。比如，倘若将树木年轮指标从多代用网络中都剔除掉，代用指标数据库的得分将显著下降。另一方面，对于数据网中包含树木年轮指标的情况来说，北半球的长期趋势则具有一定的稳健性，这说明，在多代用指标气候重建中，树木生长趋势中可能存在的偏差带来的影响并不大。利用由所有混合代用指标和长期仪器/历史指标组成的数据网所得到的重建结果具有最佳的交叉验证有效性。下文将要讨论的重建结果就是由此获得的。

温 度 重 建

这里所讨论的重建是利用所有可用指标得出的，另外还采用了前述校正实验确定出的最优特征向量子集（1780～1980 年为 11 个，1760～1779 年 9 个，1750～1759 年 8 个，1700～1749 年 5 个，1600～1699 年 4 个，1450～1599 年 2 个，1400～1449 年 1

from 1600–1699, 2 from 1450–1599, 1 from 1400–1449). To better illustrate the workings and effectiveness of the proxy pattern reconstruction procedure, we show as an example (Fig. 4) the actual, the EOF-filtered, and the reconstructed temperature patterns for a year (1941) during the calibration interval. This year was a known ENSO year, associated with a warm eastern tropical Pacific and a cold central North Pacific. Pronounced cold anomalies were also found over large parts of Eurasia. The proxy-reconstructed pattern captures these features, although in a relatively smoothed sense (describing ~30% of the full variance in that pattern), and is remarkably similar to the raw data once it has been filtered by retaining only the 11 eigenvectors (nos 1–5, 7, 9, 11, 14–16) used in pattern reconstruction. It is thus visually apparent that the multiproxy network is quite capable of resolving much of the structure resolved by the eigenvectors retained in the calibration process.

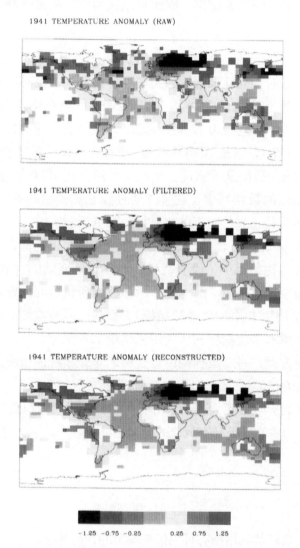

Fig. 4. Comparison of the proxy-based spatial reconstructions of the anomaly pattern for 1941 versus the raw data. Comparisons based on actual (top), EOF-filtered (middle), and proxy-reconstructed (bottom) data. Anomalies (relative to 1902–80 climatology) are indicated by the colour scale shown in °C.

个)。为了更好地阐释多代用指标分布型重建方法的运作方式及其有效性，作为示例，图 4 分别给出了校正区间内 1941 年的三种温度分布型：原始数据、EOF 过滤结果和重建结果。这一年是显著的 ENSO 年，东赤道太平洋非常暖，而中北部太平洋则较冷。欧亚地区大部也出现了显著的异常变冷现象。而根据代用指标重建出的气温分布型也恰反映出了上述特征，只是其变化趋势相对平滑一些（贡献了分布型总方差的 30% 左右）。而且，如果利用分布型重建中采用的 11 个特征向量（nos 1～5，7，9，11，14～16）对原始数据加以筛选，所剩下的数据与重建结果亦极为相似。因此，显然多数经校正过程中保留下来的特征向量解出的结构都可用多代用数据网解决。

1941　温度距平值（原始数据）

1941　温度距平值（EOF 过滤结果）

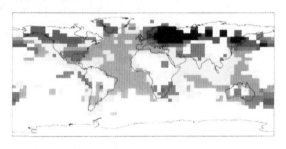

1941　温度距平值（重建结果）

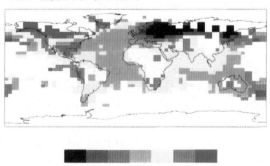

图 4. 在 1941 年的异常分布型下，根据代用指标得到的空间重建结果与原始数据的对比。图示分别为原始数据（上图）、EOF 筛选后的数据（中图）以及代用指标重建结果（下图）。距平值（相对于 1902～1980 年间的气候）以彩色图例标出，单位为℃。

We consider the temporal variations in the first five RPCs (Fig. 5a). The positive trend in RPC no. 1 during the twentieth century is clearly exceptional in the context of the long-term variability in the associated eigenvector, and indeed describes much of the unprecedented warming trend evident in the NH reconstruction. The negative trend in RPC no. 2 during the past century is also anomalous in the context of the longer-term evolution of the associated eigenvector. The recent negative trend is associated with a pattern of cooling in the eastern tropical Pacific (superimposed on warming associated with the pattern of eigenvector no. 1) which may be a modulating negative feedback on global warming[26]. RPC no. 5 shows notable multidecadal variability throughout both the modern and pre-calibration interval, associated with the wavelike trend of warming and subsequent cooling of the North Atlantic this century discussed earlier[29-33] and the longer-term multidecadal oscillations in this region detected in a previous analysis of proxy climate networks[5]. This variability may be associated with ocean–atmosphere processes related to the North Atlantic thermohaline circulation[4,34].

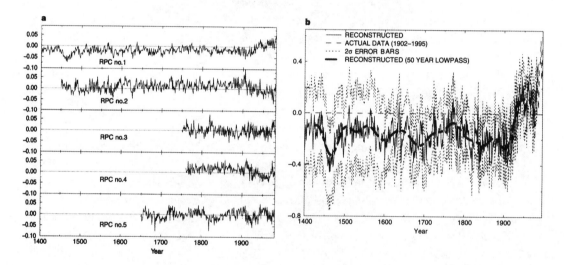

Fig. 5. Time reconstructions (solid lines) along with raw data (dashed lines). **a**, For principal components (RPCs) 1–5; **b**, for Northern Hemisphere mean temperature (NH) in °C. In both cases, the zero line corresponds to the 1902–80 calibration mean of the quantity. For **b** raw data are shown up to 1995 and positive and negative 2σ uncertainty limits are shown by the light dotted lines surrounding the solid reconstruction, calculated as described in the Methods section.

The long-term trends in the reconstructed annual mean NH series (Fig. 5b) are quite similar to those of decadal Northern Hemisphere summer temperature reconstructions[6], showing pronounced cold periods during the mid-seventeenth and nineteenth centuries, and somewhat warmer intervals during the mid-sixteenth and late eighteenth centuries, with almost all years before the twentieth century well below the twentieth-century climatological mean. Taking into account the uncertainties in our NH reconstruction (see Methods), it appears that the years 1990, 1995 and now 1997 (this value recently calculated and not shown) each show anomalies that are greater than any other year back to 1400 at 3 standard errors, or roughly a 99.7% level of certainty. We note that hemispheric mean

我们来看一下前 5 个 RPC 随时间的变化（图 5a）。在二十世纪，RPC no.1 在该特征向量长期变化的背景下的正变化趋势是异常明显的，这实际上反映了北半球重建结果中表现出的空前变暖的趋势。而上世纪 RPC no.2 在相关特征向量上的长期演化背景下的负相关变化也是异常的。最近的负相关变化趋势则可能与东赤道太平洋的变冷有关（叠加于特征向量 no.1 的变暖趋势之上），它可能是全球变暖的一种负反馈调制[26]。RPC no.5 在整个现代（器测）时期以及（器测）之前的时期都表现出显著的以数十年为周期的变化特征，这可能与本世纪北大西洋的波状变暖及之后的变冷现象相关，这一点前人已有研究[29-33]，前人根据代用指标气候网分析发现该区长期存在周期为数十年的振荡变化[5]。该变化可能反映了与北大西洋温盐循环有关的海－气相互作用过程[4,34]。

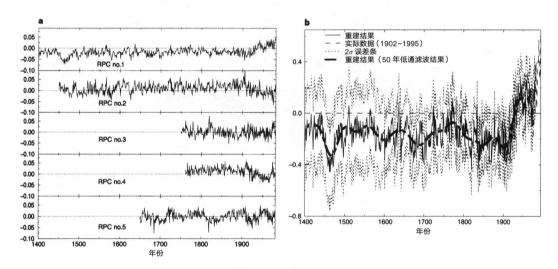

图 5. 时间序列的重建结果（实线）及原始数据（虚线），**a**，主成分因子（RPC）1～5；**b**，北半球平均温度（NH），单位℃。两图中的零线均对应 1902～1980 年间的校正平均值。**b** 图所示为直到 1995 的原始数据，其不确定界限为 ±2σ，以实线周围的浅色点虚线表示，计算过程详见方法部分。

年均北半球序列重建结果中的长期变化趋势（图 5b）与北半球夏季温度重建结果的十年平均值变化[6]相似，在十七世纪中期和十九世纪均表现为显著冷期，而十六世纪中期和十八世纪后期则有一段时间相对较暖，另外二十世纪之前的结果均低于二十世纪的气候平均值。如果考虑我们的 NH 重建结果中的不确定性（见方法），那么，1990 年、1995 年以及现在的 1997 年（该值是新计算出的，文中未列出）的距平值呈现异常，比 1400 年以来的任何年份都高，其置信水平为 99.7%（相当于三个标准差）。我们注意到，半球的平均值与全球或半球的均一趋势并不一致。图 6 上图为

values are not associated with globally or hemispherically uniform trends. An example of the global pattern for an historically documented[35] "very strong" El Niño year (1791) is shown in Fig. 6 top panel, demonstrating the classic warm eastern tropical Pacific and cold central North Pacific sea surface temperature patterns. Analysis of ENSO variability in these reconstructions is discussed in more detail elsewhere[36]. We also show the reconstructed pattern for 1816 (Fig. 6 bottom panel). Quite anomalous cold is evident throughout much of the Northern Hemisphere (even relative to this generally cold decade) but with a quadrupole pattern of warmth near Newfoundland and the Near East, and enhanced cold in the eastern United States and Europe consistent with the anomalous atmospheric circulation associated with the NAO pattern. Such a pattern is indeed observed in empirical[37] and model-based studies[38] of the atmospheric response to volcanic forcing. We infer in the 1816 temperature pattern a climatic response to the explosive Tambora eruption of April 1815 based on both the anomalous hemispheric coolness and the superimposed NAO-like pattern. Reconstructed time series RPCs nos 1–5, the NH series, the NINO3 index and reconstructions for specific grid-points can be obtained through the NOAA palaeoclimatology Web site (http://www.ngdc.noaa.gov/paleo/paleo.html).

1791 TEMPERATURE ANOMALY PATTERN

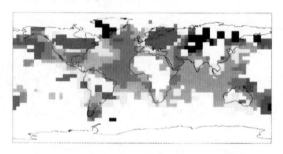

1816 TEMPERATURE ANOMALY PATTERN

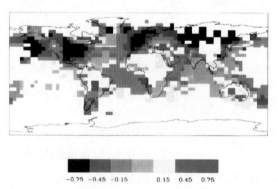

-0.75 -0.45 -0.15   0.15   0.45   0.75

Fig. 6. Reconstructed annual temperature patterns for two example years. Top, 1791; bottom, 1816. The colours indicate regions which exceeded (either positively or negatively) the threshold indicated in °C. The zero baseline is defined by the 1902–80 climatological mean for each grid-point.

前人已证实的"超强"厄尔尼诺年(1791)的全球温度分布型[35]，显示了典型的东赤道太平洋暖、中北部太平洋冷的海表温度分布型。根据上述重建分析出的 ENSO 的变化在其他文章中有更为详细的讨论[36]。我们还给出了重建的 1816 年全球温度分布型(图 6 下图)。北半球大部分地区都表现为明显的异常寒冷(即使相对于总体较冷的这个十年来说也是如此)，但在纽芬兰岛附近和近东地区则呈现四极子增暖模态，而美国东部和欧洲地区则更冷一些，这与 NAO 模态有关的大气异常环流有关。关于大气对火山强迫响应的研究，无论是经验观测[37]还是模式研究[38]，都证实了这种分布型的存在。根据北半球的异常变冷现象和叠加其上的类 NAO 模态，我们推断在 1816 年的温度分布型是对 1815 年 4 月爆发的坦博拉火山喷发的一种气候响应。重建的时间序列 RPC nos 1~5、NH 序列、NINO3 指数以及特定格点的重建结果都可从美国国家海洋和大气局(NOAA)的古气候网站找到(网址 http://www.ngdc.noaa.gov/paleo/html)。

1791　温度距平值分布型

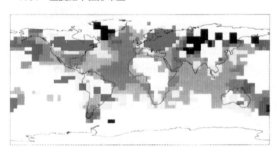

1816　温度距平值分布型

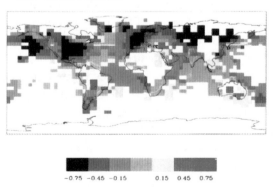

图 6. 两个典型年份上重建的年际温度分布型。上图为 1791 年；下图为 1816 年。彩色部分表示超出界限的区域(不管是正向的还是负向的)，单位℃。零基线是根据 1902~1980 年间每个格点的气候平均值得到的。

Attribution of Climate Forcings

We take an empirical approach to detecting the possible effects of external forcings on the climate. The reconstructed NH series is taken as a diagnostic of the global climate, and we examine its relationship with three candidate external forcings during the period 1610–1995 including (1) CO_2 measurements[39] as a proxy for total greenhouse-gas changes, (2) reconstructed solar irradiance variations[2] and (3) the weighted historical "dust veil index" (DVI) of explosive volcanism (see Fig. 31.1 in ref. 40) updated with recent data[41]. While we warn that historical series for these forcing agents are imperfectly known or measured, they do nonetheless represent our best estimates of the time-histories of the corresponding forcings. More detailed discussions of the estimation of, and potential sources of uncertainty or bias in, these series are available[2,39,40]. Industrial-aerosol forcing of the climate has also been suggested as an important forcing of recent climate[42,43], but its physical basis is still controversial[44], and difficult to estimate observationally. Noting that in any case, this forcing is not believed to be important before about 1940, its omission should be inconsequential in our long-term detection approach. Our empirical signal detection is complementary to that of model-based "finger-print" signal detection studies[42,43,45]; while our empirical approach relies on the faithfulness of the reconstructed forcing series and on the assumption of a linear and contemporaneous response to forcings, it does not suffer the potential weaknesses of incomplete representations of internal feedback processes[26], poorly constrained parametrizations of climatic responses[44], and underestimated natural variability[1] in model-based studies. To the extent that the response to forcing is not contemporaneous, but rather is delayed owing to the inertia of the slow-response components of the climate system (for example, the ocean and cryosphere), our detection approach will tend to underestimate the response to forcings, making the approach a relatively conservative one.

We estimate the response of the climate to the three forcings based on an evolving multivariate regression method (Fig. 7). This time-dependent correlation approach generalizes on previous studies of (fixed) correlations between long-term Northern Hemisphere temperature records and possible forcing agents[2,3]. Normalized regression (that is, correlation) coefficients r are simultaneously estimated between each of the three forcing series and the NH series from 1610 to 1995 in a 200-year moving window. The first calculated value centred at 1710 is based on data from 1610 to 1809, and the last value, centred at 1895, is based on data from 1796 to 1995—that is, the most recent 200 years. A window width of 200 yr was chosen to ensure that any given window contains enough samples to provide good signal-to-noise ratios in correlation estimates. Nonetheless, all of the important conclusions drawn below are robust to choosing other reasonable (for example, 100-year) window widths.

气候强迫的影响因素

本文采用了经验方法研究了外部强迫可能对气候产生的影响。对全球气候的诊断是采用重建的 NH 序列来进行的，同时对它与 1610~1995 年间三个候选的气候强迫之间的关系作了评估，其中包括：(1) 作为所有温室气体变化代用指标的 CO_2 的测定结果[39]，(2) 重建的太阳辐照度变化[2]，(3) 火山喷发作用的历史权重"尘幕指数"(DVI)(见文献 40 中的图 31.1)，并且已更新至最新数据[41]。但是我们还要注意一点，那就是，我们对这些强迫因素的历史序列的了解和测定都存在一定缺陷，不过尽管如此，它们仍然代表了我们对相应气候强迫的历史变化的最佳估计。前人已有关于这些序列估计值及其潜在不确定性和偏差来源的详细讨论[2,39,40]。工业气溶胶对气候的影响也被认为是近年来气候的重要强迫因子[42,43]，不过关于其物理基础仍存在争议[44]，而且也很难通过观测做出评估。不管怎样，在 1940 年以前这一强迫的影响不是那么重要，但在我们的长期探测法中漏掉它也是不合理的。我们采用的经验信号探测法是对基于模式的"指纹"信号探测研究[42,43,45]的一种补充，同时我们采用的经验方法依赖于重建的强迫序列的准确性以及这些强迫的同期响应呈线性的假定，它不受模式研究中存在的内在反馈过程[26]的不完整性、气候响应参数化的低约束性[44]以及对自然变化的低估[1]等潜在缺陷的影响。从某种程度上来说，气候对影响因素的响应并不是同期的，它会因气候体系中缓慢响应组分(如海洋以及冰冻圈等)的惯性而被延迟，因此本文采用的探测方法可能会低估气候对上述强迫的响应，也就是说这是一种相对较为保守的研究方法。

我们利用改进的多元回归法研究了气候对三个强迫因子的响应情况(图 7)。该依时相关法是根据前人关于北半球气温的长期记录与潜在影响因素之间的(固定的)相关性研究[2,3]概括而来的。同时还求出了各强迫序列与 1610 到 1995 年间 NH 序列之间的标准化回归(即相关性)系数 r，其中以 200 年作为滑动窗口。由此得到的第一个以 1710 年为中心的计算值就是根据 1610 到 1809 年之间的数据得到的，而最后一个值，以 1895 年为中心，是根据 1796 到 1995 年，即最近 200 年的数据得出的。之所以将窗口宽度选为 200 年是为了保证每个给定窗口中包含足够多的样本来为相关性研究提供最佳信噪比。不过，在选择其他合理的窗口宽度(比如 100 年)时，下文将要得出的那些重要结论是同样成立的。

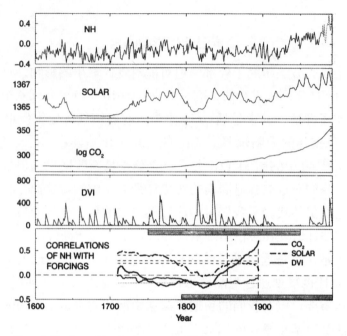

Fig. 7 Relationships of Northern Hemisphere mean (NH) temperature with three candidate forcings between 1610 and 1995. Panels, (top to bottom) as follows. "NH", reconstructed NH temperature series from 1610–1980, updated with instrumental data from 1981–95. "Solar", reconstructed solar irradiance. "log CO_2", greenhouse gases represented by atmospheric CO_2 measurements. "DVI", weighted volcanic dust veil index. Bottom panel, evolving multivariate correlation of NH series with the three forcings NH, Solar, log CO_2. The time axis denotes the centre of a 200-year moving window. One-sided (positive) 90%, 95%, 99% significance levels (see text) for correlations with CO_2 and solar irradiance are shown by horizontal dashed lines, while the one-sided (negative) 90% significance threshold for correlations with the DVI series is shown by the horizontal dotted line. The grey bars indicate two difference 200-year windows of data, with the long-dashed vertical lines indicating the centre of the corresponding window.

We test the significance of the correlation coefficients (r) relative to a null hypothesis of random correlation arising from natural climate variability, taking into account the reduced degrees of freedom in the correlations owing to substantial trends and low-frequency variability in the NH series. The reduced degrees of freedom are modelled in terms of first-order markovian "red noise" correlation structure of the data series, described by the lag-one autocorrelation coefficient ρ during a 200-year window. This parameter ranges from 0.48 in the first window (1610–1809) to 0.77 in the final window (1796–1995) of the moving correlation, the considerably larger recent value associated with the substantial global warming trend of the past century. This latter trend has been shown to be inconsistent with red noise[46] and could thus itself be argued as indicative of externally forced variability. An argument could in this sense, be made for using the smaller pre-industrial value $\rho = 0.48$ of the NH series in estimating the statistical degrees of freedom appropriate for the null hypothesis of natural variability. Nonetheless, we make the conservative choice of adopting the largest value $\rho = 0.77$ as representative of the natural serial correlation in the series. We use Monte Carlo simulations to estimate the likelihood of chance spurious correlations of such serially correlated noise with each of the three actual forcing series. The associated confidence limits are approximately constant between sliding 200-year windows. For

140

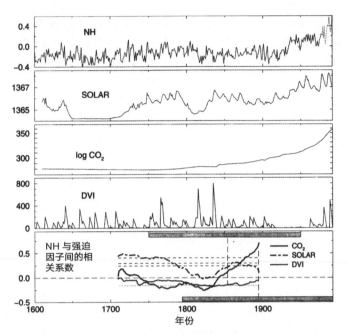

图 7. 1610～1995 年北半球平均(NH)气温与三个强迫因子之间的关系。从上到下依次为："NH"，
1610～1980 年北半球温度序列的重建结果，同时利用 1981～1995 年间的仪器测定信息得到的更新
结果；"Solar"，太阳辐照度的重建结果；"log CO₂"，以大气 CO_2 浓度的测定值表示的温室气体情况；
"DVI"，火山尘幕指数的加权平均值；底图为 NH 序列与三个强迫因子(NH、Solar、log CO₂)之间多元
回归相关系数的变化。时间轴上标出了各滑动窗口(200 年)的中心年份。单边(正向)检验条件下，气
温与 CO₂ 和太阳辐照度之间相关系数的显著性水平为 90%、95%、99%(见正文)，显著性水平在图中以
水平划线标出；而 DVI 序列单边(负向)检测条件下，气温与 CO₂ 和太阳辐照度之间相关系数的显著性
水平为 90%，显著性水平在图中以水平点线表示。灰色条带所示为两个不同的 200 年滑动窗口内的数据，
垂向长划线所示为相应滑动窗口的中心位置。

我们检验相关系数(r)相对于原(零)假设相关系数的显著性，其中原(零)假设
相关系数是由气候自然变化产生的随机序列得到的，该过程考虑了由 NH 序列的明
显趋势和低频变化所导致的相关性中自由度的降低。自由度的降低是通过数据序列
的一阶马尔可夫"红噪声"相关性结构的形式模拟出的，上述红噪声通过计算 200 年
滑动窗口中滞后 1 年的自相关系数 ρ 来描述。该参数变化范围为 0.48(滑动相关的第
一个窗口(1610～1809))到 0.77(滑动相关的最后一个窗口(1796～1995))，距今最
近的值要大得多，这与过去一个世纪中全球呈显著变暖有关。而后一种趋势与红噪
声结果[46]相矛盾，所以可以认为它本身就是衡量外部因素引起的变化的一个指标。
从这个意义上来讲，我们就可以认为利用工业化以前 NH 序列的较小值 $\rho = 0.48$ 来
估计统计自由度也是适合自然变化的无效假设的。尽管如此，我们还是保守地选择
了最大值，即 $\rho = 0.77$ 来作为序列中自然序列的相关系数。我们采用蒙特卡罗模拟
法对序列相关噪声与三个实际影响因素序列之间为偶然虚假相关的可能性作了评估。
在各个 200 年间隔的滑动窗口之间，其置信区间基本不变。对于与 CO₂ 和太阳辐照

(positive) correlations with both CO_2 and solar irradiance, the confidence levels are both approximately 0.24 (90%), 0.31 (95%), 0.41 (99%), while for the "whiter" relatively trendless, DVI index, the confidence levels for (negative) correlations are somewhat lower (-0.16, -0.20, -0.27 respectively). A one-sided significance test is used in each case because the physical nature of the forcing dictates a unique expected sign to the correlations (positive for CO_2 and solar irradiance variations, negative for the DVI fluctuations).

The correlation statistics indicate highly significant detection of solar irradiance forcing in the NH series during the "Maunder Minimum" of solar activity from the mid-seventeenth to early eighteenth century which corresponds to an especially cold period. In turn, the steady increase in solar irradiance from the early nineteenth century through to the mid-twentieth century coincides with the general warming over the period, showing peak correlation during the mid-nineteenth century. The regression against solar irradiance indicates a sensitivity to changes in the "solar constant" of ~ 0.1 K W^{-1} m^{-2}, which is consistent with recent model-based studies[42]. Greenhouse forcing, on the other hand, shows no sign of significance until a large positive correlation sharply emerges as the moving window slides into the twentieth century. The partial correlation with CO_2 indeed dominates over that of solar irradiance for the most recent 200-year interval, as increases in temperature and CO_2 simultaneously accelerate through to the end of 1995, while solar irradiance levels off after the mid-twentieth century. It is reasonable to infer that greenhouse-gas forcing is now the dominant external forcing of the climate system. Explosive volcanism exhibits the expected marginally significant negative correlation with temperature during much of 1610–1995 period, most pronounced in the 200-year window centred near 1830 which includes the most explosive volcanic events.

A variety of general circulation[42,47] and energy-balance model experiments[43,45,48] as well as statistical comparisons of twentieth-century global temperatures with forcing series[49] suggest that, although both solar and greenhouse-gas forcings play some role in explaining twentieth-century climate trends, greenhouse gases appear to play an increasingly dominant role during this century. Such a proposition is consistent with the results of this study.

As larger numbers of high-quality proxy reconstructions become available in diverse regions of the globe, it may be possible to assimilate a more globally representative multiproxy data network. Given the high level of skill possible in large-scale reconstruction back to 1400 with the present network, it is reasonable to hope that it may soon be possible to faithfully reconstruct mean global temperatures back over the entire millennium, resolving for example the enigmatic[7] medieval period. Geothermal measurements from boreholes[50] recover long-term temperature trends without many of the complications of traditional proxy indicators and, in combination with traditional multiproxy networks, may prove helpful in better resolving trends over many centuries. With a better knowledge of how the climate has varied before the twentieth century, we will be able to place even better constraints on the importance of natural and anthropogenic factors governing the climate of the past few centuries, factors which will no doubt continue to affect climate variability in the future, in addition to any anthropogenic effects.

度之间的（正）相关性，其置信水平分别约为 0.24(90%)、0.31(95%)、0.41(99%)，而对于趋势性不太明显的"更白的"尘幕指数，其（负）相关性的置信水平则低一些（分别为 -0.16、-0.20 和 -0.27）。所有情况采用的都是单边显著性检验，因为强迫因子的物理性质已决定了相关关系的唯一的期望符号（与 CO_2 和太阳辐照度的变化之间为正，与 DVI 指数的波动之间为负）。

相关性统计量表明，从十七世纪中期到十八世纪早期，即太阳活动处于"蒙德极小期"时，NH 序列中太阳辐照度强迫的影响非常明显，而此时又正是异常寒冷的一段时期。反过来，从十九世纪早期到二十世纪中期太阳辐照度的稳步增强与该时期总体变暖的趋势一致，其中十九世纪中期时两者相关性最高。太阳辐照度回归分析结果表明"太阳常数"变化约 0.1 $K \cdot W^{-1} \cdot m^{-2}$ 时较为敏感，这与近期的模式研究结果 [42] 是一致的。另一方面，温室效应强迫，只在滑动窗移至二十世纪时才突然表现出很强的正相关性，而之前没有任何明显的信号。在最近 200 年里，气候与 CO_2 的相关性实际上已经超过气候与太阳辐照度的相关性，变成了最主要的因素，主要表现为温度和 CO_2 浓度同步加速升高，直到 1995 年底仍是如此，而太阳辐照度水平则自二十世纪中期以后持续平稳。由此我们可以断定，温室气体强迫是如今气候系统中最主要的外部强迫。正如我们所料，从 1610～1995 年间的大多数时间里火山喷发与温度变化呈略显著的负相关关系，而最显著的时期是 1830 年前后的 200 年，因为多数火山喷发事件都发生在这一时期。

各种大气环流 [42,47] 与能量平衡模式实验 [43,45,48] 以及二十世纪全球温度与其强迫因子序列之间的统计比较结果 [49] 表明，虽然太阳辐射和温室气体都能在一定程度上解释二十世纪气候的变化趋势，但本世纪中温室气体的作用似乎日益显著。这一点与本文的研究结果也是一致的。

随着全球各个地区越来越多的高质量代用指标重建结果的出现，人们将可以得到一个更具有全球代表性的多代用数据网。基于目前该网得到的自 1400 年以来的大尺度重建结果的高水准性，我们有理由期待近千年来全球平均温度的可靠重建结果很快就可以实现，从而能够解开中世纪的谜团 [7]。根据钻孔 [50] 得到的地热测量结果可以恢复出长期的温度变化趋势，而无须借助复杂的传统代用指标，并且将它与传统的多代用指标网相结合，说不定还可以更好地推断出数个世纪上的变化趋势。倘若能够对二十世纪以前气候的变化情况了解得更多一些，我们将可以对过去几个世纪中影响气候变化的自然因素和人为因素所起到的重要性进行限定，除人为因素以外，未来这些因素无疑还会持续影响气候变化。

Methods

Statistics. We use as our primary diagnostic of calibration and verification reconstructive skill the conventional "resolved variance" statistic;

$$\beta = 1 - \frac{\sum(y_{\text{ref}} - \hat{y})^2}{\sum y_{\text{ref}}^2}$$

where y_{ref} is the reference series (the raw data in the case of calibration or the verification dataset in the case of verification) and $\hat{y}$ is the series being compared to it (the proxy-reconstructed data for either calibration or verification). We compute β for each grid-point, and for the NH, GLB and MULT quantities. The sum extends over the time interval of comparison, and for the multivariate case (MULT), over all gridpoints as well. We also computed a calibration β statistic for the detrended NH series (DETR) to distinguish between explanatory variance associated with the notable trend of the twentieth century, and that related to departures from the trend.

β is a quite rigorous measure of the similarity between two variables, measuring their correspondence not only in terms of the relative departures from mean values (as does the correlation coefficient r) but also in terms of the means and absolute variance of the two series. For comparison, correlation (r) and squared-correlation (r^2) statistics are also determined. The expectation value for two random series is $\beta = -1$. Negative values of β may in fact be statistically significant for sufficient temporal degrees of freedom. Nonetheless, the threshold $\beta = 0$ defines the simple "climatological" model in which a series is assigned its long-term mean. In this sense, statistically significant negative values of β might still be considered questionable in their predictive or reconstructive skill. Owing to the more rigorous "match" between two series sought by β, highly significant values of β are possible even when r^2 is only marginally significant.

Significance levels were determined for r^2 from standard one-sided tables, accounting for decreased degrees of freedom owing to serial correlation. Significance levels for β were estimated by Monte Carlo simulations, also taking serial correlation into account. Serial correlation is assumed to follow from the null model of AR(1) red noise, and degrees of freedom are estimated based on the lag-one autocorrelation coefficients (ρ) for the two series being compared. Although the values of ρ differ from grid-point to grid-point, this variation is relatively small, making it simplest to use the ensemble average values of ρ over the domain ($\rho \approx 0.2$).

Calibration. With the spatial sampling of $M = 1,082$ continuous monthly grid-point surface temperature anomaly (that is, de-seasonalized) data used (Fig. 1b), the $N = 1,128$ months of data available from 1902 to 1995 were sufficient for a unique, overdetermined eigenvector decomposition (note that $N' = 94$ years of the annual mean data would, in contrast, not be sufficient).

For each grid-point, the mean was removed, and the series was normalized by its standard deviation. A standardized data matrix $\boldsymbol{T}$ of the data is formed by weighting each grid-point by the cosine of its central latitude to ensure areally proportional contributed variance, and a conventional Principal Component Analysis (PCA) is performed,

方　法

统计量　我们把传统的"解释方差"统计量作为校正和检验重建效果的主要诊断方法：

$$\beta = 1 - \frac{\sum(y_{\text{ref}} - \hat{y})^2}{\sum y_{\text{ref}}^2}$$

其中，y_{ref} 为参考序列（校正过程中的原始数据或检验过程中的检验数据库），$\hat{y}$ 为与之相比较的序列（校正或检验时的多代用指标重建数据）。我们对每个格点以及 NH、GLB 和 MULT 序列的 β 值都做了计算。求和的范围是整个对比区间，而在多元情形（MULT）下，则对所有格点求和。我们还计算了 NH 降趋势序列（DETR）的校正 β 值，以区分与二十世纪的显著变化趋势有关的解释性方差和与之背离趋势相关的方差。

β 可以相当精确地衡量两个变量之间的相似性，它不仅可以衡量与平均值的相对偏离（如相关系数 r），还能表示出两序列之间的平均值和绝对方差。为了对比起见，我们还确定出了相关系数 r 及其平方值。两个随机序列的期望值为 $\beta = -1$。β 为负值时实际上表示它在足够的时间自由度下具有统计学意义。不管怎样，临界值 $\beta = 0$ 时，可定义出一个简单的"气候"模式，从中可指定某序列的长期平均值。从这个角度来讲，当 β 为显著负值时，所得的预测和重建结果仍值得商榷。由于根据 β 得到的两个序列之间"匹配"更精确，因此，即使当 r^2 仅略微显著时，β 值会表现出高度的显著性。

根据标准单边表确定出的 r^2 的显著性水平结果显示，自由度的下降是由序列间的相关性引起的。利用蒙特卡罗模拟法对 β 的显著性水平做了研究，同时也考虑到了序列相关性的影响。假设序列相关性服从 AR(1) 红噪声的原假设模型，其自由度根据两个对比序列滞后为 1 个单位的自相关系数（ρ）进行估算。虽然格点与格点之间的 ρ 值不同，但这个变化相对较小，因而取定义域内 ρ 的集合平均值是一种最简便的方法（$\rho \approx 0.2$）。

校正　采用 $M = 1{,}082$ 个空间格点的逐月温度距平值（消除季节性变化后）（图 1b），由此，1902～1995 年间 $N = 1{,}128$ 个月份上的可用数据足以获得唯一的超定的特征向量分解（相反，$N' = 94$ 年时的年均数据则无法满足）。

去除每个格点的平均值，并以其标准偏差对序列作标准化。以格点中心纬度的余弦值来取各格点的加权平均值从而形成标准化数据矩阵 T，以确定因面积不同而引起的变化，然后进行传统的主成分分析（PCA），其中：

$$T = \sum_{k=1}^{K} \lambda_k \mathbf{u}_k^t \mathbf{v}_k$$

decomposing the dataset into its dominant spatiotemporal eigenvectors. The M-vector or empirical orthogonal function (EOF) $\mathbf{v}_k$ describes the relative spatial pattern of the kth eigenvector, the N-vector $\mathbf{u}_k$ or principal component (PC) describes its variation over time, and the scalar λ_k describes the associated fraction of resolved (standardized and weighted) data variance.

In a given calibration exercise, we retain a specified subset of the annually averaged eigenvectors, the annually averaged PCs denoted by $\bar{u}_n^k$, where $n = 1, ..., \bar{N}$, $\bar{N} = 79$ is the number of annual averages used of the N-month length data set. In practice, only a small subset N_{eofs} of the highest-rank eigenvectors turn out to be useful in these exercises from the standpoint of verifiable reconstructive skill. An objective criterion was used to determine the particular set of eigenvectors which should be used in the calibration as follows. Preisendorfer's[25] selection rule "rule N" was applied to the multiproxy network to determine the approximate number N_{eofs} of significant independent climate patterns that are resolved by the network, taking into account the spatial correlation within the multiproxy data set. Because the ordering of various eigenvectors in terms of their prominence in the instrumental data, and their prominence as represented by the multiproxy network, need not be the same, we allowed for the selection of non-contiguous sequences of the instrumental eigenvectors. We chose the optimal group of N_{eofs} eigenvectors, from among a larger set (for example, the first 16) of the highest-rank eigenvectors, as the group of eigenvectors which maximized the calibration explained variance. It was encouraging from a consistency standpoint that this subset typically corresponded quite closely to the subset which maximized the verification explained variance statistics (see below), but the objective criterion was, as it should be, independent of the verification process. We emphasize, furthermore, that statistical significance was robustly established, as neither the measures of statistical skill nor the reconstructions themselves were highly sensitive to the precise criterion for selection. In addition to the above means of cross-validation, we also tested the network for sensitivity to the inclusion or elimination of particular trainee data (for example, instrumental/historical records, noninstrumental/historical records, or dendroclimatic proxy indicators).

These N_{eofs} eigenvectors were trained against the N_{proxy} indicators, by finding the least-squares optimal combination of the N_{eofs} PCs represented by each individual proxy indicator during the $\bar{N} = 79$ year training interval from 1902 to 1980 (the training interval is terminated at 1980 because many of the proxy series terminate at or shortly after 1980). The proxy series and PCs were formed into anomalies relative to the same 1902–80 reference period mean, and the proxy series were also normalized by their standard deviations during that period. This proxy-by-proxy calibration is well posed (that is, a unique optimal solution exists) as long as $\bar{N} > N_{\text{eofs}}$ (a limit never approached in this study) and can be expressed as the least-squares solution to the overdetermined matrix equation, $U\mathbf{x} = \mathbf{y}^{(\mathbf{p})}$, where

$$U = \begin{vmatrix} \bar{u}_1^{(1)} & \bar{u}_1^{(2)} & \cdots & \bar{u}_1^{(N_{\text{eofs}})} \\ \bar{u}_2^{(1)} & \bar{u}_2^{(2)} & \cdots & \bar{u}_2^{(N_{\text{eofs}})} \\ & \vdots & & \\ \bar{u}_{\bar{N}}^{(1)} & \bar{u}_{\bar{N}}^{(2)} & \cdots & \bar{u}_{\bar{N}}^{(N_{\text{eofs}})} \end{vmatrix}$$

is the matrix of annual PCs, and

$$T = \sum_{k=1}^{K} \lambda_k \mathbf{u}_k^\dagger \mathbf{v}_k$$

将数据库分解为主要的时空特征向量。M 向量或称经验正交函数（EOF）$\mathbf{v}_k$ 表示第 k 个特征向量的相对空间分布型，N 向量 $\mathbf{u}_k$ 或称主成分（PC），则反映了其随时间的变化，标量 λ_k 与解出（标准值和加权平均值）数据的方差有关。

在特定的校正过程中，我们保留了由特征向量年均值组成的子集，年均主成分以 $\bar{u}_n^k$ 表示，其中 $n = 1，\cdots，\bar{N}，\bar{N} = 79$ 是长度为 N 月的数据库的年平均值。从可检测的重建效果来看，在实际操作中，仅最高阶特征向量组成的较小的子集 N_{eofs} 被证明是有用的。我们采用了一项客观标准来确定特征向量的组合，如后文所示这些特征向量将用于校正过程。将普赖森多费尔[25]的选择定则"N 定则"用于多代用数据网来确定数据网所解释的显著独立气候分布型中的大致个数 N_{eofs}，同时还考虑了多代用数据库中的空间相关性。由于仪器测量数据的突出特征所反映出的特征向量的级别与多代用数据网络表示的特征之间不必完全相同，所以我们也考虑到了所选出的仪器记录特征向量为不连续序列的情况。我们从最高阶特征向量组成的较大子集（比如，前 16 个特征向量）中选出 N_{eofs} 向量的最优组合来作为校正过程中方差贡献率最大的特征向量组。令人鼓舞的是，尽管所采用的客观标准与检验过程是相互独立的（本就应该独立），从一致性角度来看，该子集与检验过程中方差贡献率最大的子集非常相近（见下文）。此外，需要强调的是，该数据的统计学意义是稳健的，统计效果的测量和重建方法本身对选择的确切标准都不是非常敏感。除上述交叉验证手段以外，我们还测试了数据网络对引入或消除特殊训练数据（例如，仪器／历史记录、非仪器／历史记录、年轮气候代用指标等）的敏感度。

利用 N_{proxy} 指标来训练 N_{eofs} 特征向量，从中找出从 1902～1980 年的 $\bar{N} = 79$ 年的训练区间上各代用指标所代表的 N_{eofs} 主成分的最小二乘方优化组合（训练区间截止于 1980 年是因为许多代用指标序列均终止于 1980 年或略晚）。从代用指标序列和各个主成分中减去 1902～1980 年间的平均值，同时利用该时段上的标准偏差对代用指标进行标准化。只要 $\bar{N} > N_{\text{eofs}}$（该界限在本研究中均未被打破）代用指标的逐个校正就是适定的（即存在唯一的优化解），可以表示为超定矩阵方程 $U\mathbf{x} = \mathbf{y}^{(\text{P})}$ 的最小二乘方解，其中

$$U = \begin{vmatrix} \bar{u}_1^{(1)} & \bar{u}_1^{(2)} & \cdots & \bar{u}_1^{(N_{\text{eofs}})} \\ \bar{u}_2^{(1)} & \bar{u}_2^{(2)} & \cdots & \bar{u}_2^{(N_{\text{eofs}})} \\ & \vdots & & \\ \bar{u}_{\bar{N}}^{(1)} & \bar{u}_{\bar{N}}^{(2)} & \cdots & \bar{u}_{\bar{N}}^{(N_{\text{eofs}})} \end{vmatrix}$$

是由各年的主成分组成的矩阵，并且

$$\mathbf{y}^{(p)} = \begin{vmatrix} y^{(p)}_1 \\ y^{(p)}_2 \\ \vdots \\ y^{(p)}_{\tilde{N}} \end{vmatrix}$$

is the time series $\tilde{N}$-vector for proxy record p.

The N_{eofs}-length solution vector $\mathbf{x} = \mathbf{G}^{(p)}$ is obtained by solving the above overdetermined optimization problem by singular value decomposition for each proxy record $p = 1, \dots, P$. This yields a matrix of coefficients relating the different proxies to their closest linear combination of the N_{eofs} PCs;

$$\boldsymbol{G} = \begin{vmatrix} G^{(1)}_1 & G^{(2)}_2 & \cdots & G^{(1)}_{N_{eofs}} \\ G^{(2)}_1 & G^{(2)}_2 & \cdots & G^{(2)}_{N_{eofs}} \\ \vdots & & & \\ G^{(P)}_1 & G^{(P)}_2 & \cdots & G^{(P)}_{N_{eofs}} \end{vmatrix}$$

This set of coefficients will not provide a single consistent solution, but rather represents an overdetermined relationship between the optimal weights on each on the N_{eofs} PCs and the multiproxy network.

Proxy-reconstructed patterns are thus obtained during the pre-calibration interval by the year-by-year solution of the overdetermined matrix equation, $\boldsymbol{G}\mathbf{z} = \mathbf{y}_{(j)}$, where $\mathbf{y}_{(j)}$ is the predictor vector of values of each of the P proxy indicators during year j. The predictand solution vector $\mathbf{z} = \hat{\mathbf{U}}$ contains the least-squares optimal values of each of the N_{eofs} PCs for a given year. This optimization is overdetermined (and thus well constrained) as long as $P > N_{eofs}$ which is always realized in this study. It is noteworthy that, unlike conventional palaeoclimate transfer function approaches, there is no specific relationship between a given proxy indicator and a given predictand (that is, reconstructed PC). Instead, the best common choice of values for the small number of N_{eofs} predictands is determined from the mutual information present in the multiproxy network during any given year. The reconstruction approach is thus relatively resistant to errors or biases specific to any small number of indicators during a given year.

This yearly reconstruction process leads to annual sequences of the optimal reconstructions of the retained PCs, which we term the reconstructed principal components or RPCs and denote by $\hat{\mathbf{u}}^k$. Once the RPCs are determined, the associated temperature patterns are readily obtained through the appropriate eigenvector expansion,

$$\hat{\boldsymbol{T}} = \sum_{k=1}^{N_{eofs}} \lambda_k \hat{\mathbf{u}}^\dagger_k \mathbf{v}_k$$

while quantities of interest (for example, NH) are calculated from the appropriate spatial averages, and appropriate calibration and verification resolved variance statistics are calculated from the raw and reconstructed data.

$$\mathbf{y}^{(p)} = \begin{vmatrix} y^{(p)}_1 \\ y^{(p)}_2 \\ \vdots \\ y^{(p)}_{\tilde{N}} \end{vmatrix}$$

是代用指标记录 p 的时间序列的 $\tilde{N}$ 维向量。

通过解答上述超定的最优化问题可得到 N_{eofs} 长度解向量 $\mathbf{x} = \mathbf{G}^{(P)}$，方法是对每个代用指标记录 $p = 1$，…，P 作奇异值分解。从而形成一个由相关系数组成的矩阵，将不同的代用指标与与之最相近的 N_{eofs} 主成分的线性组合联系起来：

$$\boldsymbol{G} = \begin{vmatrix} G^{(1)}_1 & G^{(1)}_2 & \cdots & G^{(1)}_{N_{eofs}} \\ G^{(2)}_1 & G^{(2)}_2 & \cdots & G^{(2)}_{N_{eofs}} \\ & \vdots & & \\ G^{(P)}_1 & G^{(P)}_2 & \cdots & G^{(P)}_{N_{eofs}} \end{vmatrix}$$

该组相关系数并不能解译为单个连续解，而是代表了每个 N_{eofs} 主成分的最优加权和多代用数据网之间的超定关系。

这样，利用超定矩阵方程 $\boldsymbol{G}\mathbf{z} = \mathbf{y}_{(j)}$ 在各年上的解就可获得由代用资料在器测资料校正之前的时期重建的模态，方程中 $\mathbf{y}_{(j)}$ 为 j 年每个 P 代用指标值的预测向量。预测解向量 $\mathbf{z} = \hat{\mathbf{U}}$ 包含给定年份上所有 N_{eofs} 主成分最优值的最小二乘方。只要 $P > N_{eofs}$，这种优化方法就是超定的（因而受到很好的限制），而在本研究中该条件是始终都可以保证的。值得注意的是，与传统的古气候转换函数法不同，本方法中给定的代用指标和给定的预测对象（即重建的主成分）之间并没有特殊的关系。相反，对于为数较少的 N_{eofs} 预测对象来说，一般其最优选择值是由给定年份中多代用数据网络显示的共有信息决定的。因此，该重建方法对于给定年份上少数指标带来的误差和偏差具有一定的抵抗力。

逐年重建过程可以得到每年的残余主成分理想重建结果序列，我们将其称为重建主成分，即 RPC，以 $\hat{\mathbf{u}}^k$ 表示。一旦确定出各个 RPC，就可通过适当的特征向量扩展得出相关的温度分布型：

$$\hat{\boldsymbol{T}} = \sum_{k=1}^{N_{eofs}} \lambda_k \hat{\mathbf{u}}^\dagger_k \mathbf{v}_k$$

我们所要研究的量（比如 NH）可通过适当的空间平均值计算得到，利用原始数据和重建数据可计算出合适的校正和检验方差。

Several checks were performed to ensure a reasonably unbiased calibration procedure. The histograms of calibration residuals were examined for possible heteroscedasticity, but were found to pass a χ^2 test for gaussian characteristics at reasonably high levels of significance (NH, 95% level; NINO3, 99% level). The spectra of the calibration residuals for these quantities were, furthermore, found to be approximately "white", showing little evidence for preferred or deficiently resolved timescales in the calibration process. Having established reasonably unbiased calibration residuals, we were able to calculate uncertainties in the reconstructions by assuming that the unresolved variance is gaussian distributed over time. This variance increases back in time (the increasingly sparse multiproxy network calibrates smaller fractions of variance), yielding error bars which expand back in time.

Verification. Verification resolved variance statistics (β) were determined based on two distinct verification data sets including (1) the sparse subset of the gridded data ($M' = 219$ grid-points) for which independent values are avail-able from 1854 to 1901 (see Fig. 1b) and (2) the small subset of 11 very long instrumental estimated temperature grid-point averages (10 in Eurasia, 1 in North America—see Fig. 1a) constructed from the longest available station measurements. Each of the "grid-point" series shared at least 70% of their variance with the corresponding temperature grid-point available from 1854–1980, providing verification back to at least 1820 in all cases (and back through the mid and early eighteenth century in many cases). Note that this latter verification data set is only temporally, but not spatially, independent of the multiproxy network itself, which contains these long instrumental grid-point series as a small subset of the network. In case (1), NH and GLB verification statistics are computed as well as the multivariate (MULT) grid-point level verification statistic, although these quantities represent different spatial samplings from those in the full calibration data set owing to the sparser sampling of the verification period. Case (2) provides a longer-term, albeit an even less spatially representative, multivariate verification statistic (MULTb). In this case, the spatial sampling does not permit meaningful estimates of NH or GLB mean quantities. In any of these diagnostics, a positive value of β is statistically significant at $> 99\%$ confidence as established from Monte Carlo simulations. Verification skills for the NINO3 reconstructions are estimated by other means, as the actual NINO3 index is not available far beyond the beginning of the calibration period. The (negative) correlation r of NINO3 with the SOI annual-mean from 1865 to 1901 (P. D. Jones, personal communication), and a squared congruence statistic g^2 measuring the categorical match between the distribution of warm NINO3 events and the distribution of warm episodes according to the historical[35] chronology (available back to the beginning of 1525), were used for statistical cross-validation based on one-sided tables and Monte Carlo simulations, respectively. The results of all calibration and verification experiments are available; see Supplementary Information.

(**392**, 779-787; 1998)

Michael E. Mann[*], Raymond S. Bradley[*] & Malcolm K. Hughes[†]

[*] Department of Geosciences, University of Massachusetts, Amherst, Massachusetts 01003-5820, USA
[†] Laboratory of Tree Ring Research, University of Arizona, Tucson, Arizona 85721, USA

Received 9 May 1997; accepted 27 February 1998.

为了保证校正过程的合理性和公正性，我们作了多次核对。同时对校正残差柱状图的异方差性作了检验，但结果发现，其高斯分布特征可通过 χ^2 检验，且显著性的置信水平较高（NH，95%；NINO3，99%）。另外研究还发现，这些量的校正残差的频率谱基本是白噪音，很难从中找出校正过程对时间尺度有偏好或有未充分解释的时间尺度的证据。确定出公正合理的校正残差以后，并假定未解释方差随时间呈高斯分布，那么就可从中计算出重建结果的不确定性。该方差随时间的向前追溯而逐渐变大（随着多代用数据网越来越稀疏，能校正的差异越来越小），误差范围也逐渐增大。

检验　检验解释方差 (β) 是根据两个不同的检验数据库确定出的，这两个数据库分别为：(1) 格点数据较稀疏（$M' = 219$ 个格点）的子集，从中可以得到 1854～1901 年间的独立值（见图 1b）；(2) 利用历史最悠久的站位记录构建的 11 个较长的仪器温度记录的格点平均值（欧亚地区 10 个，北美 1 个，见图 1a）。假设所有情况下检验时间均至少可回溯到 1820 年（有些情况下还可以回溯到十八世纪中期或十八世纪初），那么从 1854 到 1980 年，上述每个"格点"序列与相应的温度格点解释了至少 70% 的方差。需要注意的是，后一种检验数据库仅是时间上的而并非空间上的，与多代用数据网本身是独立的，这些长期仪器格点序列仅是多代用数据网所包含的一个小的子网。第 (1) 种情形中，计算了 NH 和 GLB 检验统计量以及多元（MULT）格点水平上的检验统计量，只是与全部校正数据库相比它们代表的空间取样间隔不同，因为检验期内的取样较稀疏。第 (2) 种情形是一种长期的多元检验统计量（MULTb），只是其空间代表性略差。在这种情形下，所用空间采样方法并不能保证 NH 或 GLB 平均量的估计值有意义。根据蒙特卡罗模拟法，在上述所有情形中，当置信水平 > 99% 时，正值 β 均具有统计学意义。NINO3 重建结果的检验效果是通过其他途径得出的，因为在校正周期开始后的很长一段时间都没有可用的实际 NINO3 指数。1865～1901 年间 NINO3 与 SOI 年均值的（负）相关系数 r（琼斯，个人交流），以及用来衡量 NINO3 变暖事件的分布与由历史 [35] 年表（可回溯到 1525 年初期）得到的变暖事件的分布情况之间的类型匹配度的平方同余数 g^2，分别根据单边检验表和蒙特卡罗模拟进行了交叉验证。校正和检验的结果见补充信息。

（齐红艳 翻译；俞永强 审稿）

References:

1. Barnett, T. P., Santer, B., Jones, P. D., Bradley, R. S. & Briffa, K. R. Estimates of low frequency natural variability in near-surface air temperature. *Holocene* **6,** 255-263 (1996).

2. Lean, J., Beer, J. & Bradley, R. S. Reconstruction of solar irradiance since 1610: Implications for climate change. *Geophys. Res. Lett.* **22,** 3195-3198 (1995).

3. Crowley, T. J. & Kim, K. Y. Comparison of proxy records of climate change and solar forcing. *Geophys. Res. Lett.* **23,** 359-362 (1996).

4. Delworth, T. D., Manabe, S. & Stouffer, R. J. Interdecadal variations of the thermohaline circulation in a coupled ocean–atmosphere model. *J. Clim.* **6,** 1993-2011 (1993).

5. Mann, M. E., Park, J. & Bradley, R. S. Global interdecadal and century-scale oscillations during the past five centuries. *Nature* **378,** 266-270 (1995).

6. Bradley, R. S. & Jones, P. D. "Little Ice Age" summer temperature variations: their nature and relevance to recent global warming trends. *Holocene* **3,** 367-376 (1993).

7. Hughes, M. K. & Diaz, H. F. Was there a "Medieval Warm Period" and if so, where and when? *Clim. Change* **26,** 109-142 (1994).

8. Diaz, H. F. & Pulwarty, R. S. An analysis of the time scales of variability in centuries-long ENSO-sensitive records in the last 1000 years. *Clim. Change* **26,** 317-342 (1994).

9. Cook, E. R., Briffa, K. R., Mehn, D. M., Graybill, D. A. & Funkhouser, G. The "segment length" curse in long tree-ring chronology development for palaeoclimatic studies. *Holocene* **5,** 229-237 (1995).

10. Jones, P. D. & Briffa, K. R. Global surface air temperature variations during the 20th century: Part 1—Spatial, temporal and seasonal details. *Holocene* **1,** 165-179 (1992).

11. Bradley, R. S. in *Climatic Variations and Forcing Mechanisms of the Last 2000 Years* (eds Jones, P. D., Bradley, R. S. & Jouzel, J.) 603-624 (Springer, Berlin, 1996).

12. Jacoby, G. C. & D'Arrigo, R. Reconstructed Northern Hemisphere annual temperature since 1671 based on high-latitude tree-ring data from North America. *Clim. Change* **14,** 39-59 (1989).

13. Jacoby, G. C., D'Arrigo, R. D. & Tsevegyn, D. Mongolian tree rings and 20th-century warming. *Science* **9,** 771-773 (1996).

14. Fritts, H. C., Blasing, T. J., Hayden, B. P. & Kutzbach, J. E. Multivariate techniques for specifying tree-growth and climate relationships and for reconstructing anomalies in paleoclimate. *J. Appl. Meteorol.* **10,** 845-864 (1971).

15. Cook, E. R., Briffa, K. R. & Jones, P. D. Spatial regression methods in dendroclimatology: a review and comparison of two techniques. *Int. J. Climatol.* **14,** 379-402 (1994).

16. Briffa, K. R., Jones, P. D. & Schweingruber, F. H. Tree-ring density reconstructions of summer temperature pattern across western North American since 1600. *J. Clim.* **5,** 735-753 (1992).

17. Schweingruber, F. H., Briffa, K. R. & Jones, P. D. Yearly maps of summer temperatures in western Europe from A.D. 1750 to 1975 and Western North American from 1600 to 1982. *Vegetatio* **92,** 5-71 (1991).

18. Fritts, H. *Reconstructing Large-scale Climatic Patterns from Tree Ring Data* (Univ. Arizona Press, Tucson, 1991).

19. Guiot, J. The combination of historical documents and biological data in the reconstruction of climate variations in space and time. *Palaeoklimaforschung* **7,** 93-104 (1988).

20. D'Arrigo, R. D., Cook, E. R., Jacoby, G. C. & Briffa, K. R. NAO and sea surface temperature signatures in tree-ring records from the North Atlantic sector. *Quat. Sci. Rev.* **12,** 431-440 (1993).

21. Stahle, D. W. & Cleaveland, M. K. Southern Oscillation extremes reconstructed from tree rings of the Sierra Madre Occidental and southern Great Plains. *J. Clim.* **6,** 129-140 (1993).

22. Smith, T. M., Reynolds, R. W., Livezey, R. E. & Stokes, D. C. Reconstruction of historical sea surface temperatures using empirical orthogonal functions. *J. Clim.* **9,** 1403-1420 (1996).

23. Kaplan, A. *et al.* Analyses of global sea surface temperature 1856-1991. *J. Geophys. Res.* (in the press).

24. Halpert, M. S. & Ropelewski, C. F. Surface temperature patterns associated with the Southern Oscillation. *J. Clim.* **5,** 577-593 (1992).

25. Preisendorfer, R. W. *Principal Component Analysis in Meteorology and Oceanography* (Elsevier, Amsterdam, 1988).

26. Cane, M. *et al.* Twentieth-century sea surface temperature trends. *Science* **275,** 957-960 (1997).

27. Hurrell, J. W. Decadal trends in the North Atlantic Oscillation, regional temperatures and precipitation. *Science* **269,** 676-679 (1995).

28. Chang, P., Ji, L. & Li, H. A decadal climate variation in the tropical Atlantic Ocean from thermodynamic air–sea interactions. *Nature* **385,** 516-518 (1997).

29. Folland, C. K., Parker, D. E. & Kates, F. E. Worldwide marine temperature fluctuations 1856-1981. *Nature* **310,** 670-673 (1984).

30. Kushnir, Y. Interdecadal variations in North Atlantic sea surface temperature and associated atmospheric conditions. *J. Clim.* **7,** 141-157 (1994).

31. Schlesinger, M. E. & Ramankutty, N. An oscillation in the global climate system of period 65-70 years. *Nature* **367,** 723-726 (1994).

32. Mann, M. E. & Park, J. Global-scale modes of surface temperature variability on interannual to century timescales. *J. Geophys. Res.* **99,** 25819-25833 (1994).

33. Mann, M. E. & Park, J. Joint spatiotemporal modes of surface temperature and sea level pressure variability in the northern hemisphere during the last century. *J. Clim.* **9,** 2137-2162 (1996).

34. Delworth, T. D., Manabe, S. & Stouffer, R. J. Multidecadal climate variability in the Greenland Sea and surrounding regions: a coupled model simulation. *Geophys. Res. Lett.* **24,** 257-260 (1997).

35. Quinn, W. H. & Neal, V. T. in *Climate Since A.D. 1500* (eds Bradley, R. S. & Jones, P. D.) 623-648 (Routledge, London, 1992).

36. Mann, M. E., Bradley, R. S. & Hughes, M. K. in *El Niño and the Southern Oscillation: Multiscale Variability and its Impacts on Natural Ecosystems and Society* (eds Diaz, H. F. & Markgraf, V.) (Cambridge Univ. Press, in the press).

37. Kelly, P. M., Jones, P. D. & Pengqun, J. The spatial response of the climate system to explosive volcanic eruptions. *Int. J. Climatol.* **16,** 537-550 (1996).

38. Kirchner, I. & Graf, H. F. Volcanoes and El Niño: signal separation in Northern Hemisphere winter. *Clim. Dyn.* **11,** 341-358 (1995).

39. Raynaud, D., Barnola, J. M., Chappellaz, J. & Martinerie, P. in *Climatic Variations and Forcing Mechanisms of the Last 2000 Years* (eds Jones, P. D., Bradley, R. S. & Jouzel, J.) 547-562 (Springer, Berlin, 1996).

40. Bradley, R. S. & Jones, P. D. in *Climate Since A.D. 1500* (eds Bradley, R. S. & Jones, P. D.) 606-622 (Routledge, London, 1992).

41. Robock, A. & Free, M. P. Ice cores as an index of global volcanism from 1850 to the present. *J. Geophys. Res.* **100,** 11549-11567 (1995).

42. Hegerl, G. C. *et al.* Multi-fingerprint detection and attribution analysis of greenhouse gas, greenhouse gas-plus-aerosol and solar forced climate change. *Clim. Dyn.* **13,** 613-634 (1997).

43. North, G. R. & Stevens, M. J. Detecting climate signals in the surface temperature record. *J. Clim.* (in the press).

44. Hansen, J., Sato, M. & Ruedy, R. The missing climate forcing. *Phil. Trans. R. Soc. Lond. B* **352,** 231-240 (1997).

45. Stevens, M. J. & North, G. R. Detection of the climate response to the solar cycle. *J. Clim.* (in the press).

46. Mann, M. E. & Lees, J. Robust estimation of background noise and signal detection in climatic time series. *Clim. Change* **33,** 409-445 (199?)

47. Rind, D. & Overpeck, J. Hypothesized causes of decadal-to-century climate variability: climate model results. *Quat. Sci. R*

48. Wigley, T. M. L. & Raper, S. C. B. Natural variability of the climate system and detection of the greenh

49. Thomson, D. J. Dependence of global temperatures on atmospheric CO

50. Pollack, H. N. & Chap

Supplementary Information is available on *Nature*'s World-Wide Web site (http://www.nature.com) or as paper copy from Mary Sheehan at the London editorial office of *Nature*.

Acknowledgements. This work benefited from discussions with M. Cane, E. Cook, M. Evans, A. Kaplan and collaborators at the Lamont Doherty Earth Observatory. We acknowledge discussions with K. Briffa, T. Crowley, P. Jones, S. Manabe, R. Saravanan and K. Trenberth, as well as the comments of G. Hegerl. We thank R. D'Arrigo, D. Fisher, G. Jacoby, J. Lean, A. Robock, D. Stahle, C. Stockton, E. Vaganov, R. Villalba and the numerous contributors to the International Tree-Ring Data Bank and other palaeoclimate researchers who have made their data available to us for use in this study; we also thank F. Keimig, M. Munro, R. Holmes and C. Aramann for their technical assistance. This work was supported by the NSF and the US Department of Energy. M.E.M. acknowledges support through the Alexander Hollaender Distinguished Postdoctoral Research Fellowship program of the Department of Energy. This work is a contribution to the NSF- and NOAA-sponsored Analysis of Rapid and Recent Climatic Change (ARRCC) project.

Correspondence and requests for materials should be addressed to M.E.M. (e-mail: mann@snow.geo.umass.edu).

Identification of a Host Galaxy at Redshift z=3.42 for the γ-ray Burst of 14 December 1997

S. R. Kulkarni *et al.*

Editor's Note

Gamma-ray bursts were first discovered by a satellite launched to look for detonations of nuclear weapons. The first counterparts at other wavelengths were found in 1997 by the BeppoSAX satellite, after which it became clear that the bursts were happening in distant and faint galaxies. But the energy release could not be calibrated without knowing just how far away they are. Here Shrinivas Kulkarni at the California Institute of Technology and coworkers determine that a gamma-ray burst of 14 December 1997 happened in a galaxy at a redshift of 3.42. This meant that if it was radiated uniformly, the energy released as gamma rays alone would be more than 10^{53} erg, or about 1,000 times greater than a supernova. Astronomers now believe that the burst is beamed through a small angle, reducing the total energy to about 10^{51} erg, or about ten times that of a normal supernova.

Knowledge of the properties of γ-ray bursts has increased substantially following recent detections of counterparts at X-ray, optical and radio wavelengths. But the nature of the underlying physical mechanism that powers these sources remains unclear. In this context, an important question is the total energy in the burst, for which an accurate estimate of the distance is required. Possible host galaxies have been identified for the first two optical counterparts discovered, and a lower limit obtained for the redshift of one of them, indicating that the bursts lie at cosmological distances. A host galaxy of the third optically detected burst has now been identified and its redshift determined to be $z = 3.42$. When combined with the measured flux of γ-rays from the burst, this large redshift implies an energy of 3×10^{53} erg in the γ-rays alone, if the emission is isotropic. This is much larger than the energies hitherto considered, and it poses a challenge for theoretical models of the bursts.

EVER since their discovery nearly three decades ago[1], it was understood that progress in solving the puzzle of γ-ray bursts (GRBs) depends on their identification at other— preferably optical—wavelengths, so that the distances could be measured using standard spectroscopic techniques. From distances and flux measurements one can then infer luminosities and other physical parameters, which can then be used to test theoretical models of the bursts and their origins.

A recent breakthrough in this field was the precise localization of bursts by the BeppoSAX

1997 年 12 月 14 日的伽马射线暴在红移 3.42 处的寄主星系的证认

库尔卡尼等

编者按

伽马射线暴最早是被一个探测核武器爆炸的卫星所发现的。1997 年，贝波 X 射线天文卫星首次探测到伽马射线暴在其他波段上的对应体。自那以后，人们逐渐意识到这些暴发生于遥远的暗弱星系当中。然而，如果我们不知道它们到底离我们有多远，我们就没办法确定它们释放的能量。这里，根据加州理工学院的什里尼沃斯·库尔卡尼与他的合作者的研究结果得知，1997 年 12 月 14 日发现的伽马射线暴产生于一个红移为 3.42 的星系当中。这意味着如果它各向均匀地向外辐射，那么单以 γ 射线形式释放的能量就将超过 10^{53} erg，大约是一个超新星能量的 1,000 倍。当前天文学家相信伽马射线暴是在一个小角度范围内定向爆发，这样使得该伽马射线暴所释放的能量降至大约 10^{51} erg，对应一个普通超新星能量的十倍左右。

随着近来对伽马射线暴在 X 射线、光学以及射电波段对应体的观测，我们对这种天体特征的认知大大地获得增加。然而我们对于提供暴能源的底层物理机制仍不清楚。这其中，一个重要的问题就是，要确定整个暴的总能量，需要精确地测量它相对我们的距离。到目前为止，人们已经辨认出了最先发现的两个光学对应体的可能寄主星系，并定出了其中一个星系的红移下限，这表明伽马射线暴发生在宇宙学距离上。在这里，我们也证认出了第三个从光学上探测到对应体的暴的寄主星系，并定出其红移为 3.42。根据实际探测到的 γ 射线流量，如果辐射是各向同性的，那么如此高的红移就意味着单 γ 射线上释放的能量就达到 3×10^{53} erg。这比先前考虑的能量都要大得多，因此对伽马射线暴的理论模型提出了挑战。

自三十年前发现伽马射线暴[1]起，人们就知道，要解决伽马射线暴(GRB)的谜团需要证认出其他波段——尤其是可见光波段的对应体，这样才能通过标准的光谱方法来确定它们的距离。由距离和流量我们就能推算出它的光度和其他物理量，用以验证伽马射线暴机制和起源的理论模型。

这个领域最新的突破是 BeppoSAX 卫星对暴的精确定位[2]，它帮助我们首次在

satellite[2], which has led to the first identifications of GRBs at other wavelengths: X-rays[3], optical[4] and radio[5]. This has further led to the determination of the distance scale of GRBs, with the detection of intergalactic absorption lines[6] in the optical transient[7,8] (OT) of GRB970508 (refs 9, 10). Apparent host galaxies have been detected for the first two optical afterglows found[11-14].

Here we report follow-up studies of the OT[15] of a relatively bright burst, GRB971214 (refs 16–18). As the OT faded away, we found an extended object with a red-band magnitude $R = 25.6 \pm 0.15$ at the position of the OT. Based on the excellent positional coincidence, 0.06 ± 0.06 arcsec, we argue here that this is the host galaxy of GRB971214. Spectroscopic observations show that the host is a typical star-forming galaxy[19-21] at a redshift $z = 3.418$.

Given this high redshift, the γ-ray energy release of this burst is unexpectedly large, about 3×10^{53} erg, assuming isotropic emission, corresponding to about 16% of the rest-mass energy of our Sun. Energy released in other forms of radiation, for example, neutrinos or gravity waves, is not included in this energy budget. Nonetheless, the inferred energy release in γ-rays alone is so substantial that it may present difficulties for some of the currently popular theoretical models for the origin of the bursts (coalescence of neutron stars). We may be forced to consider even more energetic possibilities[22,23] or to find ways of extracting more electromagnetic energy in coalescence models.

The Optical Transient

Our follow-up at the Keck Observatory began on the night of 1997 December 15 UT, approximately 13 h after the burst. The only available imager was a "guide" camera with a small field-of-view and relatively low sensitivity; consequently, these data were not used in the analysis described below. For the subsequent nights we used the Low Resolution Imaging Spectrograph[24] (LRIS) mounted at the Cassegrain focus of the Keck II 10-m telescope. Our early observations were made in the I band. This choice was driven by the presence of a bright Moon in the sky. Later observations were done under darker sky conditions and with a R-band filter (OG570+KG3). In Table 1 we give a summary of our observations as well as other reported measurements[25-29].

Figure 1 shows an I-band image obtained on 1997 December 16.52 UT, when the OT[26] was still bright. I-band and R-band magnitudes of the OT are given in Table 1 and shown in Fig. 2. In order to establish consistency between our measurements and the earlier measurements reported by others, we measured instrumental magnitudes of a number of "secondary" stars in the general vicinity of the OT and then used them to define the zero-point of our instrumental magnitudes for all of our data sets. We calibrated the instrumental magnitudes by using observations of the standard star PG0231+051 (for which we assume I-band magnitude $I = 16.639$ mag; ref. 30) obtained on 1997 December 22 UT. We then derived the magnitudes for stars H1 and H2 (identified in Fig. 1) as 16.1 ± 0.1 mag and 18.6 ± 0.2 mag, respectively. For these two stars, Halpern *et al.*[15] obtained 15.93 mag and 18.46 mag.

其他波段(X 射线 [3]、光学波段 [4]、射电波段 [5])证认出 GRB 的对应体。据此,我们可以进一步确定 GRB 的距离尺度,比如对 GRB970508 的光学暂现源 [7,8](OT) 的星系际吸收线 [6] 进行探测以确定 GRB970508(参考文献 9 和 10)的距离。到目前为止,人们已经探测到最早发现的两个光学余辉所对应的寄主星系 [11-14]。

这里我们要报道的是对一个较亮的 GRB971214(参考文献 16 ~ 18) 的 OT[15] 的后续研究。在 OT 消失以后,我们在原来的位置上发现了一个 R 波段星等为 25.6±0.15 的延展天体。由于位置上惊人的一致——相差仅 0.06±0.06 角秒,我们推测这就是 GRB971214 的寄主星系。光谱分析表明这个星系是一个典型的恒星形成星系 [19-21],红移为 3.418。

如此之高的红移意味着这个伽马射线暴所释放的 γ 射线的能量相当大,如果辐射是各向同性的,那么释放的能量约为 3×10^{53} erg,相当于 16% 的太阳静质能。这其中还不包括其他形式的辐射,比如中微子和引力波等。然而,单是这推算出的 γ 射线波段的能量如此之大就已经给目前流行的伽马射线暴爆发起源理论模型(中子星并合)带来了挑战。我们不得不考虑其他可以产生更高能量的理论模型 [22,23],或者在并合模型中找到可以提取更多电磁能量的方式。

光学暂现源

我们于世界时 1997 年 12 月 15 日夜里(大约是伽马射线暴爆发后的 13 小时)在凯克观测站开始进行后续观测。唯一可用的成像设备是一个小视场、相对低灵敏度的导星相机,因此这批数据没有用于下面的分析。接下来的几个夜里我们使用装在 10 米凯克望远镜 II 卡塞格林焦点上的低分辨率成像摄谱仪 [24](LRIS)。前期的观测由于有月光干扰,因此只在 I 波段进行。在天空背景变暗后我们采用 R 波段滤光片 (OG570+KG3)进行后续的观测。在表 1 中我们对我们的观测条件以及其他已有的测量结果 [25-29] 进行了简单的总结。

图 1 显示的是世界时 1997 年 12 月 16.52 日获得的 I 波段图像,图像中 OT[26] 还非常明亮,我们将 OT I 波段和 R 波段的星等列于表 1 中并标示于图 2 中。为了使所得数据格式同前人的测量结果相一致,我们测量了 OT 附近一系列二级标准星的仪器星等,用以确定我们所有数据集的仪器星等的零点。然后我们通过世界时 1997 年 12 月 22 日对标准星 PG0231+051(取其 I 波段星等为 16.639 星等;参考文献 30) 的观测对仪器星等进行定标。这样,我们得到恒星 H1 和 H2(在图 1 中标出)的绝对星等分别为 16.1±0.1 星等和 18.6±0.2 星等。相应地,哈尔彭等人 [15] 给出的结果是 15.93 星等和 18.46 星等,两者在误差范围内一致,于是我们可以将我们 I 波段的测

Within errors, these two sets of measurements agree and thus allow us to link our I-band magnitudes to those of Halpern *et al.*[26]. The zero point for our R-band magnitudes was set by assuming $R = 20.1 \pm 0.1$ for object 2 of Henden *et al.*[29]; see also Fig. 1. Our R-band magnitude is in excellent agreement with that obtained by Diercks *et al.*[28] at about the same epoch.

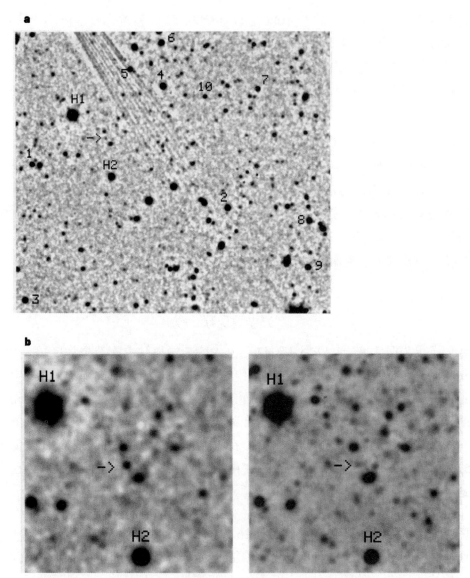

Fig. 1. Images of the field of the optical transient of GRB971214 and the associated host galaxy. **a**, I-band image of the field of the optical transient (OT) of GRB971214. The image is 3.5 arcmin by 3.0 arcmin and has been smoothed by a gaussian of full-width at half-maximum (FWHM) of 0.53 arcsec. The numbered stars are "secondary" stars used for achieving consistency within our various data sets. Stars H1 and H2 allow us to link our photometry with that of Halpern *et al.*[26] and Henden *et al.*[29]. The "rays" are light from the bright star (6.7 mag) SAO15663 diffracted by the telescope structure; this bright star is ~1 arcmin northeast of the OT. The rays rotate as the telescope tracks the source. In **b**, the left panel is an I-band image of the OT obtained from data taken on 1997 December 16 UT. The image is a square of side 1

量结果与他们的结果 [26] 联合起来使用。R 波段的零点是在假定亨登等人 [29] 的 2 号天体 R 波段星等为 20.1±0.1 的情况下而确定的；参见图 1。这样，我们获得的 R 波段数据与迪克斯等人 [28] 在同时期的测量结果相当吻合。

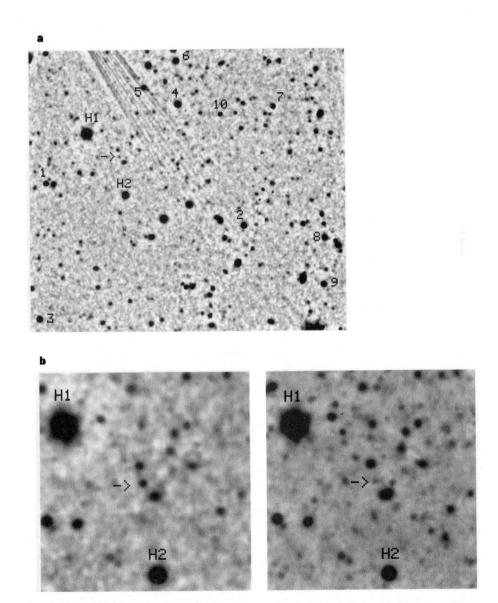

图 1. GRB971214 的 OT 及其寄主星系所在视场的图像。**a**. GRB971214 的 OT 所在视场的 I 波段图像。图像为 3.5 角分 ×3.0 角分，已做半峰全宽为 0.53 角秒的高斯平滑处理。已编号的恒星是用以同其他批次数据进行定标的二级标准星。恒星 H1 和 H2 使我们的测光数据得以同哈尔彭等人 [26] 和亨登等人 [29] 的数据进行比较。图中的"射线状特征"是被望远镜结构衍射的亮星 SAO15663(6.7 星等) 的星光；这颗亮星位于 OT 东北约 1 角分处。随着望远镜跟踪目标时的视场旋转，射线也相应转动。**b**. 左图是世界时 1997 年 12 月 16 日时拍摄的 OT 的 I 波段图像。图像大小为 1 角分见方，经半峰全宽为 0.53 角秒的

arcmin, and has been smoothed by a guassian with FWHM of 0.53 arcsec. The OT is marked. The right panel is an R-band image of the field of OT obtained from data taken on 1998 January 10 UT. The image is a square of side 1 arcmin, and has been smoothed by a gaussian with FWHM of 0.53 arcsec. An extended object is seen at the position of the OT. We suggest that this is the host galaxy of GRB971214. For all three images, North is to the top and East to the left. In all three images the optical transient is marked by an arrow. We have also obtained the spectrum of the galaxy ~4.5 arcsec to the northeast of the OT. The spectrum shows a prominent [O II] 3,727 emission line and the usual absorption features at a redshift of 0.5023.

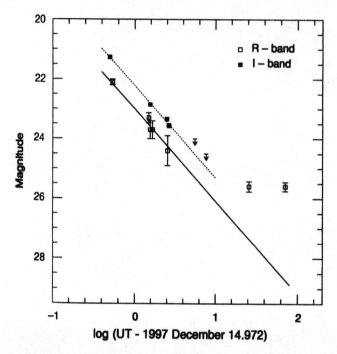

Fig. 2. The R- and I-band light curve of the OT of GRB971214. The OT is identified in Fig.1. The x-axis is $\log(t - t_0)$ where t is the UT date of the observation and t_0 is the UT date of the GRB. All times are measured in units of days. The GRB took place on 1997 December 14.9727 (ref.16). At I band the OT is well detected for the first three nights. Upper limits exist for the nights of UT 1997 December 20 and December 22. The dotted line is a linear least-squares fit to the I-band data but restricted to these first three nights. The assumed model is $I(t - t_0) = I_1 + s \log(t - t_0)$; here I is the magnitude at time t, s is the slope and I_1 the offset. The fits yield $I_1 = 22.22 \pm 0.18$ and $s = 3.09 \pm 0.52$. For reasons discussed in the text it is reasonable to assume that the R-band light curve has the same slope, s. The solid line is parallel to the dotted line but goes through our R-band magnitude of 1997 December 16.63 UT. This line describes adequately the R-band light curve for the first three nights. Clearly, the December 15.50 point of Henden et al.[29] is not well accounted for by this model. The R-band measurements of 1998 January 10 and February 24 lie well above the extrapolation of the dotted line. We attribute this excess over the decaying optical transient as arising from galaxy K shown in Fig. 1. In both the January 10 and February 24 images, the full width at half maximum (FWHM) of K is 1.07 arcsec which is larger than the same estimated from stars in the vicinity of K (Table 1). The rough size of K is thus ~0.65 arcsec.

The decay of previous OTs associated with GRBs appears to be well characterized by a power law: $S(t) \propto t^{-a}$, where $S(t)$ is the flux of the OT and t is the time since the γ-ray burst. Accepting this parametric form and considering only the time interval December 15–17,

高斯平滑处理。图中标出了 OT 的位置。右图是世界时 1998 年 1 月 10 日获得的 OT 所在视场的 R 波段图像，也是 1 角分见方，同样经半峰全宽为 0.53 角秒的高斯平滑处理。在 OT 原来的位置上可以看见一个延展天体，我们认为那就是 GRB971214 的寄主星系。所有的三幅图片均为上北左东，OT 由箭头标出。我们也获得了 OT 东北约 4.5 角秒处的星系的光谱，光谱中有明显的 [O $_{II}$] 3,727 发射线和红移 0.5023 处常见的吸收线。

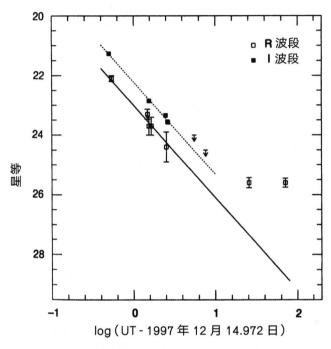

图 2. GRB971214 光学暂现源 R 和 I 波段的光变曲线。OT 已在图 1 中标出。x 轴是 $\log(t-t_0)$，其中 t 是观测的世界时，t_0 是 GRB 的世界时，均以天为单位。GRB 发生于 1997 年 12 月 14.9727 日（参考文献 16）。在 I 波段前三个晚上的探测都可得到很确切的结果，在 12 月 20 日和 22 日都只能得到上限。图中的虚线是 I 波段数据的线性最小二乘法拟合，不过仅限于对前三天的观测进行拟合。假定的模型是 $I(t-t_0)=I_1+s\log(t-t_0)$，其中 I 是时间 t 时的星等，s 是斜率，I_1 是偏置。拟合的结果为 $I_1=22.22\pm0.18$，$s=3.09\pm0.52$。由于正文中提到的原因，有理由相信 R 波段光变曲线有着同样的斜率 s。图中实线与虚线平行，且过世界时 1997 年 12 月 16.63 日测得的 R 波段星等，同前三天的 R 波段光变数据基本吻合。亨登等人 [29]12 月 15.50 日的数据点显然不满足这个模型。1998 年 1 月 10 日和 2 月 24 日的 R 波段测量结果都在虚线的外推之上，我们认为这种对衰减的光学暂现源的亮度超出是由图 1 中显示的星系 K 引起的，在 1 月 10 日和 2 月 24 日的图像中 K 的半峰全宽为 1.07 角秒，这个值比它邻近的恒星的估值大得多（见表 1）。由此估计 K 的角直径约为 0.65 角秒。

之前提到的与 GRB 相关的 OT 的光度衰减可以按幂律形式进行描述，即 $S(t)\propto t^a$，其中 $S(t)$ 为 OT 的流量，t 是从伽马射线暴爆发算起的时间。基于这个参数表达式且仅考虑 12 月 15~17 日的数据，我们可以得到 $I(t)=(22.22\pm0.18)+(3.09\pm0.52)\log(t)$，

we obtain $I(t) = (22.22 \pm 0.18) + (3.09 \pm 0.52) \log (t)$: here t is in days. As magnitudes are defined to be $-2.5 \log(S) + $ constant, the value of a is 1.22 ± 0.2—remarkably similar to those reported for previous OTs[31-33].

From the data in Fig. 2 we see that initially the R-band flux appears to decrease in approximately the same fashion as the I-band flux. Such broad-band decay has been noted in previous OTs[32,33] and indeed is an expectation of simple fireball models[31,34]. Accepting that all optical bands decay with the same a, we obtain $R(t) = (23.0 \pm 0.22) + (3.09 \pm 0.52) \log(t)$. In arriving at this equation we have used our 16 December R-band measurement (Table 1) to set the zero point.

Identifying the Host Galaxy

Using the above decay formula we predict $R = 27.4 \pm 0.8$ mag on 1998 January 10 UT. In contrast, we find at the same epoch an object at the position of the OT with $R = 25.6 \pm 0.17$ mag (see Fig. 1). Furthermore, this object, which we will call "K", is extended in comparison to stars in the same field. We suggest that K is the host galaxy of the GRB and has become apparent now that the OT has dimmed.

It is important to assess how well K coincides with the OT. To this end, we determined the location of the OT relative to 15 "tertiary" stars in the vicinity of the OT in the December 16.52 I-band image, and likewise for K in the January 10 R-band image. The error in the measured angular difference between the OT and K is determined by two factors, as follows. (1) The error in determining the centroid of K, which primarily arises from the small number of signal photons and the sky background. We estimate this to be 0.20 pixel (r.m.s.); each LRIS pixel is a square of side 0.211 arcsec. The similar error for the OT is negligible because the OT is much brighter than K in the December 16 image that we used. (2) Errors due to coordinate transformation between the two images which are of different orientations. We determined a coordinate transformation between the frames using the measurements of the tertiary stars, and then transformed the centroid of the OT into the frame of January 10 image. This error turns out to be negligible compared to the error discussed in (1). The difference between the position of the OT in the transformed image and that of K is $\Delta X = 0.24 \pm 0.20$ pixel $= 0.05 \pm 0.04$ arcsec and $\Delta Y = 0.14 \pm 0.20$ pixel $= 0.03 \pm 0.04$ arcsec, corresponding to a radial separation of 0.06 ± 0.06 arcsec.

One may argue that the density of galaxies as faint as K is sufficiently high that the coincidence of the OT with a background galaxy is not improbable. The cumulative surface density[35] of galaxies with $R \leqslant 25.5$ mag is 3.9×10^5 per degree^2. Considering angular offsets up to the apparent radius of the galaxy, 0.35 arcsec (see Fig. 1), the chance coincidence probability is reasonably small ($\sim 10^{-3}$). We therefore suggest that galaxy K is the host of the OT and thus is the host of GRB971214.

162

其中 t 的单位是天。由于星等值的定义可以表示为 $-2.5 \log(S) +$ 常数，因此 a 的值为 1.22 ± 0.2，这个结果同以前 OT 的研究结果 [31-33] 很接近。

从图 2 中的数据可以看出初始阶段 R 波段的流量衰减特征几乎与 I 波段的情况相同，这一宽波段的衰减现象也曾在以前对 OT 的观测 [32,33] 中被人们注意到，与此同时这个现象也符合简单火球模型的理论预言 [31,34]。如果所有光学波段的衰减的幂指数都对应相同的 a 值，那么 $R(t) = (23.0 \pm 0.22) + (3.09 \pm 0.52) \log(t)$。此处我们采用了 12 月 16 日 R 波段的测量结果（见表 1）来标定星等零点。

寄主星系证认

利用上面的衰减公式我们估计出世界时 1998 年 1 月 10 日的 OT R 波段星等为 27.4 ± 0.8，然而相比之下，我们在同一时期 OT 的位置上发现一个 R 波段星等为 25.6 ± 0.17 的天体（见图 1）。此外，该天体（称为"K"）相比于同视场的恒星而言是一个展源。因此我们推断 K 可能是这个 GRB 的寄主星系，正是在 OT 变得暗淡之后才得以显现出来。

明确天体 K 与 OT 的一一对应的关系是非常重要的。为此，我们在 12 月 16.52 日的 I 波段图像上确定了 OT 相对于它附近 15 个三级标准星的相对位置，并在 1 月 10 日的 R 波段图像上对天体 K 进行同样的操作。这里有两个因素决定 OT 和 K 之间角度差异的误差。(1) K 天体形心的定位误差，这个误差主要来源于过少数量的信号光子以及天空背景。我们估计其误差大小为 0.20 个像素（均方根），而 LRIS 望远镜的每个像素为长 0.211 角秒的方格。相比之下，OT 的此类误差则可以忽略，因为在选取的 12 月 16 日的图像中 OT 比 K 亮很多。(2) 将方位角不同的两幅图像进行坐标变换所引入的误差。我们通过测量三级标准星来对两个图像系统进行坐标变换，然后依此把 OT 的形心位置变换到 1 月 10 日的图像中去。相比于前面讨论的因素(1)，这个误差可以忽略不计。转换后的图像中 OT 与 K 的位置差别为 $\Delta X = 0.24 \pm 0.20$ 像素 $= 0.05 \pm 0.04$ 角秒，$\Delta Y = 0.14 \pm 0.20$ 像素 $= 0.03 \pm 0.04$ 角秒，这对应它们的径向距离为 0.06 ± 0.06 角秒。

有人也许会怀疑，像 K 这样暗淡的星系其分布密度相当高，OT 有可能刚好与一个背景星系位置相重合。但是所有 R 波段星等小于等于 25.5 的星系累积面密度 [35] 为每平方度 3.9×10^5 个。即使考虑到这个星系视半径的角度偏移（也只有 0.35 角秒，见图 1），这种意外重合的可能性也相当小（约为 10^{-3}）。因此我们认为星系 K 是 OT 的寄主星系，也是 GRB971214 的寄主星系。

Determining the Redshift

Spectroscopic observations of the OT and its host galaxy were obtained using LRIS[24]. Our initial attempt to obtain the spectrum of the OT on 1997 December 17 UT was unsuccessful, largely due to the bright moonlight. Subsequent observations were obtained under much better sky conditions, and after the OT had faded and object K dominated the light. Almost all spectra were obtained with the slit position angle close to the parallactic angle, and the wavelength-dependent slit losses are not important for the discussion below. The log and the technical details of the spectroscopic observations can be found in Table 2.

The spectrum, shown in Fig. 3, shows a prominent emission line at 5,382.1 Å with a clear continuum drop immediately on the blue (short-wavelength) side of the line. No other emission line is detected. The continuum rises slightly towards the red and is undetectable at about 4,000 Å.

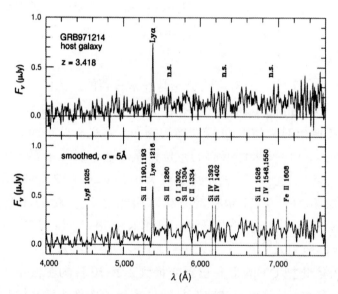

Fig. 3. The composite Keck spectrum of the host galaxy of GRB971214. The top panel shows the original data, with the locations of the prominent night sky (n.s.) emission lines indicated. The bottom panel shows the same spectrum smoothed with a gaussian with σ = 5 Å, which is approximately equal to the effective instrumental resolution. Locations of several absorption features commonly seen in the spectra of $z \approx 3$ galaxies are indicated. The redshift, $z = 3.418$, has been derived from the mean point of the Lyα emission and absorption features, as described in the text. The log of the observations and the details of the spectrograph settings can be found in Table 2. The "lower" resolution data were reduced completely independently by two of the authors (S.G.D. and K.L.A.), using independent reduction packages. The results are in an excellent mutual agreement. The "higher" resolution data are also fully consistent with them, showing essentially the same spectroscopic features. The useful wavelength range spanned by the lower-resolution data is ~4,000–7,600 Å, and the higher-resolution data spans ~4,900–7,300 Å. All of the spectra have been averaged with appropriate signal-to-noise weighting, after suitable resampling.

确 定 红 移

我们通过 LRIS[24] 获得了 OT 及其寄主星系的光谱观测结果。在世界时 1997 年 12 月 17 日，我们初次尝试拍摄 OT 光谱，但主要由于月光太亮而没能成功。后续观测的天气状况较好，且当时 OT 已减弱，亮度主要由天体 K 主导。在拍摄所有光谱时基本上狭缝位置角都接近于星位角，从而波长依赖的狭缝损失在下文中的讨论并不重要。我们在表 2 中给出了一些光谱观测日志和技术细节。

在图 3 显示的光谱中，我们在波长 5,382.1 Å 处发现了一条很强的发射线，并在紧挨该发射线的蓝（短波）端发现连续谱有明显的下降。同时并没有探测到其他的发射线。另外，连续谱朝着红端有轻微上升，并在大约 4,000 Å 处已不能被测到。

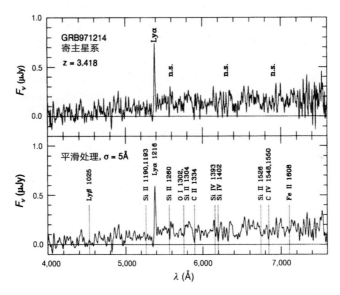

图 3. GRB971214 寄主星系的凯克望远镜合成光谱。上部分是原始数据，标出了明显的夜晚天光 (ns) 发射线的位置。下部分是经 σ = 5 Å 的高斯平滑处理之后所得到的光谱，与仪器的有效分辨率接近。图中标出了红移在 3 左右的星系中常见的吸收线，如正文中所述，通过 Lyα 发射线的中心和吸收线特征得出该星系的红移 z = 3.418。观测日志和光谱细节可以在表 2 中看到。"较低"分辨率数据由作者中的两位 (乔尔戈夫斯基和阿德尔贝格尔) 用完全不同的测光包各自独立处理得出，结果相当一致。"较高"分辨率的数据也与之吻合，显现出相同的光谱特征。低分辨率数据的有效波长范围约为 4,000 ~ 7,600 Å，高分辨率数据约为 4,900 ~ 7,300 Å。全部光谱都经适当重采样后对信噪比加权进行了平均操作。

There are only two plausible interpretations of this prominent emission line. All other choices would require a presence of other, stronger emission lines which are not seen in our spectrum. First, the line could be [O II] 3,727 at $z = 0.444$. We would then expect to see comparably strong [O III] 5,007 and 4,959 lines at 7,230 Å and 7,160 Å, Hβ 4,861 Å at 7,020 Å, and perhaps also other Balmer lines. Yet none of these lines are seen in our data, even though they fall in a reasonably clean part of the spectrum. It would be also hard to understand the abrupt drop in the continuum blueward of the emission line.

The alternative interpretation, and the interpretation which we favour, is that the emission line is Lyα 1,215.7 Å at a redshift $z_{em} = 3.428$. The overall appearance of the spectrum is typical of the known, star-forming galaxies at comparable redshifts[19-21]. In the absence of an active nucleus, no other strong emission lines would be expected within the wavelength range probed by our spectrum.

Additional arguments in favour of the large-redshift interpretation include:

(1) The drop across the Lyα line, due to the absorption by intervening hydrogen, is characteristic of objects at such high redshifts. The measured drop amplitude[36] $D_A = 0.35 \pm 0.05$ is exactly the mean value seen in the spectra of quasars at this redshift[37].

(2) We expect no flux blueward of 4,030 Å, the redshifted Lyman continuum break. Additionally, we expect a relatively flat, star-formation-powered continuum redward of the Lyα line. Our spectrum is consistent with these expectations. The expected suppression of blue light is also supported by our imaging observations in the B band (Table 1); this band covers the range 3,900–4,800 Å. Our 3σ upper limit on the B magnitude of galaxy K is 26.8 mag, corresponding to a B-band flux of < 84 nJy. We find that the red portion of the spectrum can be approximated by $F_\nu = 174(\nu/\nu_R)^\alpha$ nJy with $\alpha = -0.7 \pm 0.2$; here F_ν is the spectral density at frequency ν and $\nu_R = 4.7 \times 10^{14}$ Hz, the centre frequency of the R band. The extrapolation of this spectrum to the B band ($\nu_B = 6.8 \times 10^{14}$ Hz) is 134 nJy, above our upper limit.

(3) We plot in Fig. 3 the expected locations of some of the commonly observed interstellar gas absorption lines seen in the spectra of star-forming galaxies at high redshifts[19-21]. We note several coincidences with dips in the observed spectrum.

Table 1. Summary of R- and I-band measurements

Epoch (UT)	Band	Magnitude	Notes and refs*
1997 Dec.15.47	I	21.27 ± 0.30	Ref. 26
1997 Dec.16.52	I	22.85 ± 0.33	9 × 120 s, 0.9″, R.G.
1997 Dec.17.45	I	23.34 ± 0.31	9 × 120 s, 1.1″, R.G. †
1997 Dec.17.60	I	23.55 ± 0.26	11 × 120 s, 0.8″, R.G.
1997 Dec.17.63	I	23.57 ± 0.40	11 × 120 s, 0.63″, R.G.‡

对于这条显著的发射线的解释只有两种可能，因为所有其他的解释都还要求存在其他更强的发射线，但它们未在我们光谱中出现。首先一种可能是这条谱线是红移为 0.444 的 [O II] 3,727。如果是这样，我们还将期待在 7,230 Å 和 7,160 Å 处看到差不多强的 [O III] 5,007 和 4,959 线，以及在 7,020 Å 处的 Hβ 4,861 Å，也许还有其他的巴耳末谱线。但是我们的数据中却并没有出现这些谱线，即使它们落在光谱中相对干净的一块区域。与此同时，这种解释也很难解释发射线蓝端连续谱的突降。

因此我们更倾向于另一种可能，即这条显著的发射线其实是红移 3.428 的 Lyα 1,215.7 Å。整个光谱的特征同具有差不多红移的已知恒星形成星系的典型光谱特征相一致 [19-21]。因为没有活跃的星系核，在我们所观测的光谱范围内不会看到其他更强的谱线。

支持高红移的其他理由还包括：

(1) 由传播过程中的氢吸收所导致的跨 Lyα 线的流量陡降是这类高红移天体的普遍特征。测量到的下降幅度 [36] $D_A = 0.35 \pm 0.05$，正是这个红移处观测类星体光谱所得到的平均值 [37]。

(2) 我们预期在 4,030 Å（即红移后的莱曼连续谱跳变）的更蓝端测不到流量。另外，我们期待在 Lyα 线的红端看到一个相对平坦的由恒星形成活动所贡献的连续谱。我们的光谱结果与这些预期相吻合。我们所预期的蓝光的压低也在 B 波段（3,900 ~ 4,800 Å，见表 1）的成像观测中得到了证实。天体 K 的 B 波段 3σ 置信度的星等上限为 26.8，对应的 B 波段流量小于 84 nJy。我们发现光谱红端部分的谱密度近似满足公式 $F_\nu = 174(\nu/\nu_R)^\alpha$ nJy，其中 F_ν 为频率 ν 处的谱密度，$\alpha = -0.7 \pm 0.2$，$\nu_R = 4.7 \times 10^{14}$ Hz 为 R 波段的中央频率；将此光谱外推到 B 波段（$\nu_B = 6.8 \times 10^{14}$ Hz），那么对应的流量值为 134 nJy，这将高于我们测得的流量上限。

(3) 我们在图 3 中标出了通常能在高红移的恒星形成星系中观测到的星际气体吸收线的位置 [19-21]，我们注意到它们与观测到的光谱中的几处凹陷区域相对应。

表 1. R 和 I 波段测量数据汇总

时期（世界时）	波段	星等	注释和参考文献 *
1997 年 12 月 15.47 日	I	21.27 ± 0.30	参考文献 26
1997 年 12 月 16.52 日	I	22.85 ± 0.33	9 × 120 s，0.9″，古德里奇
1997 年 12 月 17.45 日	I	23.34 ± 0.31	9 × 120 s，1.1″，古德里奇 †
1997 年 12 月 17.60 日	I	23.55 ± 0.26	11 × 120 s，0.8″，古德里奇
1997 年 12 月 17.63 日	I	23.57 ± 0.40	11 × 120 s，0.63″，古德里奇 ‡

Continued

Epoch (UT)	Band	Magnitude	Notes and refs*
1997 Dec. 20.53	I	> 24	10 × 360 s, A.V.F.1.4″ §
1997 Dec. 22.63	I	> 24.5	2 × 400 s, 0.9″, M.D. ‖
1997 Dec.15.50	R	21.7 ± 0.10	Ref. 29
1997 Dec.15.51	R	22.1 ± 0.10	Ref. 28
1997 Dec.16.46	R	23.3 ± 0.17	6 × 180 s, 0.86″, R.G.
1997 Dec.16.63	R	23.7 ± 0.10	3 × 180 s, 0.56″, R.G. ‡
1997 Dec.16.52	R	23.7 ± 0.30	Ref. 28
1997 Dec.17.51	R	24.4 ± 0.50	Ref. 25
1998 Jan.10.62	R	25.6 ± 0.17	12 × 300 s, 0.86″, J. Aycock¶
1998 Feb. 24.52	R	25.6 ± 0.15	10 × 300 s, 0.87″, S.R.K.
1998 Feb. 24.58	B	> 26.8	6 × 300, 3 × 600,1.00″, S.R.K.#

* The entries in the last column are: (for Keck observations) the number of frames obtained, the integration time, the seeing specified as full-width at half-maximum (FWHM), the observer or the reference (for observations reported in the literature). Additional notes are indicated by a letter. The log of the December 15 "guide" camera observations is not included here as we did not use those data in any quantitative analysis. (The OT was detected in those data, though.) Successive frames are displaced by 5–15 arcsec and the final image is obtained by registering the frames and then adding. All the imaging analysis was carried out with IRAF, a software package supplied by the National Optical Astronomy Observatories.

† Rhoads[27] reports $I = 22.9 ± 0.4$ on 1997 Dec 17.37 UT. This is consistent, within errors, with our measurement of the same night. However, due to the large error bars, this is not included in our compilation (above).

‡ This observation was affected by light diffracted by the telescope structure contaminating the region in the vicinity of the OT (see Fig.1). The diffracted pattern was ~15 pixels wide and long. An image was formed by assigning each pixel a value equal to the median of a box 75×5 pixels, aligned along the diffraction spikes. All features smaller than this box disappear in the resultant image, leaving only the diffracted rays. This median image was then subtracted from the original image leaving an image free from the diffracted pattern. Photometry of the OT, reported in this table, was then performed on this subtracted image.

§ The data were taken in the polarimetric mode of LRIS and hence the light is split into two orthogonal polarization components—we call this the top and bottom channels. The images from each of the channels were reduced separately, averaged together and the two resultant images were co-added to give the final image. The seeing as estimated by observations of the standard star SA 104 440 (ref. 30) was exceptionally bad, ~1.4 arcsec. No object was found in the final image at the position of the OT. We do not have a good estimate of the true point-spread function in the final image as the useful field is only 25 arcsec across and there are no bright stars within this. Based on noise statistics within an aperture of radius 1.5 arcsec we set a 3σ detection limit of $I = 24$ mag for this data. Analysis of artificial stars with FWHM of 1.4 arcsec using the IRAF task DAOFIND led to a detection limit of 23 mag. However, visual inspection of the artificial stars knowing the position of the object results in a more stringent limit, closer to the $I = 24$ mag obtained above.

‖ 3σ upper limit estimated from the detection statistics of artificial stars which have the same point-spread function as that of nearby stars and embedded in the general vicinity of the OT. This observation suffered from high sky brightness (19.1 mag per square arcsec) and also lack of numerous exposures.

¶ The first four exposures suffered from brighter sky as the Moon was still up in the sky. The image shown in Fig. 1 is the sum of the next eight exposures.

3σ upper limit. See footnote ‖ for details of the procedure of deriving the upper limit.

If this interpretation is correct, a better estimate of the galaxy's redshift can be obtained as follows. The Lyα emission line tends to be shifted systematically to the red in such galaxies, due to resonant scattering and absorption in the ambient gas. In any given case,

续表

时期（世界时）	波段	星等	注释和参考文献 *
1997 年 12 月 20.53 日	I	> 24	10×360 s，菲力片科 1.4″§
1997 年 12 月 22.63 日	I	> 24.5	2×400 s，0.9″，戴维斯‖
1997 年 12 月 15.50 日	R	21.7 ± 0.10	参考文献 29
1997 年 12 月 15.51 日	R	22.1 ± 0.10	参考文献 28
1997 年 12 月 16.46 日	R	23.3 ± 0.17	6×180 s，0.86″，古德里奇
1997 年 12 月 16.63 日	R	23.7 ± 0.10	3×180 s，0.56″，古德里奇‡
1997 年 12 月 16.52 日	R	23.7 ± 0.30	参考文献 28
1997 年 12 月 17.51 日	R	24.4 ± 0.50	参考文献 25
1998 年 1 月 10.62 日	R	25.6 ± 0.17	12×300 s，0.86″，艾科克¶
1998 年 2 月 24.52 日	R	25.6 ± 0.15	10×300 s，0.87″，库尔卡尼
1998 年 2 月 24.58 日	B	> 26.8	6×300，3×600，1.00″，库尔卡尼 #

* 最后一列的内容分别是：（对于凯克的观测数据来说）观测帧数、积分时间、以半峰全宽表示的视宁度、观测者或参考文献（对于文章报道的观测）。额外的注释由符号标出。12 月 15 日的导星照相机的观测数据没有用于任何定量分析，因此这里没有给出日志（尽管其中观测数据也已探测到 OT）。相邻各帧之间位移了 5～15 角秒，通过对齐和叠加来获得最终的图像。所有的图像分析工作都使用美国国家光学天文台开发的 IRAF 软件包完成。

† 罗兹 [27] 给出世界时 1997 年 12 月 17.37 日 $I = 22.9 \pm 0.4$。这与我们同天的结果在误差范围内一致。但是这个结果由于误差棒过大，所以没有列入上面的列表中。

‡ 望远镜结构造成的衍射光污染了 OT 邻近视场（见图 1），观测结果受到一定影响。衍射花纹长宽各大概 15 像素。我们给每一个像素赋值，该值取为 75×5 像素的方框区域内的中值，而该方框沿着衍射峰成直线。如此得到的图像中所有小于此方框的特征都会消失，只剩衍射线。将此图像从原始图像中减去，便可除去衍射花纹。表中所给出的 OT 的测光值，便是在相减后的图像上得到的。

§ 这些数据是由 LRIS 在偏振模式下获得的，因此光被分解为两个正交的极化分量，分别称为上、下通道。各通道的图像分开进行处理，求平均，然后将得到的两幅图像进行叠加得到最后的图像。视宁度是通过观测标准星 SA 104 440（参考文献 30）估计出来的，结果异常地差，约为 1.4 角秒。在最后的图像中 OT 的位置上并未出现任何天体。由于有效视场跨度只有 25 角秒，而且其中没有亮星，我们无法较好地估计最后的图像中的点扩展函数。根据孔径 1.5 角秒内的噪声统计，我们估计数据在 3σ 内的探测极限为 $I = 24$ 星等。用软件 IRAF 的 DAOFIND 函数包分析半峰全宽为 1.4 角秒的虚拟假星，所获得的探测极限可达 23 星等。不过若考虑假星的位置已知，按上述方法我们可以得到一个更加严格的限制，更接近上面得到的 24 星等。

‖ 我们对与邻近恒星具有同样点扩展函数且落在 OT 附近的假星进行探测统计所获得的 3σ 上限。此观测受到明亮天光的影响（每平方角秒 19.1 星等），且曝光次数也不够。

¶ 由于有月亮在天上，前四次曝光受到明亮天光的影响。图 1 中的图像是由随后八次曝光叠加而成。

3σ 上限。处理细节参看脚注 ‖。

　　如果这个解释正确，就可以用下面的方法更好地估计星系的红移值。在此类星系中 Lyα 发射线会因为环境气体的共振散射和吸收而系统地向红端移动。在任何特

the bias thus introduced in the centring of the peak of the Lyα emission depends on the exact geometry and kinematics of the gas and dust, which are not known; but on average the effect is to shift the peak of the Lyα emission to the red, relative to the systemic redshift of the galaxy[19-21]. We measure the wavelength of the apparent absorption dip immediately on the high-frequency side of the emission line as: $\lambda_{obs,air} = 5,356.1$ Å. Taking the mean of this wavelength and that of the emission line, and applying the standard air-to-vacuum correction, we obtain for the systemic redshift of this galaxy, $z_K = 3.418 \pm 0.010$.

As discussed above the continuum slope of the observed spectrum redward of the emission line is $F_\nu \propto \nu^a$ with $a = -0.7 \pm 0.2$. Unobscured star-forming galaxies have $\alpha \approx 0.0$ to -0.5, depending on the initial mass function and the history of star formation. The slightly steeper slope of our spectrum suggests a modest amount of rest-frame extinction for the galaxy as a whole. This of course does not constrain the extinction along any particular line of sight, such as the direction to the OT itself. We note that similar extinctions are inferred for star-forming galaxies at comparable redshift[19].

Two corrections must be taken into account before a quantitative interpretation of the spectrum can be done. First, the Galactic reddening in this direction is estimated[33] to be $E_{B-V} \approx 0.016$ mag, implying a Galactic extinction correction to the observed fluxes of about 6% at the wavelength of the emission line, and about 4% in the R band. Second, by comparing the spectrum to our measured R-band photometry of the galaxy, we estimate the loss due to the finite size of the slit to be a factor of 1.48.

The Nature of the Host Galaxy

We now discuss the physical properties and nature of this galaxy. Assuming a standard Friedman model cosmology with $H_0 = 65$ km s^{-1} Mpc^{-1} and $\Omega_0 = 0.3$, we derive a luminosity distance (d_L) of 9.7×10^{28} cm (changing the cosmological parameters to $H_0 = 50$ km s^{-1} Mpc^{-1} and $\Omega_0 = 1$ yields $d_L = 8.6 \times 10^{28}$ cm). (H_0, the Hubble constant, is the expansion rate of the Universe at the present time. Ω_0 is the ratio of the mean density of the Universe to the closure density.) Assuming that the intrinsic spectrum F_ν is $\propto \nu^{-0.7}$ and scaling from the observed R-band magnitude corrected for the Galactic extinction, we find the restframe B-band flux to be ~ 0.45 µJy. The corresponding absolute B magnitude is $M_B \approx -20.9$, and is about equal to the absolute magnitude of a typical, L_* galaxy today. Given its high redshift, this galaxy probably has yet to produce most of its stars, and even with some evolutionary fading of its present-day luminosity it may evolve to an L_* galaxy.

The observed Lyα emission line flux is $F_{Ly\alpha} = (3.8 \pm 0.4) \times 10^{-18}$ erg cm^{-2} s^{-1}. Correcting for the flux calibration zero-point and the Galactic extinction, this becomes $F_{Ly\alpha} = (6.2 \pm 0.7) \times 10^{-18}$ erg cm^{-2} s^{-1}. For our assumed cosmology, the implied Lyα line luminosity is $L_{Ly\alpha} = (7.3 \pm 0.8) \times 10^{41}$ erg s^{-1}. Estimates of conversion of the Lyα line luminosity to the implied unobscured star-formation rate (SFR) are in the range $L_{Ly\alpha} = (7 \pm 4) \times 10^{41}$ erg s^{-1} for SFR $= 1$ $M_\odot$ yr^{-1} (where $M_\odot$ is the solar mass), depending on the stellar initial mass

定情况下，定位 Lyα 发射线的峰所引入的偏差都与具体的然而未知的气体和尘埃几何和动力学特征有关，但是平均来讲这些效应都会使 Lyα 发射线的峰相对星系的系统性红移更偏向红端 [19-21]。我们测量了紧挨着发射线的高频端的明显吸收凹陷的波长为 5,356.1 Å。取此值和发射线波长的平均值，再应用标准大气－真空修正关系，我们可以得到此星系的系统性红移为 $z_K = 3.418 \pm 0.010$。

正如前文提到过的，观测上发射线红端的连续谱斜率满足 $F_v \propto v^\alpha$，其中 $\alpha = -0.7 \pm 0.2$。对于没有被遮挡的恒星形成星系，α 约为 0.0～-0.5，这依赖于星系的初始质量函数和恒星形成历史。我们光谱数据中略陡的斜率表明该星系从总体来说受到了中等程度的静止参照系消光。这当然不能限制任何特定方向上的消光，比如 OT 的方向上。我们也注意到具有差不多红移的其他恒星形成星系也存在类似的消光 [19]。

在对光谱进行定量解释之前还要做两个修正。首先，该方向上的银河系红化可以近似估计 [33] 为 $E_{B-V} \approx 0.016$ 星等，由此可推算出在发射线波长处应对流量进行 6% 的银河系消光修正，而在 R 波段进行 4% 的修正。其次，通过把光谱和我们得到的 R 波段的星系测光结果相比较，我们估计由狭缝尺度有限导致的光损失因子为 1.48。

寄主星系性质

下面来讨论这个星系的物理特征和性质。在 $H_0 = 65$ km·s⁻¹·Mpc⁻¹（H_0 是哈勃常数，指宇宙现在的膨胀速率），$\Omega_0 = 0.3$（Ω_0 是宇宙平均密度与闭合密度的比值）的标准弗里德曼宇宙学模型下，我们可以得到该星系的光度距离为 9.7×10^{28} cm（如果将宇宙学参量改变为 $H_0 = 50$ km·s⁻¹·Mpc⁻¹，$\Omega_0 = 1$，则 $d_L = 8.6 \times 10^{28}$ cm）。假定本征光谱满足 $F_v \propto v^{-0.7}$，并由改正了银河系消光后的 R 波段星等进行推算，我们求得静止参照系中 B 波段的流量约为 0.45 μJy，对应的 B 波段绝对星等 $M_B \approx -20.9$，基本和目前典型的 L_* 型星系的绝对星等相同。考虑到它是一个高红移的星系，大部分恒星应该还没有产生，甚至它现在的光度受一些演化过程影响而变暗，它可能正向一个 L_* 型星系演化。

观测到的 Lyα 发射线的流量为 $F_{Ly\alpha} = (3.8 \pm 0.4) \times 10^{-18}$ erg·cm⁻²·s⁻¹，经零点修正和银河系消光修正后变为 $F_{Ly\alpha} = (6.2 \pm 0.7) \times 10^{-18}$ erg·cm⁻²·s⁻¹。在我们所采用的宇宙学模型框架下，我们可以推算出 Lyα 谱线的光度为 $L_{Ly\alpha} = (7.3 \pm 0.8) \times 10^{41}$ erg·s⁻¹。从谱线光度和暗示的未遮挡恒星形成率（SFR）之间的转换关系来看，对应 SFR = 1 $M_\odot$·yr⁻¹（其中 $M_\odot$ 为太阳质量）的区间是 $L_{Ly\alpha} = (7 \pm 4) \times 10^{41}$ erg·s⁻¹，具体依赖于恒星的初始质

function (IMF)[39,40]. We thus estimate the unobscured star-formation rate in this galaxy to be approximately $(1.0 \pm 0.5)\, M_\odot\, \mathrm{yr}^{-1}$.

Table 2. Summary of spectroscopic observations

Epoch (UT)	Grating (lines per mm)	Slit width (arcsec)	Int. time (s)	Notes
1997 Dec. 28	300	1.5	2×1800 s	T.K.
1998 Feb. 3	300	1.0	5×1800 s	S.R.K
1998 Feb. 22	600	1.0	2×1800 s	S.R.K.
1998 Feb. 23	600	1.0	6×1800 s	S.R.K.

Entries (from left to right) are: the UT date of the observation, the grating used, the slit width in arcsec, the number of exposures and the integration time per exposure, and the name of the principal observer. The 300 lines per mm grating is blazed to 5,000 Å and the blaze wavelength of the 600 lines per mm grating is 7,500 Å. The centre wavelength of the 300 lines per mm spectra (hereafter "lower" resolution data) is $\sim$6,400 Å and the dispersion is $\sim$2.45 Å per pixel. The centre wavelength of the 600 lines per mm spectra, the "higher resolution" data, is 6,100 Å and the dispersion is 1.25 Å per pixel. In both cases, the spectral resolution as parametrized by the FWHM is $\sim$5 pixels. The spectrum of the standard star HZ 44 (ref. 45) obtained on February 3 was used to flux calibrate the lower-resolution spectra, and the spectrum of GD 153 (ref. 46), observed on February 22, was used to calibrate the higher-resolution spectra. The zero-point uncertainty of the flux calibration is estimated to be $\sim$10%, judging by the internal agreement, but the slit losses are likely to be higher. Wavelength calibration was derived from arc lamp spectra taken immediately after the observations of the target. The random errors in the wavelength calibration are 0.3 Å and the estimated systematic errors due to instrument flexure are of the same order. Both these uncertainties are unimportant for our redshift determination discussed in the text.

However, the Lyα emission line is probably attenuated by resonant scattering in neutral hydrogen and by absorption by dust. An independent, and perhaps more robust, estimate of the star-formation rate can be obtained from the rest-frame continuum luminosity at 1,500 Å (ref. 41). We note that the observed flux at a wavelength of $1,500(1 + z_K)$ Å is 0.22 μJy. This translates to a star-formation rate, under the usual assumption of a Salpeter IMF, of $5.2\, M_\odot\, \mathrm{yr}^{-1}$. These numbers should be regarded as lower limits given the unknown extinction in the rest-frame of galaxy K.

On the whole, the properties of galaxy K are typical of the known systems at comparable redshifts[19-21], thus giving us some confidence that our redshift interpretation is indeed correct.

The fluence of this GRB[17] above the observed photon energy of 20 keV is $F = 1.1 \times 10^{-5}$ erg cm^{-2}. The γ-ray fluence as observed by the Gamma-ray Burst Monitor (GRBM) on board BeppoSAX is $(0.9 \pm 0.9) \times 10^{-5}$ erg cm^{-2} s^{-1} (above 40 keV). The burst was also observed by the All Sky Monitor on board the X-ray satellite XTE (ref. 42). D. A. Smith (personal communication) estimates the fluence in the 2–12 keV band to be $1.8(\pm 0.03) \times 10^{-7}$ erg cm^{-2}. Thus the isotropic energy loss in γ-rays alone at the distance to the host is $4\pi d_L^2 F/(1 + z) \approx 3 \times 10^{53}$ erg.

The currently favoured model[43] for GRBs is coalescence of neutron stars. The coalescence is expected to release most of the energy in neutrinos (as in type II supernovae) and about

量函数 (IMF)[39,40]。我们据此估计该星系的未遮挡恒星形成率约为 $(1.0 \pm 0.5)\, M_\odot \cdot \mathrm{yr}^{-1}$。

表 2. 光谱观测数据汇总

时期 (世界时)	光栅 (每毫米线数)	狭缝宽度 (角秒)	积分时间 (秒)	注释
1997 年 12 月 28 日	300	1.5	2×1800 秒	昆迪茨
1998 年 2 月 3 日	300	1.0	5×1800 秒	库尔卡尼
1998 年 2 月 22 日	600	1.0	2×1800 秒	库尔卡尼
1998 年 2 月 23 日	600	1.0	6×1800 秒	库尔卡尼

表头从左到右依次为：观测的标准世界时，所用光栅，缝宽 (角秒为单位)，曝光次数和每次曝光时间，主要观测者的名字。每毫米 300 线光栅的闪耀波长为 5,000 Å，每毫米 600 线的光栅闪耀波长则为 7,500 Å。300 线光栅光谱 (以下称低分辨率数据) 的中央波长约为 6,400 Å，色散约为每像素 2.45 Å。600 线光栅光谱 (高分辨率数据) 的中央波长为 6,100 Å，色散为每像素 1.25 Å。两种情况下由半峰全宽参数化表示的光谱分辨率约为 5 像素，2 月 3 日获得的标准星 HZ 44 的光谱 (参考文献 45) 用于低分辨率光谱流量校准，而 2 月 22 日获得的 GD 153 的光谱 (参考文献 46) 用于高分辨率光谱流量校准。流量校准时零点的不确定性，通过内部符合判断，约为 10%，但狭缝损失率可能会更高。波长修正是从观测完目标后立即拍摄的弧光灯光谱中得到的，随机误差为 0.3 Å，仪器弯沉也会带来同量级的系统误差。所有这些误差对文中所讨论的红移的确定并不重要。

但是，Lyα 发射线的强度可能被中性氢的共振散射和尘埃吸收所减弱。利用静止参考系下连续谱在 1,500 Å 处的光度可以对恒星形成率给出另一个独立的、更稳健的估计 (参考文献 41)。我们注意到在波长 $1,500(1+z_K)$ Å 处的观测流量为 0.22 μJy，如果采用萨尔皮特初始质量函数的通常假定，得出的恒星形成率为 5.2 $M_\odot \cdot \mathrm{yr}^{-1}$。鉴于星系 K 在静止参考系中的消光情况未知，这些结果应视为恒星形成率的下限。

总的来说，星系 K 的性质在已知的同等红移的星系 [19-21] 中非常典型，这使我们更加确信自己的红移推算是正确的。

这个 GRB[17] 在 20 keV 的光子能量以上的能流为 $F = 1.1 \times 10^{-5}\ \mathrm{erg} \cdot \mathrm{cm}^{-2}$。由装在 BeppoSAX 卫星上的伽马射线暴监测器 (GRBM) 观测到的 γ 射线流量 (40 keV 以上) 则为 $(0.9 \pm 0.9) \times 10^{-5}\ \mathrm{erg} \cdot \mathrm{cm}^{-2} \cdot \mathrm{s}^{-1}$。X 射线时变探测器 (参考文献 42) 上的全天探测器同样观测到这次暴，史密斯 (个人交流) 估计该暴在 2～12 keV 能段的能流为 $1.8(\pm 0.03) \times 10^{-7}\ \mathrm{erg} \cdot \mathrm{cm}^{-2}$。因此在这个距离上各向同性释放的 γ 射线能量为 $4\pi d_L^2 F/(1+z) \approx 3 \times 10^{53}\ \mathrm{erg}$。

目前倾向于用中子星并合模型 [43] 来解释 GRB。并合过程中大部分能量应以中微子的形式放出 (与 II 型超新星相同)，有大约 10^{51} erg 的能量以电磁能的形式释放。

10^{51} erg is supposed to be in the form of electromagnetic energy. The measured fluence of GRB971214 when combined with our proposed redshift for the GRB appears to be inconsistent with the expectations of the neutron-star merger model in its simplest form; however, it is possible that more elaborate versions of this model could be made to fit these observations. Non-spherical emission could reduce the strain on the energy budget but will not alter the fact that this fairly bright burst originated from such a large redshift.

The most significant implication of our hypothesis that K is the host galaxy of GRB971214 is the implied extreme energetics. The energy budget for GRBs goes up from the traditional 10^{51} erg to perhaps the entire energy available in the coalescence of neutron-star mergers[43]. Other energetic models, ranging from the death of an extremely massive star[23] to coalescence of black-hole neutron star binaries[22] may then become more attractive. In the latter models, GRBs are directly related to the formation rate of massive stars. If that is the case, then the typical redshift of GRBs is approximately 2 (ref. 44), consistent with the suggested high redshift for GRB971214. Moreover, this burst evidently did not originate in a highly obscured star-forming region in its host galaxy.

Regardless of the details of their genesis, GRBs appear to be the brightest known objects in the Universe, albeit over the limited duration of the burst. The term "hypernova"[23] can be justifiably used to describe these most extreme events, especially their after-glow. GRB971214 was not a particularly faint event, and thus statistically we expect many fainter events to arise from larger redshifts. The high brightness of GRBs and their optical transients offer us exciting and new opportunities to probe the Universe at high redshifts.

(**393**, 35-39; 1998)

S. R. Kulkarni*, S. G. Djorgovski*, A. N. Ramaprakash*†, R. Goodrich‡, J. S. Bloom*, K. L. Adelberger*, T. Kundic*, L. Lubin,*, D. A. Frail§, F. Frontera‖#, M. Feroci¶, L. Nicastro*, A. J. Barth**, M. Davis**, A. V. Filippenko** & J. Newman**

* Palomar Observatory 105-24, California Institute of Technology, Pasadena, California 91125, USA

† Inter-University Centre for Astronomy and Astrophysics, Ganeshkhind, Pune 411 007, India

‡ W. M. Keck Observatory, 65-0120 Mamalahoa Highway, Kamuela, Hawaii 96743, USA

§ National Radio Astronomy Observatory, Socorro, New Mexico 8801, USA

‖ Istituto Tecnnologie Studio delle Radiazioni Extraterrestri, CNR, via Gobetti 101, Bologna I-40129, Italy

Dipartimento di Fisica, Universita Ferrara, Via Paradiso 12, I-44100, Italy

¶ Istituto di Astrofisica Spaziale, CNR, via Fosso del Cavaliere, Roma I-00133, Italy

* Istituto Fisica Cosmica App. Info., CNR, via U. La Malfa 153, Palermo I-90146, Italy

** Department of Astronomy, University of California, Berkeley, California 94720, USA

Received 18 March; accepted 14 April 1998.

References:

1. Klebesadel, R. W., Strong, I. B. & Olson, R. A. Observations of gamma-ray bursts of cosmic origin. *Astrophys. J.* **182**, L85-L88 (1973).

2. Boella, G. *et al.* BeppoSAX, the wide band mission for x-ray astronomy. *Astron. Astrophys. Suppl. Ser.* **122**, 299-399 (1997).

3. Costa, E. *et al.* Discovery of an X-ray afterglow associated with the γ-ray burst of 28 February 1997. *Nature* **387**, 783-785 (1997).

4. van Paradijs, J. *et al.* Transient optical emission from the error box of the γ-ray burst of 28 February 1997. *Nature* **386**, 686-689 (1997).

5. Frail, D. A., Kulkarni, S. R., Nicastro, L., Feroci, M. & Taylor, G. B. The radio afterglow from the γ-ray burst of 8 May 1997. *Nature* **389**, 261-263 (1997).

而在 GRB971214 中，结合推算出的红移以及测量到的能量，我们发现结果与最简单的中子星并合模型预言不符，然而对模型作更细致的调整也许能使模型与观测数据相吻合。非球对称的辐射也会缓解释放能量不足的问题，但是仍很难解释在这样高的红移处有如此亮的爆发。

在 K 是 GRB971214 的寄主星系的前提下，最富意义的暗示就是所推算出的巨大能量。GRB 的能量需求从传统的 10^{51} erg 增加到了大概中子星并合事件所能释放的全部能量 [43]。诸如极大质量恒星死亡 [23]，黑洞–中子星并合事件 [22] 等其他模型因此变得更有吸引力。在后面的这两个模型中，GRB 将与大质量恒星的形成率直接相关，如果是这样，GRB 的典型红移应为 2 左右（参考文献 44），这与 GRB971214 给出的高红移的结果相一致。此外，这个暴很明显地并没有发生在该寄主星系被严重遮蔽的恒星形成区中。

不论它们的具体起源是什么，GRB 是宇宙中已知的最亮的天体，尽管只是在有限的持续时间之内。术语"极超新星"[23] 可以用来形容这类极端事件，特别是它们的余辉。GRB971214 并不是一个特别暗的事件，因此我们可以期望统计上在更高红移处会探测到很多更暗的事件。GRB 及其光学暂现源的巨大亮度为我们探索高红移处的宇宙提供了令人兴奋的新机会。

（余恒 翻译；黎卓 审稿）

6. Metzger, M. R. *et al.* Spectral constraints on the redshift of the optical counterpart to the γ-ray burst of 8 May 1997. *Nature* **387,** 878-880 (1997).

7. Bond, H. E. *et al. IAU Circ.* No. 6665 (1997).

8. Djorgovski, S. G. *et al.* The optical counterpart to theγ-ray burst 970508. *Nature* **387,** 876-878 (1997).

9. Costa, E. *et al. IAU Circ.* No. 6649 (1997).

10. Piro, L. *et al. IAU Circ.* No. 6656 (1997).

11. Bloom, J. S., Kulkarni, S. R., Djorgovski, S. G. & Frail, D. A. *GCN Note* No. 30 (1998).

12. Zharikov, S. V., Sokolov, V. V. & Baryshev, Y. V. *GCN Note* No. 31 (1998.).

13. Galama, T. J. *et al.* Optical followup of GRB 970508. *Astrophys. J.* (submitted).

14. Sahu, K. C. *et al.* Observations of GRB 970228 and GRB 970508 and the neutron star merger mode. *Astrophys. J.* **489,** L127-L131 (1997).

15. Halpern, J. P., Thorstensen, J. R., Helfand, D. J. & Costa, E. Optical afterglow of the γ-ray burst of 14 December 1997 *Nature* **393,** 41-43 (1998).

16. Heise, J. *et al. IAU Circ.* No. 6787 (1997).

17. Kippen, R. M. *et al. IAU Circ.* No. 6789 (1997).

18. Antonelli, L. A. *et al. IAU Circ.* No. 6792 (1997).

19. Steidel, C. C., Giavalisco, M., Pettini, M., Dickinson, M. & Adelberger, K. Spectroscopic confirmation of a population of normal star-forming galaxies at redshifts $z > 3$. *Astrophys. J.* **462,** L17-L21 (1996).

20. Steidel, C. C., Giavalisco, M., Dickinson, M. & Adelberger, K. L. Spectroscopy of Lyman break galaxies in the Hubble Deep Field. *Astron. J.* **112,** 352-358 (1996).

21. Steidel, C. C. *et al.* A large structure of galaxies at redshift $z \sim 3$ and its cosmological implications. *Astrophys. J.* **492,** 428-438 (1998).

22. Meszáros, P. & Rees, M. J. Poynting jets from black holes and cosmological gamma-ray bursts. *Astrophys. J.* **482,** L29-L31 (1997).

23. Paczyński, B. Are gamma-ray bursts in star-forming regions? *Astrophys. J.* **492,** L45-L48 (1998).

24. Oke, J. B. *et al.* The Keck low-resolution imaging spectrometer. *Publ. Astron. Soc. Pacif.* **107,** 375-385 (1995).

25. Castander, F. J. *et al. GCN Note* No. 11 (1997).

26. Halpern, J., Thorstensen, J., Helfand, D. & Costa, E. *IAU Circ.* No. 6788 (1997).

27. Rhoads, J. *IAU Circ.* No. 6793 (1997).

28. Diercks, A. *et al. IAU Circ.* No. 67921 (1997).

29. Henden, A. A., Luginbuhl, C. B. & Vrba, F. J. *GCN Note.* No. 16 (1997).

30. Landolt, A. U. UBVRI photometric standard stars in the magnitude range $11.5 < V < 16.0$ around the celestial equator. *Astron. J.* **104,** 340-371 (1992).

31. Wijers, R. A. M. J., Rees, M. J. & Meszáros, P. Shocked by GRB 970228: the afterglow of a cosmological fireball. *Mon. Not. R. Astron Soc.* **288,** L51-L56 (1997).

32. Sokolov, V. V. *et al.* BVR$_C$I$_C$ photometry of GRB 970508 optical remnant: May–August, 1997. *Astron. Astrophys.* (in the press); preprint http://xxx.lanl.gov, astro-ph/0902341 (1998).

33. Galama, T. J. *et al.* Optical follow-up of GRB 970508. *Astrophys. J.* (in the press); preprint http:// xxx.lanl.gov, astro-ph/9802160 (1998).

34. Waxman, E. Gamma-ray-burst afterglow: supporting the cosmological fireball model, constraining parameters, and making predictions. *Astrophys. J.* **485,** L5-L8 (1997).

35. Hogg, D. W. *et al.* Counts and colors of faint galaxies in the U and R bands. *Mon. Not. R. Astron. Soc.* **288,** 404-410 (1997).

36. Oke, J. B. & Korycansky, D. Absolute spectrophotometry of very large redshift quasars. *Astrophys. J.* **255,** 11-19 (1996).

37. Kennefick, J. D., Djorgovski, S. G. & de Carvalho, R. R. The luminosity function of $z > 4$ quasars from the second Palomar sky survey. *Astron. J.* **110,** 2553-2565 (1995).

38. Schlegel, D. J., Finkbeiner, D. P. & Davis, M. Maps of dust IR emission for use in estimation of reddening and CMBR foregrounds. *Astrophys. J.* (in the press); preprint http://xxx.lanl.gov, astro-ph/0910327.

39. Thompson, D., Djorgovski, S. & Trauger, J. A narrow-band imaging survey for primeval galaxies. *Astron. J.* **110,** 963-981 (1995).

40. Charlot, S. & Fall, S. M. Lyman-alpha emission from galaxies. *Astrophys. J.* **415,** 580-588 (1993).

41. Leitherer, C., Robert, C. & Heckman, T. M. Atlas of synthetic ultraviolet-spectra of massive star populations. *Astrophys. J. Suppl.* **99,** 173-187 (1995).

42. Doty, J. P. The All Sky Monitor for the X-ray Timing Explorer. *Proc. SPIE* **982,** 164-172 (1988).

43. Narayan, R., Pacsyński, B. & Piran, T. Gamma-ray bursts as the death throes of massive binary stars. *Astrophys. J.* **395,** L83-L86 (1992).

44. Wijers, R. A. M. J., Bloom, J., Bagla, J. S. & Natarajan, P. Gamma-ray bursts from stellar remnants: probing the universe at high redshift. *Mon. Not. R. Astron. Soc.* **294,** L13-L17 (1998).

45. Massey, P., Strobel, K., Barnes, J. & Anderson, E. Spectrophotometric standards. *Astrophys. J.* **328,** 315-333 (1988).

46. Bohlin, R., Colina, L. & Finley, D. White dwarf standard stars: G191-B2B, GD 71, GD 153, HZ 43. *Astron. J.* **110,** 1316-1325 (1995).

Acknowledgements. The observations reported here were obtained at the W. M. Keck Observatory, which is operated by the California Association for Research in Astronomy, a scientific partnership among California Institute of Technology, the University of California and NASA. It was made possible by the financial support from W. M. Keck Foundation. We thank W. Sargent, Director of the Palomar Observatory, F. Chaffee, Director of the Keck Observatory and our colleagues for continued support of our GRB program. We thank J. C. Clemens and M. H. van Kerkwijk for help with observations and exchange of dark time. S.R.K.'s research is supported by the NSF and NASA. S.G.D. acknowledges partial support from the Bressler Foundation. A.N.R. is grateful to the International Astronomical Union for a travel grant.

Correspondence and requests for materials should be addressed to S.R.K. (e-mail: srk:surya.caltech.edu).

A Silicon-based Nuclear Spin Quantum Computer

B. E. Kane

Editor's Note

In the 1980s, physicists began trying to build computing devices that exploit quantum effects, which could be much more powerful than classical devices, at least for certain computational tasks. The practical development of such machines was immensely challenging because of their inherent sensitivity to disruption from the environment. Here Australian physicist Bruce Kane proposes a scheme for implementing a quantum computer by storing its information in the relatively isolated spins of nuclei of dopant atoms inserted into silicon electronic devices. Logical operations, he suggests, could be performed on these spins with external electric fields, and measurements made with currents of spin-polarized electrons. Practical quantum computers are now being developed, and this strategy remains one of many being explored.

Quantum computers promise to exceed the computational efficiency of ordinary classical machines because quantum algorithms allow the execution of certain tasks in fewer steps. But practical implementation of these machines poses a formidable challenge. Here I present a scheme for implementing a quantum-mechanical computer. Information is encoded onto the nuclear spins of donor atoms in doped silicon electronic devices. Logical operations on individual spins are performed using externally applied electric fields, and spin measurements are made using currents of spin-polarized electrons. The realization of such a computer is dependent on future refinements of conventional silicon electronics.

ALTHOUGH the concept of information underlying all modern computer technology is essentially classical, physicists know that nature obeys the laws of quantum mechanics. The idea of a quantum computer has been developed theoretically over several decades to elucidate fundamental questions concerning the capabilities and limitations of machines in which information is treated quantum mechanically[1,2]. Specifically, in quantum computers the ones and zeros of classical digital computers are replaced by the quantum state of a two-level system (a qubit). Logical operations carried out on the qubits and their measurement to determine the result of the computation must obey quantum-mechanical laws. Quantum computation can in principle only occur in systems that are almost completely isolated from their environment and which consequently must dissipate no energy during the process of computation, conditions that are extraordinarily difficult to fulfil in practice.

Interest in quantum computation has increased dramatically in the past four years because of two important insights: first, quantum algorithms (most notably for prime factorization[3,4]

178

硅基核自旋量子计算机

凯恩

编者按

20世纪80年代，物理学家开始尝试构建利用量子效应的计算设备，至少对于某些计算任务而言，它可能比传统设备性能强大得多。由于其对来自环境的干扰具有内在的敏感性，这类机器的实际研发极具挑战性。本文中，澳大利亚物理学家布鲁斯·凯恩提出了一种实现量子计算机的方案，将其信息存储在硅电子器件中内嵌掺杂原子的相对孤立的核自旋中。他建议，逻辑运算可以在外部电场辅助下用这些自旋进行，并用自旋极化电子的电流进行测量。可行的量子计算机正在开发中，该方案仍然是许多正在被探索的策略之一。

量子计算机的计算效率可能会超过传统经典计算机，因为量子算法允许用更少的步骤来执行某些任务。但是这些机器的实际研发仍然是一个艰巨的挑战。这里我将提出一种构建基于量子力学的计算机的方法。计算信息被编码在掺杂硅的电子器件中施主原子的核自旋上。单个自旋的逻辑运算用外部施加电场来实现，而自旋测量基于自旋极化电子的电流。这样一个计算机的实现依赖于传统硅器件未来的精密程度。

尽管所有现代计算机所依赖的信息概念本质上都是基于经典理论的，然而物理学家知道自然界遵从量子力学定律。量子计算机的概念在理论上已经发展了几十年，旨在阐明使用量子力学来处理信息的计算机的能力和极限等基本问题[1,2]。具体而言，在量子计算机中，经典计算机中的0和1被一个二能级系统的量子态（量子比特）所取代。在量子比特上进行的逻辑运算及用来决定计算结果的测量必须遵守量子力学定律。量子计算理论上只能发生在与外界环境几乎完全隔离的系统中，这样在计算过程中不会存在能量的耗散，这样的苛刻条件在现实中很难满足。

在过去四年中，两个重大的突破使人们对量子计算的兴趣与日俱增。第一个是量子算法（尤其值得一提的是质因子分解[3,4]和穷举搜索[5]）的开发，而且这些算法

and for exhaustive search[5]) have been developed that outperform the best known algorithms doing the same tasks on a classical computer. These algorithms require that the internal state of the quantum computer be controlled with extraordinary precision, so that the coherent quantum state upon which the quantum algorithms rely is not destroyed. Because completely preventing decoherence (uncontrolled interaction of a quantum system with its surrounding environment) is impossible, the existence of quantum algorithms does not prove that they can ever be implemented in a real machine.

The second critical insight has been the discovery of quantum error-correcting codes that enable quantum computers to operate despite some degree of decoherence and which may make quantum computers experimentally realizable[6,7]. The tasks that lie ahead to create an actual quantum computer are formidable: Preskill[8] has estimated that a quantum computer operating on 10^6 qubits with a 10^{-6} probability of error in each operation would exceed the capabilities of contemporary conventional computers on the prime factorization problem. To make use of error-correcting codes, logical operations and measurement must be able to proceed in parallel on qubits throughout the computer.

The states of spin 1/2 particles are two-level systems that can potentially be used for quantum computation. Nuclear spins have been incorporated into several quantum computer proposals[9-12] because they are extremely well isolated from their environment and so operations on nuclear spin qubits could have low error rates. The primary challenge in using nuclear spins in quantum computers lies in measuring the spins. The bulk spin resonance approach to quantum computation[11,12] circumvents the single-spin detection problem essentially by performing quantum calculations in parallel in a large number of molecules and determining the result from macroscopic magnetization measurements. The measurable signal decreases with the number of qubits, however, and scaling this approach above about ten qubits will be technically demanding[37].

To attain the goal of a 10^6 qubit quantum computer, it has been suggested that a "solid state" approach[13] might eventually replicate the enormous success of modern electronics fabrication technology. An attractive alternative approach to nuclear spin quantum computation is to incorporate nuclear spins into an electronic device and to detect the spins and control their interactions electronically[14]. Electron and nuclear spins are coupled by the hyperfine interaction[15]. Under appropriate circumstances, polarization is transferred between the two spin systems and nuclear spin polarization is detectable by its effect on the electronic properties of a sample[16,17]. Electronic devices for both generating and detecting nuclear spin polarization, implemented at low temperatures in $GaAs/Al_xGa_{1-x}As$ heterostructures, have been developed[18], and similar devices have been incorporated into nanostructures[19,20]. Although the number of spins probed in the nanostructure experiments is still large ($\sim 10^{11}$; ref. 19), sensitivity will improve in optimized devices and in systems with larger hyperfine interactions.

在执行相同任务时优于经典计算机上最著名的算法。这些算法要求量子计算机的内部状态被准确无误地控制，这样量子算法所依赖的相干量子态就不会被破坏。因为完全阻止量子退相干（量子系统和外界环境之间非受控的相互作用）是不可能的，所以量子算法的存在并不能保证它们可以在真实的计算机上面得以实现。

第二个重大的突破是量子纠错码的发现，它可以使量子计算机运行时容忍一定程度的量子退相干，这使得量子计算机在实验上的实现成为可能[6,7]。眼下制造一台真正的量子计算机的任务是艰巨的：据普雷斯基尔[8]估计，一台使用 10^6 个量子比特且每次运算误码率为 10^{-6} 的量子计算机可以在质因子分解问题上超过同时代的传统计算机。为了利用纠错码，逻辑运算和测量必须能够在整个计算机的量子比特上并行进行。

自旋为 1/2 的粒子的状态是二能级系统，具备应用于量子计算的潜质。核自旋已经被纳入数个量子计算机提案[9-12]中，因为它们与外界环境间被很好地隔离，因此利用核自旋比特来运算，误码率可能很低。在量子计算机中使用核自旋的主要挑战在于测量自旋。量子计算的系综自旋共振方法[11,12]基本上是通过在大量分子上的并行量子计算并由宏观磁化测量确定结果来绕开单个自旋的测量问题。然而可测量信号随着量子比特数目的增多而减少，因此将这种方法扩展到 10 个量子比特以上将会对技术要求极为苛刻[37]。

为了实现 10^6 量子比特计算机，有人建议使用"固态"方法[13]，这种方法也许可以最终复制现代电子器件微加工技术的巨大成功。核自旋量子计算的一种有吸引力的替代方法是将核自旋融合到电子器件中，并且通过电学方式来测量自旋以及控制其之间的相互作用[14]。电子和核自旋被超精细相互作用耦合在一起[15]。在适当情况下，极化可以在两个自旋系统中转移，核自旋极化可以通过其对样品的电子性质的影响来测量[16,17]。用来产生和探测核自旋极化的电子器件也已经在低温砷化镓（GaAs）/铝镓砷（$Al_xGa_{1-x}As$）异质结构中实现[18]，类似的器件也已经被整合到纳米结构中[19,20]。尽管在纳米结构试验中研究的自旋数目仍然很大（约为 10^{11}；文献 19），但在经过优化的器件中以及更强的超精细相互作用系统中，灵敏度会得到进一步的改善。

Here I present a scheme for implementing a quantum computer on an array of nuclear spins located on donors in silicon, the semiconductor used in most conventional computer electronics. Logical operations and measurements can in principle be performed independently and in parallel on each spin in the array. I describe specific electronic devices for the manipulation and measurement of nuclear spins, fabrication of which will require significant advances in the rapidly moving field of nanotechnology. Although it is likely that scaling the devices proposed here into a computer of the size envisaged by Preskill[8] will be an extraordinary challenge, a silicon-based quantum computer is in a unique position to benefit from the resources and ingenuity being directed towards making conventional electronics of ever smaller size and greater complexity.

Quantum Computation with a ^{31}P Array in Silicon

The strength of the hyperfine interaction is proportional to the probability density of the electron wavefunction at the nucleus. In semiconductors, the electron wavefunction extends over large distances through the crystal lattice. Two nuclear spins can consequently interact with the same electron, leading to electron-mediated or indirect nuclear spin coupling[15]. Because the electron is sensitive to externally applied electric fields, the hyperfine interaction and electron-mediated nuclear spin interaction can be controlled by voltages applied to metallic gates in a semiconductor device, enabling the external manipulation of nuclear spin dynamics that is necessary for quantum computation.

The conditions required for electron-coupled nuclear spin computation and single nuclear spin detection can arise if the nuclear spin is located on a positively charged donor in a semiconductor host. The electron wavefunction is then concentrated at the donor nucleus (for s orbitals and energy bands composed primarily of them), yielding a large hyperfine interaction energy. For shallow-level donors, however, the electron wavefunction extends tens or hundreds of ångströms away from the donor nucleus, allowing electron-mediated nuclear spin coupling to occur over comparable distances. The quantum computer proposed here comprises an array of such donors positioned beneath the surface of a semiconductor host (Fig. 1). A quantum mechanical calculation proceeds by the precise control of three external parameters: (1) gates above the donors control the strength of the hyperfine interactions and hence the resonance frequency of the nuclear spins beneath them; (2) gates between the donors turn on and off electron-mediated coupling between the nuclear spins[13]; (3) a globally applied a.c. magnetic field B_{ac} flips nuclear spins at resonance. Custom adjustment of the coupling of each spin to its neighbours and to B_{ac} enables different operations to be performed on each of the spins simultaneously. Finally, measurements are performed by transferring nuclear spin polarization to the electrons and determining the electron spin state by its effect on the orbital wavefunction of the electrons, which can be probed using capacitance measurements between adjacent gates.

　　这里我将介绍一种利用位于硅材料施主上面的核自旋阵列来实现量子计算机的方法，硅材料是大多数传统计算机电子学中使用的半导体。理论上逻辑运算和测量可以在阵列中的各个自旋上独立且并行执行。我将介绍特定的电子器件来实现核自旋的控制和测量，其制造需要迅猛发展的纳米科技领域的重大突破。把这里提出的器件缩放到普雷斯基尔[8]设想的尺寸的计算机中，这也许是一个巨大的挑战，然而，由于人们为了将传统电子器件做得更小且更加复杂的过程中积累了资源和聪明才智，硅基量子计算机仍然具有独特的地位。

使用硅中 ^{31}P 阵列的量子计算

　　超精细相互作用的强度正比于原子核处电子波函数的概率密度。在半导体中，电子波函数在晶格上延展很长的距离。两个核自旋因此可以和同一个电子相互作用，这就导致了电子介导的或者间接的核自旋耦合[15]。由于电子对于外加电场的敏感，超精细相互作用以及电子介导的核自旋相互作用可以通过调节施加在半导体器件中金属栅极上的电压来控制，这使量子计算所要求的核自旋动力学的外部调制成为了可能。

　　如果核自旋位于半导体宿主中带正电的施主上面，电子耦合核自旋计算以及单核自旋检测所需要的条件会被满足。电子波函数将会集中于施主的原子核（对于 s 轨道和主要由它们组成的能带），产生大的超精细相互作用能。然而对于浅能级施主来说，电子波函数从原子核向外扩展了几十甚至几百埃，从而允许电子介导核自旋耦合在相对远的距离发生。本文提出的量子计算机包括在半导体宿主表面下方的一系列这样的施主（如图 1）。量子力学计算通过精确控制三个外部参数来进行：（1）施主上方的栅极控制超精细相互作用的强度，因此也就控制了栅极下方的核自旋的共振频率；（2）施主之间的栅极可以用来开通和关闭核自旋之间的电子介导耦合[13]；（3）一个对全体施加的交变磁场 B_{ac} 用来翻转与磁场共振的核自旋。单独调节每个自旋之间的耦合以及自旋与外加磁场 B_{ac} 的耦合，使每个自旋上的不同运算可以同时进行。最后，通过转移核自旋极化到电子身上进而影响电子轨道波函数，最终通过测试其对相邻的栅极之间的电容的改变来测得自旋态。

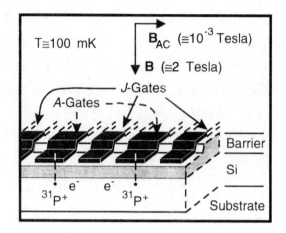

Fig. 1. Illustration of two cells in a one-dimensional array containing ^{31}P donors and electrons in a Si host, separated by a barrier from metal gates on the surface. "*A* gates" control the resonance frequency of the nuclear spin qubits; "*J* gates" control the electron-mediated coupling between adjacent nuclear spins. The ledge over which the gates cross localizes the gate electric field in the vicinity of the donors.

An important requirement for a quantum computer is to isolate the qubits from any degrees of freedom that may lead to decoherence. If the qubits are spins on a donor in a semiconductor, nuclear spins in the host are a large reservoir with which the donor spins can interact. Consequently, the host should contain only nuclei with spin $I = 0$. This simple requirement unfortunately eliminates all III–V semiconductors as host candidates, because none of their constituent elements possesses stable $I = 0$ isotopes[21]. Group IV semiconductors are composed primarily $I = 0$ isotopes and can in principle be purified to contain only $I = 0$ isotopes. Because of the advanced state of Si materials technology and the tremendous effort currently underway in Si nanofabrication, Si is the obvious choice for the semiconductor host.

The only $I = 1/2$ shallow (group V) donor in Si is [31]P. The Si:[31]P system was exhaustively studied 40 years ago in the first electron–nuclear double-resonance experiments[22,23]. At sufficiently low ^{31}P concentrations at temperature $T = 1.5$ K, the electron spin relaxation time is thousands of seconds and the ^{31}P nuclear spin relaxation time exceeds 10 hours. It is likely that at millikelvin temperatures the phonon limited ^{31}P relaxation time is of the order of 10^{18} seconds (ref. 24), making this system ideal for quantum computation.

The purpose of the electrons in the computer is to mediate nuclear spin interactions and to facilitate measurement of the nuclear spins. Irreversible interactions between electron and nuclear spins must not occur as the computation proceeds: the electrons must be in a non-degenerate ground state throughout the computation. At sufficiently low temperatures, electrons only occupy the lowest energy-bound state at the donor, whose twofold spin degeneracy is broken by an applied magnetic field B. (The valley degeneracy of the Si conduction band is broken in the vicinity of the donor[25]. The lowest donor excited state is approximately 15 meV above the ground state[23].) The electrons will only occupy the lowest energy spin level when $2\mu_B B \gg kT$, where μ_B is the Bohr magneton. (In Si, the Landé

184

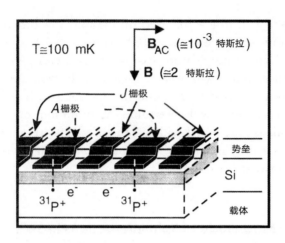

图 1. 硅宿主中 [31]P 施主和电子的一维阵列中的两个单元的示意图，图中的势垒层将该阵列与表面的金属栅极隔离开。"A 栅极"用来控制核自旋量子比特的共振频率。"J 栅极"用来控制相邻核自旋的电子介导耦合。栅极末端的台阶状设计使得加在施主附近的电场局域化。

对于量子计算机来说，一个重要的要求是量子比特应该与可能导致退相干的任何自由度隔离开来。如果量子比特是在半导体中的施主上的自旋，施主自旋可以和宿主材料中其他大量核自旋构成的群体相互作用。因此宿主应该只含有自旋 $I = 0$ 的原子核。这个简单的要求遗憾地排除了所有 III–V 半导体作为宿主的可能性，因为他们中的任何元素都没有稳定的 $I = 0$ 的同位素[21]。而 IV 半导体主要包含 $I = 0$ 同位素，而且理论上可以被提纯到只含有 $I = 0$ 的同位素。由于硅材料技术的领先状态以及现阶段人们对硅纳米加工方面的巨大投入，硅理所当然地成为半导体宿主的选择。

硅中唯一的 $I = 1/2$ 浅施主（第五主族）是 [31]P。在 40 年以前人们发现的第一个电子–原子核双共振的试验中[22,23]Si:[31]P 系统已经得到了深入的研究。温度 $T = 1.5$ K 时，当 [31]P 的浓度足够低时，电子自旋的弛豫时间是几千秒，[31]P 的核自旋弛豫时间超过 10 小时。这样的话，在毫开尔文的温度下，声子散射受限的 [31]P 的弛豫时间在 10^{18} 秒的量级（文献 24），使得这种系统成为量子计算的理想选择。

该计算机利用电子来促成核自旋之间的相互作用以及核自旋的测量。电子和核自旋之间不可逆的相互作用必须在计算过程中避免：电子必须在整个计算过程中处于非简并基态。在足够低的温度下，通过外加磁场 B 打破电子基态的双重自旋简并度，这样电子只占据施主的最低能量束缚态。（硅的导带的谷简并在施主附近被破坏了[25]。施主的最低激发态大概是基态上方 15 meV[23]。）当 $2\mu_B B \gg kT$ 时，电子仅占据最低自旋能级，其中 μ_B 是波尔磁子。（在硅中，朗德 g 因子非常接近 +2，所以在我

g-factor is very close to $+2$, so $g = 2$ is used throughout this discussion.) The electrons will be completely spin-polarized $(n_\uparrow/n_\downarrow < 10^{-6})$ when $T \leqslant 100$ mK and $B \geqslant 2$ tesla. A quantum-mechanical computer is non-dissipative and can consequently operate at low temperatures. Dissipation will arise external to the computer from gate biasing and from eddy currents caused by B_{ac}, and during polarization and measurement of the nuclear spins. These effects will determine the minimum operable temperature of the computer. For this discussion, I will assume $T = 100$ mK and $B = 2$ T. Note that these conditions do not fully polarize the nuclear spins, which are instead aligned by interactions with the polarized electrons.

Magnitude of Spin Interactions in Si:^{31}P

The size of the interactions between spins determines both the time required to do elementary operations on the qubits and the separation necessary between donors in the array. The hamiltonian for a nuclear spin–electron system in Si, applicable for an $I = 1/2$ donor nucleus and with $B\|z$ is $H_{en} = \mu_B B \sigma_z^e - g_n \mu_n B \sigma_z^n + A\sigma^e \cdot \sigma^n$, where σ are the Pauli spin matrices (with eigenvalues ± 1), μ_n is the nuclear magneton, g_n is the nuclear g-factor (1.13 for ^{31}P; ref. 21), and $A = \frac{8}{3}\pi\mu_B g_n \mu_n |\Psi(0)|^2$ is the contact hyperfine interaction energy, with $|\Psi(0)|^2$, the probability density of the electron wavefunction, evaluated at the nucleus. If the electron is in its ground state, the frequency separation of the nuclear levels is, to second order

$$hv_A = 2g_n\mu_n B + 2A + \frac{2A^2}{\mu_B B} \tag{1}$$

In Si:^{31}P, $2A/h = 58$ MHz, and the second term in equation (1) exceeds the first term for $B < 3.5$ T.

An electric field applied to the electron–donor system shifts the electron wavefunction envelope away from the nucleus and reduces the hyperfine interaction. The size of this shift, following estimates of Kohn[25] of shallow donor Stark shifts in Si, is shown in Fig. 2 for a donor 200 Å beneath a gate. A donor nuclear spin–electron system close to an "*A* gate" functions as a voltage-controlled oscillator: the precession frequency of the nuclear spin is controllable externally, and spins can be selectively brought into resonance with B_{ac}, allowing arbitrary rotations to be performed on each nuclear spin.

们的讨论中使用 $g = 2$。) 当 $T \leqslant 100$ mK 且 $B \geqslant 2$ T 的时候，电子将会被完全自旋极化 ($n_\uparrow/n_\downarrow < 10^{-6}$)。量子力学计算机是无耗散的，因此可以在低温状态下运行。然而当栅极偏压出现以及 B_{ac} 引起的涡电流出现时，计算机外部的功耗会上升，这在核自旋的极化和测量时也会出现。这些效应决定了计算机的最低工作温度。为了讨论这些问题，我们假设 $T = 100$ mK 且 $B = 2$ T。注意这些条件并不是完全极化了核自旋，核自旋是通过和极化电子的作用来对齐的。

在 Si:^{31}P 中的自旋相互作用的量级

自旋相互作用的大小决定了量子比特基本运算所需要的时间以及阵列中施主之间需要的间隔。硅中的核自旋–电子系统的哈密顿量，在施主原子核 $I = 1/2$、磁场 B 方向平行于 z 方向时，为 $H_{en} = \mu_B B \sigma e_z - g_n \mu_n B \sigma e_z + A \sigma^e \cdot \sigma^n$，这里 σ 是泡利自旋矩阵（本征值为 ± 1），μ_n 是核磁子，g_n 是原子核的 g 因子（对于 ^{31}P 而言，g 因子为 1.13；文献 21），$A = \frac{8}{3}\pi\mu_B g_n\mu_n|\Psi(0)|^2$ 是接触超精细相互作用能，其中 $|\Psi(0)|^2$ 是电子波函数在原子核的概率密度。如果电子处于基态，那么原子核能级的频率间隔（取到二阶）是，

$$hv_A = 2g_n\mu_n B + 2A + \frac{2A^2}{\mu_B B} \tag{1}$$

在 Si:^{31}P 中，$2A/h = 58$ MHz；在 $B < 3.5$ T 的时候，等式（1）中的第二项大于第一项。

一个施加于电子–施主系统的外加电场会使得电子波函数的包络线向远离原子核的方向平移从而减少超精细相互作用。根据科恩[25]对硅中浅施主斯塔克位移的估测，图 2 给出了栅极下方 200 Å 处施主上述位移的大小。靠近"A 栅极"的施主核自旋–电子系统起到电压控制振荡器的作用：这样核自旋的进动频率就可以外部控制了，自旋也可以选择性地与 B_{ac} 共振，这就使得每一个核自旋可以任意地旋转。

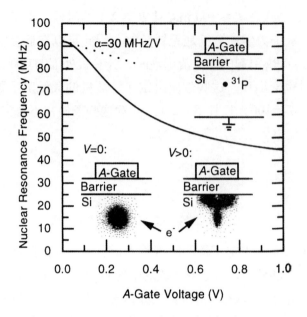

Fig. 2. An electric field applied to an A gate pulls the electron wavefunction away from the donor and towards the barrier, reducing the hyperfine interaction and the resonance frequency of the nucleus. The donor nucleus–electron system is a voltage-controlled oscillator with a tuning parameter α of the order of 30 MHz V^{-1}.

Quantum mechanical computation requires, in addition to single spin rotations, the two-qubit "controlled rotation" operation, which rotates the spin of a target qubit through a prescribed angle if, and only if, the control qubit is oriented in a specified direction, and leaves the orientation of the control qubit unchanged[26,27]. Performing the controlled rotation operation requires nuclear-spin exchange between two donor nucleus–electron spin systems[13], which will arise from electron-mediated interactions when the donors are sufficiently close to each other. The hamiltonian of two coupled donor nucleus–electron systems, valid at energy scales small compared to the donor–electron binding energy, is $H = H(B) + A_1 \sigma^{1n} \cdot \sigma^{2e} + A_2 \sigma^{2n} \cdot \sigma^{2e} + J \sigma^{1e} \cdot \sigma^{2e}$, where $H(B)$ are the magnetic field interaction terms for the spins. A_1 and A_2 are the hyperfine interaction energies of the respective nucleus–electron systems. $4J$, the exchange energy, depends on the overlap of the electron wavefunctions. For well separated donors[28]

$$4J(r) \cong 1.6 \, \frac{e^2}{\epsilon a_B} \left(\frac{r}{a_B} \right)^{\frac{5}{2}} \exp\left(\frac{-2r}{a_B} \right) \tag{2}$$

where r is the distance between donors, ϵ is the dielectric constant of the semiconductor, and a_B is the semiconductor Bohr radius. This function, with values appropriate for Si, is plotted in Fig. 3. Equation (2), originally derived for H atoms, is complicated in Si by its valley degenerate anisotropic band structure[29]. Exchange coupling terms from each valley interfere, leading to oscillatory behaviour of $J(r)$. In this discussion, the complications introduced by Si band structure will be neglected. In determining $J(r)$ in Fig. 3, the

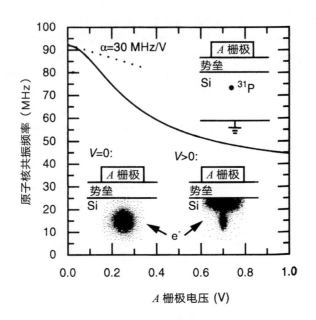

图 2. 施加于 A 栅极的电场使得电子波函数远离施主而朝向势垒，这就减少了超精细相互作用以及原子核的共振频率。施主的原子核–电子系统是一个电压调节的振荡器，调谐参数 α 量级为 $30\ MHz \cdot V^{-1}$。

　　除了单个自旋的旋转，量子力学计算需要双量子比特的"控制旋转"操作，当且仅当控制量子比特的取向在特定方向时，该操作将会使目标量子比特的自旋按照指定的角度旋转，从而使得控制量子比特的取向不变 [26,27]。要实现这种控制旋转操作需要两个施主原子核–电子自旋系统中的核自旋交换 [13]，当两个施主足够接近的时候，这将会由于电子介导相互作用而发生。当能量尺度与施主–电子结合能相比较小的时候，两个耦合的施主原子核–电子系统的哈密顿量可以表示为 $H = H(B)+A_1\sigma^{1n} \cdot \sigma^{2e}+A_2\sigma^{2n} \cdot \sigma^{2e}+J\sigma^{1e} \cdot \sigma^{2e}$。其中 $H(B)$ 为自旋的磁场相互作用项。A_1 和 A_2 为两个原子核–电子系统中各自的超精细相互作用能量。$4J$ 为依赖于电子波函数重叠的交换能。对于足够隔离的施主 [28] 而言，

$$4J(r)\cong 1.6\frac{e^2}{\epsilon a_{\mathrm{B}}}\left(\frac{r}{a_{\mathrm{B}}}\right)^{\frac{5}{2}}\exp\left(\frac{-2r}{a_{\mathrm{B}}}\right) \tag{2}$$

其中 r 是施主之间的距离，ϵ 是半导体的介电常数，a_{B} 是半导体的玻尔半径。根据硅来选取适当的参数，这个函数的图像如图 3 所示。起初由氢原子推导出来的等式 (2) 在硅中是非常复杂的，这是因为硅的谷简并各向异性能带结构 [29]。每一个谷干涉的交换耦合项都会导致 $J(r)$ 的振荡行为。在这里的讨论中，我们将忽略由于硅的能带结构导致的复杂问题。为了在图 3 中确定 $J(r)$，采用硅的横质量（$\cong 0.2m_e$），

transverse mass for Si ($\cong 0.2m_e$) has been used, and $a_B = 30$ Å. Because J is proportional to the electron wave function overlap, it can be varied by an electrostatic potential imposed by a "J-gate" positioned between the donors[13]. As shall be seen below, significant coupling between nuclei will occur when $4J \approx \mu_B B$, and this condition approximates the necessary separation between donors of 100–200 Å. Whereas actual separations may be considerably larger than this value because the J gate can be biased positively to reduce the barrier between donors, the gate sizes required for the quantum computer are near the limit of current electronics fabrication technology.

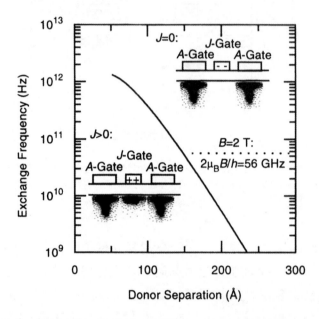

Fig. 3. J gates vary the electrostatic potential barrier V between donors to enhance or reduce exchange coupling, proportional to the electron wavefunction overlap. The exchange frequency ($4J/h$) when $V = 0$ is plotted for Si.

For two-electron systems, the exchange interaction lowers the electron singlet ($|\uparrow\downarrow - \downarrow\uparrow\rangle$) energy with respect to the triplets[30]. (The $|\uparrow\downarrow\rangle$ notation is used here to represent the electron spin state, and the $|01\rangle$ notation the nuclear state; in the $|\downarrow\downarrow 11\rangle$ state, all spins point in the same direction. For simplicity, normalization constants are omitted.) In a magnetic field, however, $|\downarrow\downarrow\rangle$ will be the ground state if $J < \mu_B B/2$ (Fig. 4a). In the $|\downarrow\downarrow\rangle$ state, the energies of the nuclear states can be calculated to second order in A using perturbation theory. When $A_1 = A_2 = A$, the $|10 - 01\rangle$ state is lowered in energy with respect to $|10 + 01\rangle$ by:

$$h\nu_J = 2A^2 \left(\frac{1}{\mu_B B - 2J} - \frac{1}{\mu_B B} \right) \qquad (3)$$

The $|11\rangle$ state is above the $|10 + 01\rangle$ state and the $|00\rangle$ state below the $|10 - 01\rangle$ state by an energy $h\nu_A$, given in equation (1). For the Si:^{31}P system at $B = 2$ T and for $4J/h = 30$ GHz, equation (3) yields $\nu_J = 75$ kHz. This nuclear spin exchange frequency approximates the rate at which binary operations can be performed on the computer (ν_J can be increased

$a_B = 30$ Å。因为 J 正比于电子波函数的重叠，它会通过施主之间的"J 栅极"所产生的静电势而变化[13]。如下所示，当 $4J \approx \mu_B B$ 的时候，核之间会产生显著的耦合，这个已经逼近了施主之间的最小要求间距 $100 \sim 200$ Å。然而实际上的间距可能远大于这个值，因为 J 栅极可以被施加正偏压来减少施主之间的势垒，量子计算机所需要的栅极的尺寸已经接近了当今电子学微加工技术的极限。

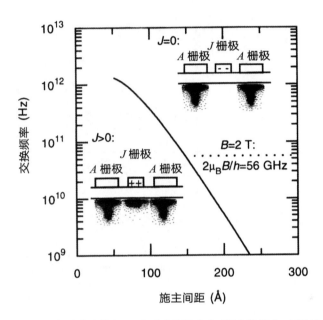

图 3. J 栅极可以改变施主之间的静电势垒 V，以此来增强或者减弱交换耦合，正比于电子波函数的重叠。当 $V=0$ 时，硅的交换频率$(4J/h)$如图所示。

对于双电子系统来说，交换作用使得电子的单重态$(|\uparrow\downarrow - \downarrow\uparrow\rangle)$能量相对于三重态能量降低[30]。$(|\uparrow\downarrow\rangle$ 符号被用来代表电子自旋态，$|01\rangle$ 符号表示核自旋态；在 $|\downarrow\downarrow 11\rangle$ 状态下，所有自旋在同一方向。为了简化，归一化的常数被省略)。然而在磁场存在时，当 $J < \mu_B B/2$ 时(如图 4a 所示)，$|\downarrow\downarrow\rangle$ 将会是电子的基态。在 $|\downarrow\downarrow\rangle$ 态中，核态的能量可以用微扰理论取到 A 的二阶项来计算。当 $A_1 = A_2 = A$ 时，$|10-01\rangle$ 态的能量相比 $|10+01\rangle$ 态的能量小了：

$$h\nu_J = 2A^2 \left(\frac{1}{\mu_B B - 2J} - \frac{1}{\mu_B B} \right) \tag{3}$$

$|11\rangle$ 态在 $|10+01\rangle$ 态上方距离能量 $h\nu_A$ 处，而 $|00\rangle$ 在 $|10-01\rangle$ 下方距离能量 $h\nu_A$ 处，由等式(1)给出。对于 Si:[31]P 系统，在 $B = 2$ T 且 $4J/h = 30$ GHz 时，等式(3)给出解为 $\nu_J = 75$ kHz。这个核自旋交换频率近似于计算机进行二进制运算的频率(ν_J 可以

by increasing J, but at the expense of also increasing the relaxation rate of the coupled nuclear–electron spin excitations). The speed of single spin operations is determined by the size of B_{ac} and is comparable to 75 kHz when $B_{ac} = 10^{-3}$ T.

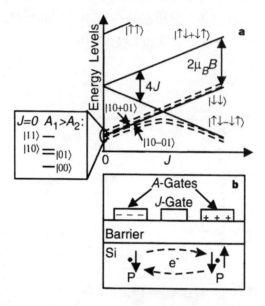

Fig. 4. Two qubit quantum logic and spin measurement. **a**, Electron (solid lines) and lowest energy-coupled electron–nuclear (dashed lines) energy levels as a function of J. When $J < \mu_B B/2$, two qubit computations are performed by controlling the $|10-01\rangle - |10+01\rangle$ level splitting with a J gate. Above $J = \mu_B B/2$, the states of the coupled system evolve into states of differing electron polarization. The state of the nucleus at $J = 0$ with the larger energy splitting (controllable by the A gate bias) determines the final electron spin state after an adiabatic increase in J. **b**, Only $|\uparrow\downarrow - \downarrow\uparrow\rangle$ electrons can make transitions into states in which electrons are bound to the same donor (D^- states). Electron current during these transitions is measurable using capacitive techniques, enabling the underlying spin states of the electrons and nuclei to be determined.

Spin Measurements

Measurement of nuclear spins in the proposed quantum computer is accomplished in a two-step process: distinct nuclear spin states are adiabatically converted into states with different electron polarization, and the electron spin is determined by its effect on the symmetry of the orbital wavefunction of an exchange-coupled two-electron system. A procedure for accomplishing this conversion is shown in Fig. 4. While computation is done when $J < \mu_B B/2$ and the electrons are fully polarized, measurements are made when $J > \mu_B B/2$, and $|\uparrow\uparrow - \downarrow\uparrow\rangle$ states have the lowest energy (Fig. 4a). As the electron levels cross, the $|\downarrow\downarrow\downarrow\rangle$ and $|\uparrow\downarrow - \downarrow\uparrow\rangle$ states are coupled by hyperfine interactions with the nuclei. During an adiabatic increase in J, the two lower-energy nuclear spin states at $J = 0$ evolve into $|\uparrow\downarrow - \downarrow\uparrow\rangle$ states when $J > \mu_B B/2$, whereas the two higher-energy nuclear states remain $|\downarrow\downarrow\downarrow\rangle$. If, at $J = 0$, $A_1 > A_2$, the orientation of nuclear spin 1 alone will determine whether the system evolves into the $|\uparrow\downarrow - \downarrow\uparrow\rangle$ or the $|\downarrow\downarrow\downarrow\rangle$ state during an adiabatic increase in J.

通过增加 J 来增加，但是也以增加了耦合原子核–电子自旋激发的弛豫速率为代价）。单个自旋的运算速度是由 B_{ac} 的大小来决定的，当 $B_{ac} = 10^{-3}$ T 时相当于 75 kHz。

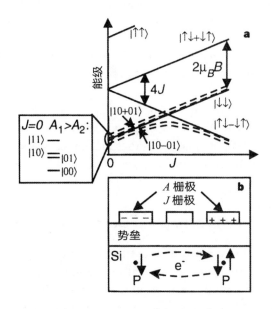

图 4. 两个量子比特的逻辑和自旋测量。**a**，电子（实线）和最低能量耦合电子–核自旋（虚线）能级随 J 的变化。当 $J < \mu_B B/2$，两个量子比特计算通过使用 J 栅极控制 $|10-01\rangle - |10+01\rangle$ 能级分裂来实现。在 $J = \mu_B B/2$ 上方，耦合系统的态演化为不同电子极化的态。原子核的具有很大能量分裂的（受 A 栅极偏压的控制）在 $J = 0$ 的态决定了在 J 绝热增加后，电子最终的自旋态。**b**，只有 $|\uparrow\downarrow-\downarrow\uparrow\rangle$ 电子才可以转换成电子与同一个施主结合的态（D^- 态）。这个转换过程的电子电流是可以通过电容技术来测量的，这就使得电子和原子核的基础自旋态可以被探测。

自 旋 测 量

在本文提出的量子计算机中，核自旋测量是分两步来实现的：不同的原子核自旋态绝热转化为不同电子极化的态，且电子自旋是由其对交换耦合双电子系统的轨道波函数对称性的影响来决定的。图 4 展示了如何实现这一转换的过程。当计算完成时 $J < \mu_B B/2$ 且电子被充分地极化，测量在 $J > \mu_B B/2$ 时进行，$|\uparrow\downarrow-\downarrow\uparrow\rangle$ 态有最低的能量（图 4a）。当电子能级交叉时，$|\downarrow\downarrow\rangle$ 态和 $|\uparrow\downarrow-\downarrow\uparrow\rangle$ 态就会通过与原子核产生超精细相互作用耦合。在 J 绝热增加过程中，当 $J > \mu_B B/2$ 时，两个处于 $J = 0$ 的低能量核自旋态演化为 $|\uparrow\downarrow-\downarrow\uparrow\rangle$ 态，然而两个高能级的核自旋态仍然保持 $|\downarrow\downarrow\rangle$ 态。如果在 $J = 0$ 时，$A_1 > A_2$，1 号原子核自旋的取向将独自决定系统在 J 的绝热增加过程中是演化为 $|\uparrow\downarrow-\downarrow\uparrow\rangle$ 态或者是 $|\downarrow\downarrow\rangle$ 态。

A method to detect the electron spin state by using electronic means is shown in Fig. 4b. Both electrons can become bound to the same donor (a D^- state) if the A gates above the donors are biased appropriately. In Si:P, the D^- state is always a singlet with a second electron binding energy of 1.7 meV (refs 31, 32). Consequently, a differential voltage applied to the A gates can result in charge motion between the donors that only occurs if the electrons are in a singlet state. This charge motion is measurable using sensitive single-electron capacitance techniques[33]. This approach to spin measurement produces a signal that persists until the electron spin relaxes, a time that, as noted above, can be thousands of seconds in Si:P.

The spin measurement process can also be used to prepare nuclear spins in a prescribed state by first determining the state of a spin and flipping it if necessary so that it ends up in the desired spin state. As with the spin computation procedures already discussed, spin measurement and preparation can in principle be performed in parallel throughout the computer.

Initializing the Computer

Before any computation, the computer must be initialized by calibrating the A gates and the J gates. Fluctuations from cell to cell in the gate biases necessary to perform logical operations are an inevitable consequence of variations in the positions of the donors and in the sizes of the gates. The parameters of each cell, however, can be determined individually using the measurement capabilities of the computer, because the measurement technique discussed here does not require precise knowledge of the J and A couplings. The A-gate voltage at which the underlying nuclear spin is resonant with an applied B_{ac} can be determined using the technique of adiabatic fast passage[34]: when $B_{ac} = 0$, the nuclear spin is measured and the A gate is biased at a voltage known to be off resonance. B_{ac} is then switched on, and the A gate bias is swept through a prescribed voltage interval. B_{ac} is then switched off and the nuclear spin is measured again. The spin will have flipped if, and only if, resonance occurred within the prescribed A-gate voltage range. Testing for spin flips in increasingly small voltage ranges leads to the determination of the resonance voltage. Once adjacent A gates have been calibrated, the J gates can be calibrated in a similar manner by sweeping J-gate biases across resonances of two coupled cells.

This calibration procedure can be performed in parallel on many cells, so calibration is not a fundamental impediment to scaling the computer to large sizes. Calibration voltages can be stored on capacitors located on the Si chip adjacent to the quantum computer. External controlling circuitry would thus need to control only the timing of gate biases, and not their magnitudes.

Spin Decoherence Introduced by Gates

In the quantum computer architecture outlined above, biasing of A gates and J gates

一种用来探测电子自旋态的方法是使用电子手段，如图 4b 所示。如果施主上方的 A 栅极被施加适当的偏压，两个电子可以都结合到同一个施主（D^- 态）。在 Si:P 中 D^- 态总是具有 1.7 meV 第二电子结合能的单重态 [31,32]。因此，施加一个差分电压于栅极 A 上可以导致只有处于单重态的电子可以在施主间产生电荷移动。这种电荷移动可以通过灵敏的单电子电容技术 [33] 来测量。这种测量自旋的方法产生的信号持续到电子自旋发生弛豫为止，如前所述，在 Si:P 系统中这种过程可以有几千秒的时间。

自旋的测量过程也可以用来将核自旋初始化到指定的态上：首先确定自旋的态，如果需要的话就翻转它使它停在想要的自旋态上。正如我们已经讨论的自旋计算过程，自旋的测量和准备原则上可以在计算机中并行执行。

初始化计算机

在进行任何计算之前，计算机必须通过定标 A 栅极和 J 栅极来实现初始化。施主位置以及栅极大小的不均匀必然造成不同单元执行逻辑运算所必需的栅极偏压的涨落，不过每个单元的参数都可以通过计算机的测量功能来独自地确定，因为这里讨论的测量技术不需要非常准确地了解 J 栅极和 A 栅极的耦合。A 栅极下面核自旋与外加 B_{ac} 共振时，所需要施加的 A 栅极电压可以通过绝热快速通过技术 [34] 来确定：当 $B_{ac} = 0$ 时，核自旋被测量，且 A 栅极被施加一个能消除共振的偏压。B_{ac} 然后被启动，在指定的电压区间内扫描 A 栅极偏压。然后关闭 B_{ac} 并再次测量核自旋。当且仅当共振在指定的 A 栅极电压范围内出现的时候，自旋才会被翻转。在越来越小的电压范围内测试自旋的翻转将能确定出共振的电压。当邻近的 A 栅极被定标之后，J 栅极可以用同样的方法在两个耦合单元的共振区扫描 J 栅极偏压来定标。

这种定标程序可以在多个单元上面并行进行，所以定标问题不是将计算机扩展到大尺寸的主要阻碍。定标电压可以存储在连接到量子计算机上的硅芯片的电容中。因此，外部控制电路只需要控制栅极偏压的时间，而不是他们的大小。

栅极导致的自旋退相干

在上面概述的量子计算机的设计结构中，对 A 类和 J 类栅极施加偏压可以自

enables custom control of the qubits and their mutual interactions. The presence of the gates, however, will lead to decoherence of the spins if the gate biases fluctuate away from their desired values. These effects need to be considered to evaluate the performance of any gate-controlled quantum computer. During the computation, the largest source of decoherence is likely to arise from voltage fluctuations on the A gates. (When $J < \mu_B B/2$, modulation of the state energies by the J gates is much smaller than by the A gates. J exceeds $\mu_B B/2$ only during the measurement process, when decoherence will inevitably occur.) The precession frequencies of two spins in phase at $t = 0$ depends on the potentials on their respective A gates. Differential fluctuations of the potentials produce differences in the precession frequency. At some later time $t = t_\phi$, the spins will be 180° out of phase; t_ϕ can be estimated by determining the transition rate between $|10+01\rangle$ (spins in phase) and $|10-01\rangle$ (spins 180° out of phase) of a two-spin system. The hamiltonian that couples these states is $H_\phi = \frac{1}{4} h\Delta(\sigma_z^{1n} - \sigma_z^{2n})$, where Δ is the fluctuating differential precession frequency of the spins. Standard treatment of fluctuating hamiltonians[34] predicts: $t_\phi^{-1} = \pi^2 S_\Delta(\nu_{st})$, where S_Δ is the spectral density of the frequency fluctuations, and ν_{st} is the frequency difference between the $|10-01\rangle$ and $|10+01\rangle$ states. At a particular bias voltage, the A gates have a frequency tuning parameter $\alpha = d\Delta/dV$. Thus:

$$t_\phi^{-1} = \pi^2 \alpha^2(V) S_V(\nu_{st}) \tag{4}$$

where S_V is the spectral density of the gate voltage fluctuations.

S_V for good room temperature electronics is of order 10^{-18} V^2/Hz, comparable to the room temperature Johnson noise of a 50-Ω resistor. The value of α, estimated from Fig. 2, is 10–100 MHz V^{-1}, yielding $t_\phi = 10$–1,000 s; α is determined by the size of the donor array cells and cannot readily be reduced (to increase t_ϕ) without reducing the exchange interaction between cells. Because α is a function of the gate bias (Fig. 2), t_ϕ can be increased by minimizing the voltage applied to the A gates.

Although equation (4) is valid for white noise, at low frequencies it is likely that materials-dependent fluctuations ($1/f$ noise) will be the dominant cause of spin dephasing. Consequently, it is difficult to give hard estimates of t_ϕ for the computer. Charge fluctuations within the computer (arising from fluctuating occupancies of traps and surface states, for example) are likely to be particularly important, and minimizing them will place great demands on computer fabrication.

Although materials-dependent fluctuations are difficult to estimate, the low-temperature operations of the computer and the dissipationless nature of quantum computing mean that, in principle, fluctuations can be kept extremely small: using low-temperature electronics to bias the gates (for instance, by using on chip capacitors as discussed above) could produce $t_\phi \approx 10^6$ s. Electronically controlled nuclear spin quantum computers thus have the theoretical capability to perform at least 10^5 to perhaps 10^{10} logical operations during t_ϕ, and can probably meet Preskill's criterion[8] for an error probability of 10^{-6} per qubit operation.

定义地控制量子比特以及他们之间的相互作用。然而如果栅极的偏压振荡出期望的范围，栅极的存在将导致自旋退相干。这些效应应该在任何通过栅极控制来实现的量子计算机的性能评价中被考虑到。在计算过程中，导致退相干的主要来源一般是 A 栅极上的电压振荡。（当 $J < \mu_B B/2$ 时，态能量被 J 栅极调制的幅度远小于被 A 栅极调制的幅度。J 超过 $\mu_B B/2$ 只会发生在测量过程，那时退相干是不可避免的。）在 $t = 0$ 时刻同相位中的两个自旋的进动频率取决于他们各自 A 栅极上的电势。这两个电势的差值波动引起进动频率的相对变化。在稍后的 $t = t_\phi$ 时刻，这两个自旋的相位将会相差 180 度（反相位）；t_ϕ 可以通过这个双自旋系统在 $|10+01\rangle$（自旋同相位）和 $|10-01\rangle$（自旋 180 度反相位）两种状态间的转换速率来确定。这些态之间耦合的哈密顿量为 $H_\phi = \frac{1}{4}h\Delta(\sigma_z^{1n} - \sigma_z^{2n})$，其中 Δ 是不同自旋进动频率差值的振荡。对该振荡哈密顿量进行标准处理 [34] 导出：$t_\phi^{-1} = \pi^2 S_\Delta(\nu_{st})$，其中 S_Δ 是频率振荡的频谱密度，ν_{st} 是 $|10-01\rangle$ 和 $|10+01\rangle$ 态之间的频率差。在一个固定的偏压下，A 类栅极有一个频率调谐参数 $\alpha = d\Delta/dV$。因此

$$t_\phi^{-1} = \pi^2 \alpha^2(V) S_V(\nu_{st}) \tag{4}$$

其中 S_V 是栅极电压涨落的频谱密度。

良好的室温电子元件的 S_V 的量级是 10^{-18} V²/Hz，这和一个 50 Ω 的电阻的室温约翰逊噪音是接近的。从图 2 中估算出的 α 值大概为 $10 \sim 100$ MHz·V⁻¹，导出 $t_\phi = 10 \sim 1,000$ s；α 可以由施主阵列单元的大小来确定，且不能在不减小单元之间的相互作用的情况下被轻易地减小（来增加 t_ϕ）。因为 α 是栅极偏压的函数（图 2），t_ϕ 可以通过最小化 A 类栅极偏压来增加。

虽然等式 4 对白噪声来说是正确的，在低频率条件下，导致自旋退相的主要原因很可能是依赖于材料的波动（1/f 噪音）。因此给出计算机一个准确的预期值 t_ϕ 是非常困难的。计算机内部的电荷涨落（例如由陷阱和表面态的占位数涨落引起）很可能是非常重要的，这就要求在计算机制造中最小化这种电荷涨落。

虽然依赖材料的波动很难预期，在低温条件下的计算机运算以及量子计算无耗散的特征意味着，原则上波动可以被保持得很小：使用低温电子器件来给栅极施加偏压（例如之前讨论中提到的使用芯片电容）可以使得 t_ϕ 约等于 10^6 s。电子器件控制的核自旋量子计算机因此理论上具备了在 t_ϕ 时间内执行 10^5 到 10^{10} 逻辑运算的能力，这样基本上就满足了普雷斯基尔的每个量子比特运算误码率为 10^{-6} 的准则 [8]。

Constructing the Computer

Building the computer presented here will obviously be an extraordinary challenge: the materials must be almost completely free of spin ($I \neq 0$ isotopes) and charge impurities to prevent dephasing fluctuations from arising within the computer. Donors must be introduced into the material in an ordered array hundreds of Å beneath the surface. Finally, gates with lateral dimensions and separations ~100 Å must be patterned on the surface, registered to the donors beneath them. Although it is possible that the computer can use SiO_2 as the barrier material (the standard MOS technology used in most current conventional electronics), the need to reduce disorder and fluctuations to a minimum means that heteroepitaxial materials, such as Si/SiGe, may ultimately be preferable to Si/SiO_2.

The most obvious obstacle to building to the quantum computer presented above is the incorporation of the donor array into the Si layer beneath the barrier layer. Currently, semiconductor structures are deposited layer by layer. The δ-doping technique produces donors lying on a plane in the material, with the donors randomly distributed within the plane. The quantum computer envisaged here requires that the donors be placed into an ordered one- or two-dimensional array; furthermore, precisely one donor must be placed into each array cell, making it extremely difficult to create the array by using lithography and ion implantation or by focused deposition. Methods currently under development to place single atoms on surfaces using ultra-high-vacuum scanning tunnelling microscopy[35] or atom optics techniques[36] are likely candidates to be used to position the donor array. A challenge will be to grow high-quality Si layers on the surface subsequent to placement of the donors.

Fabricating large arrays of donors may prove to be difficult, but two-spin devices, which can be used to test the logical operations and measurement techniques presented here, can be made using random doping techniques. Although only a small fraction of such devices will work properly, adjacent conventional Si electronic multiplexing circuitry can be used to examine many devices separately. The relative ease of fabricating such "hybrid"(quantum-conventional) circuits is a particularly attractive feature of Si-based quantum computation.

In a Si-based nuclear spin quantum computer, the highly coherent quantum states necessary for quantum computation are incorporated into a material in which the ability to implement complex computer architectures is well established. The substantial challenges facing the realization of the computer, particularly in fabricating 100-Å-scale gated devices, are similar to those facing the next generation of conventional electronics; consequently, new manufacturing technologies being developed for conventional electronics will bear directly on efforts to develop a quantum computer in Si. Quantum computers sufficiently complex that they can achieve their theoretical potential may thus one day be built using the same technology that is used to produce conventional computers.

(**393**, 133-137; 1998)

构造计算机

构造一个这里讨论的计算机显然将是非常巨大的挑战：选取的材料必须完全没有自旋（$I \neq 0$ 同位素）和电荷杂质，以避免计算机中出现退相位扰动。施主必须作为一个有序的阵列导入材料表面几百埃以下。最后，横向尺寸和间距为 100 Å 左右的栅极被加工在表面，与其下方的施主对应。虽然使用二氧化硅（大多数当前传统电子器件所使用的标准 MOS 技术）作为势垒层材料是可行的，但是由于需要最小化无序和扰动，意味着异质外延材料如硅/硅锗（Si/SiGe）最终可能优于硅/二氧化硅（Si/SiO_2）。

构造上文描述的量子计算机最明显的困难是如何将施主阵列注入势垒层下方的硅层中。现阶段半导体结构是逐层沉积制造的。δ 掺杂技术得到的施主位于材料中的单层中，其中施主是随机分布在这一层中的。我们这里设想的量子计算机要求施主被放置于有序的一维或者二维的阵列中；另外，施主必须被准确地放置在每个阵列单元中，这就使得通过光刻技术和离子注入或者通过聚焦沉积方法都很难制造出这种阵列。目前正在发展的将单原子放在表面的技术是利用超高真空扫描隧道显微镜 [35] 或者原子光学技术 [36] 实现的，这些是可能被用来放置施主阵列的备选方法。接着的一个挑战将是在施主放置处表面上生长高质量的硅层。

制造大的施主阵列也许是非常困难的，但是用来检验此处介绍的逻辑计算和测量技术的双自旋器件可以使用随机掺杂技术来制造。虽然只有一小部分这样的器件可以正常工作，但是相关的传统硅电子学中的复用电路可以用来独立地检测许多器件。制造这种"混合"（量子–传统）电路的相对简易性是硅基量子计算机的独特魅力所在。

在基于硅的核自旋量子计算机中，量子计算所必需的高度相干的量子态被整合到能够实现复杂计算机结构的材料中。实现量子计算机面临的重要挑战，尤其是 100 Å 尺度的栅极器件制造，与下一代传统电子器件面临的挑战类似；因此，正为传统电子器件研发的新制造技术将与开发硅量子计算机的努力直接相关。足够复杂以至于可以实现其理论潜力的量子计算机，也许会在将来的某一天使用传统计算机类似的技术搭建起来。

（姜克 翻译；杜江峰 审稿）

B. E. Kane

Semiconductor Nanofabrication Facility, School of Physics, University of New South Wales, Sydney 2052, Australia

Received 10 November 1997; accepted 24 February 1998.

References:

1. Steane, A. Quantum computing. *Rep. Prog. Phys.* **61,** 117-173 (1998).

2. Bennett, C. H. Quantum information and computation. *Physics Today* 24-30 (Oct. 1995).

3. Shor, P. W. in *Proc. 35th Annu. Symp. Foundations of Computer Science* (ed. Goldwasser, S.) 124-134 (IEEE Computer Society, Los Alamitos, CA, 1994).

4. Ekert, A. & Jozsa, R. Quantum computation and Shor's factoring algorithm. *Rev. Mod. Phys.* **68,** 733-753 (1996).

5. Grover, L. K. Quantum mechanics helps in searching for a needle in a haystack. *Phys. Rev. Lett.* **79,** 325-328 (1997).

6. Calderbank, A. R. & Shor, P. W. Good quantum error correcting codes exist. *Phys. Rev. A* **54,** 1098-1105 (1996).

7. Steane, A. M. Error correcting codes in quantum theory. *Phys. Rev. Lett.* **77,** 793-797 (1996).

8. Preskill, J. Reliable quantum computers. *Proc. R. Soc. Lond. A* **454,** 385-410 (1998).

9. Lloyd, S. A potentially realizable quantum computer. *Science* **261,** 1569-1571 (1993).

10. DiVincenzo, D. P. Quantum computation. *Science* **270,** 255-261 (1995).

11. Gershenfeld, N. A. & Chuang, I. L. Bulk spin-resonance quantum computation. *Science* **275,** 350-356 (1997).

12. Cory, D. G., Fahmy, A. F. & Havel, T. F. Ensemble quantum computing by NMR spectroscopy. *Proc. Natl Acad. Sci. USA* **94,** 1634-1639 (1997).

13. Loss, D. & DiVincenzo, D. P. Quantum computation with quantum dots. *Phys. Rev. A* **57,** 120-126 (1998).

14. Privman, V., Vagner, I. D. & Kventsel, G. Quantum computation in quantum Hall systems. *Phys. Lett. A* **239,** 141-146 (1998).

15. Slichter, C. P. *Principles of Magnetic Resonance* 3rd edn, Ch 4 (Springer, Berlin, 1990).

16. Dobers, M., Klitzing, K. v., Schneider, J., Weimann, G. & Ploog, K. Electrical detection of nuclear magnetic resonance in GaAs-Al$_x$Ga$_{1-x}$As heterostructures. *Phys. Rev. Lett.* **61,** 1650-1653 (1988).

17. Stich, B., Greulich-Weber, S. & Spaeth, J.-M. Electrical detection of electron nuclear double resonance in silicon. *Appl. Phys. Lett.* **68,** 1102-1104 (1996).

18. Kane, B. E., Pfeiffer, L. N. & West, K. W. Evidence for an electric-field-induced phase transition in a spin-polarized two-dimensional electron gas. *Phys. Rev. B* **46,** 7264-7267 (1992).

19. Wald, K. W., Kouwenhoven, L. P., McEuen, P. L., van der Vaart, N. C. & Foxon, C. T. Local dynamic nuclear polarization using quantum point contacts. *Phys. Rev. Lett.* **73,** 1011-1014 (1994).

20. Dixon, D. C., Wald, K. R., McEuen, P. L. & Melloch, M. R. Dynamic polarization at the edge of a two-dimensional electron gas. *Phys. Rev. B* **56,** 4743-4750 (1997).

21. *CRC Handbook of Chemistry and Physics* 77th edn 11-38 (CRC Press, Boca Raton, Florida, 1996).

22. Feher, G. Electron spin resonance on donors in silicon. I. Electronic structure of donors by the electron nuclear double resonance technique. *Phys. Rev.* **114,** 1219-1244 (1959).

23. Wilson, D. K. & Feher, G. Electron spin resonance experiments on donors in silicon. III. Investigation of excited states by the application of uniaxial stress and their importance in relaxation processes. *Phys. Rev.* **124,** 1068-1083 (1961).

24. Waugh, J. S. & Slichter, C. P. Mechanism of nuclear spin-lattice relaxation in insulators at very low temperatures. *Phys. Rev. B* **37,** 4337-4339 (1988).

25. Kohn, W. *Solid State Physics* Vol. 5 (eds Seitz, F. & Turnbull, D.) 257-320 (Academic, New York, 1957).

26. DiVincenzo, D. P. Two-bit gates are universal for quantum computation. *Phys. Rev. A* **51,** 1015-1021 (1995).

27. Lloyd, S. Almost any quantum logic gate is universal. *Phys. Rev. Lett.* **75,** 346-349 (1995).

28. Herring, C. & Flicker, M. Asymptotic exchange coupling of two hydrogen atoms. *Phys. Rev.* **134,** A362–A366 (1964).

29. Andres, K., Bhatt, R. N., Goalwin, P., Rice, T. M. & Walstedt, R. E. Low-temperature magnetic susceptibility of Si:P in the nonmetallic region. *Phys. Rev. B* **24,** 244-260 (1981).

30. Ashcroft, N. W. & Mermin, N. D. in *Solid State Physics* Ch. 32 (Saunders College, Philadelphia, 1976).

31. Larsen, D. M. Stress dependence of the binding energy of D^- centers in Si. *Phys. Rev. B* **23,** 5521-5526 (1981).

32. Larsen, D. M. & McCann, S. Y. Variational studies of two- and three-dimensional D^- centers in magnetic fields. *Phys. Rev. B* **46,** 3966-3970 (1992).

33. Ashoori, R. C. Electrons in artificial atoms. *Nature* **379,** 413-419 (1996).

34. Abragam, A. *Principles of Nuclear Magnetism* (Oxford Univ. Press, London, 1961).

35. Lyding, J. W. UHV STM nanofabrication: progress, technology spin-offs, and challenges. *Proc. IEEE* **85,** 589-600 (1997).

36. Adams, C. S., Sigel, J. & Mlynek, J. Atom optics. *Phys. Rep.* **240,** 143-210 (1994).

37. Warren, W. S. The usefulness of NMR quantum computing. *Science* **277,** 1688-1690 (1997).

Acknowledgements. This work has been supported by the Australian Research Council. I thank R. G. Clark for encouragement and E. Hellman for suggesting that the work in ref. 18 could be relevant to quantum computation.

Correspondence should be addressed to the author (e-mail: kane@newt.phys.unsw.edu.au).

Collective Dynamics of "Small-world" Networks

D. J. Watts and S. H. Strogatz

Editor's Note

We live in a world of networks, from human social networks and biological food webs to neural systems, transportation networks and the modern Internet. In this paper, Duncan Watts and Steven Strogatz demonstrate that these networks often share surprising architectural similarities. Most notably, they are "small worlds", in that only a few steps are needed to move along links from any one point to another, even in networks containing an enormous number of elements. Such networks also tend to be highly "clustered", with many redundant paths, making them connected wholes that are resilient to the removal of links. This paper stimulated an explosion of work on complex networks by providing a new conceptual framework for analysing them.

Networks of coupled dynamical systems have been used to model biological oscillators[1-4], Josephson junction arrays[5,6], excitable media[7], neural networks[8-10], spatial games[11], genetic control networks[12] and many other self-organizing systems. Ordinarily, the connection topology is assumed to be either completely regular or completely random. But many biological, technological and social networks lie somewhere between these two extremes. Here we explore simple models of networks that can be tuned through this middle ground: regular networks "rewired" to introduce increasing amounts of disorder. We find that these systems can be highly clustered, like regular lattices, yet have small characteristic path lengths, like random graphs. We call them "small-world" networks, by analogy with the small-world phenomenon[13,14] (popularly known as six degrees of separation[15]). The neural network of the worm *Caenorhabditis elegans*, the power grid of the western United States, and the collaboration graph of film actors are shown to be small-world networks. Models of dynamical systems with small-world coupling display enhanced signal-propagation speed, computational power, and synchronizability. In particular, infectious diseases spread more easily in small-world networks than in regular lattices.

TO interpolate between regular and random networks, we consider the following random rewiring procedure (Fig. 1). Starting from a ring lattice with n vertices and k edges per vertex, we rewire each edge at random with probability p. This construction allows us to "tune" the graph between regularity ($p = 0$) and disorder ($p = 1$), and thereby to probe the intermediate region $0 < p < 1$, about which little is known.

"小世界"网络的集体动力学

沃茨，斯特罗加茨

编者按

我们生活在一个网络世界里：从人类社会网络和生物食物网到神经系统、运输网络和现代互联网。本文中，邓肯·沃茨和史蒂文·斯特罗加茨证明了这些网络通常具有惊人的结构相似性。最值得注意的是，它们是"小世界"，因为即使在包含大量元素的网络中，也只需要几个步骤就可以沿着链路从一个点移动到另一个点。这种网络也倾向于高度"集群化"，许多冗余路径使得它们连接在一起成为一个整体，从而对链路的移除具有恢复力。本文提供了一个新的概念框架来分析复杂网络，促进了复杂网络研究工作的爆炸式增长。

耦合动力系统网络已被广泛应用于生物振荡 [1-4]、约瑟夫森结阵列 [5,6]、可激发介质 [7]、神经网络 [8-10]、空间博弈 [11]、基因控制网络 [12] 以及多种其他自组织系统的建模。通常情况下，其拓扑结构被认为或是完全规则的，或是完全随机的。但是许多生物、技术和社会的网络介于这两种极端状况之间。本文阐述了一种处于这种中间态的简单的网络模型：将规则网络"重新连接"来引入更多无序性。我们发现这些系统可以像有序晶格一样被高度集聚，同时具有很小的特征路径长度，类似随机图。通过与小世界现象 [13,14]（人们所熟知的六度分隔 [15]）的类比，我们称它们为"小世界"网络。秀丽隐杆线虫的神经网络、美国西部的电力网络以及电影演员的合作图都表现为小世界网络。具有小世界耦合的动力系统模型显示了更强的信号传播速度、计算能力和同步性。特别需要指出的是，传染病在小世界网络中传播要比在规则网络中传播更容易。

为了实现规则网络和随机网络之间的中间状态，我们来考虑以下的随机重连过程（图 1）。从一个具有 n 个顶点，每个顶点 k 条边的环形晶格网络开始，我们以概率 p 随机地将每条边重新连接。这种构造允许我们在规则（$p=0$）和无序（$p=1$）之间"调谐"，从而探索目前被了解得很少的 $0<p<1$ 的中间区域。

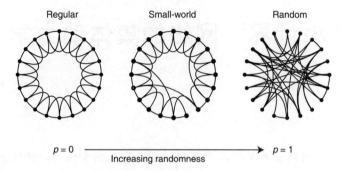

Fig. 1. Random rewiring procedure for interpolating between a regular ring lattice and a random network, without altering the number of vertices or edges in the graph. We start with a ring of n vertices, each connected to its k nearest neighbours by undirected edges. (For clarity, $n = 20$ and $k = 4$ in the schematic examples shown here, but much larger n and k are used in the rest of this Letter.) We choose a vertex and the edge that connects it to its nearest neighbour in a clockwise sense. With probability p, we reconnect this edge to a vertex chosen uniformly at random over the entire ring, with duplicate edges forbidden; otherwise we leave the edge in place. We repeat this process by moving clockwise around the ring, considering each vertex in turn until one lap is completed. Next, we consider the edges that connect vertices to their second-nearest neighbours clockwise. As before, we randomly rewire each of these edges with probability p, and continue this process, circulating around the ring and proceeding outward to more distant neighbours after each lap, until each edge in the original lattice has been considered once. (As there are $nk/2$ edges in the entire graph, the rewiring process stops after $k/2$ laps.) Three realizations of this process are shown, for different values of p. For $p = 0$, the original ring is unchanged; as p increases, the graph becomes increasingly disordered until for $p = 1$, all edges are rewired randomly. One of our main results is that for intermediate values of p, the graph is a small-world network: highly clustered like a regular graph, yet with small characteristic path length, like a random graph. (See Fig. 2.)

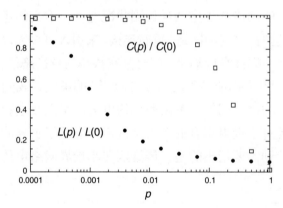

Fig. 2. Characteristic path length $L(p)$ and clustering coefficient $C(p)$ for the family of randomly rewired graphs described in Fig. 1. Here L is defined as the number of edges in the shortest path between two vertices, averaged over all pairs of vertices. The clustering coefficient $C(p)$ is defined as follows. Suppose that a vertex v has k_v neighbours; then at most $k_v(k_v-1)/2$ edges can exist between them (this occurs when every neighbour of v is connected to every other neighbour of v). Let C_v denote the fraction of these allowable edges that actually exist. Define C as the average of C_v over all v. For friendship networks, these statistics have intuitive meanings: L is the average number of friendships in the shortest chain connecting two people; C_v reflects the extent to which friends of v are also friends of each other; and thus C measures the cliquishness of a typical friendship circle. The data shown in the figure are averages over 20 random realizations of the rewiring process described in Fig. 1, and have been normalized by the values $L(0)$, $C(0)$ for a regular lattice. All the graphs have $n = 1,000$ vertices and an average degree of $k = 10$ edges per vertex. We note that a logarithmic horizontal scale has been used to resolve the rapid drop in $L(p)$, corresponding

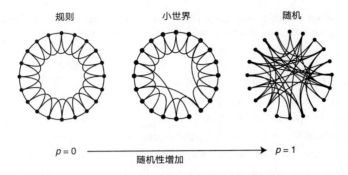

规则　　　　　小世界　　　　　随机

$p = 0$ ———————————→ $p = 1$
随机性增加

图 1. 随机重新连接来制造处于规则环形晶格网络和随机网络之间的状态，而顶点和边的数量不变。我们从一个具有 n 个顶点的环开始，每个顶点都与最近的 k 个相邻顶点通过无向边相连。（为了表述清晰，图中的例子取 $n = 20$，$k = 4$，但在文中的其他部分我们取了更大的 n 和 k 值。）我们选取一个顶点以及顺时针方向上连接该顶点和最近邻点的边。以概率 p 将这条边重新连接到整个环上随机选取的任一顶点，不允许有重复连接的边；否则保持这条边不动。我们沿环的顺时针方向重复这个过程，按顺序考虑到每个顶点直到一圈完成。接下来，我们考虑连接顶点与其顺时针方向第二近邻点的边。像上次操作一样，我们以概率 p 随机地重新连接这些边，然后我们重复这个过程，沿环循环，并在每一圈之后对连接更远近邻点的边继续该操作，直到每条原始网络中的边都被考虑到为止。（因为整个图中共有 $nk/2$ 条边，所以重新连接过程在 $k/2$ 圈后结束。）图中表示了不同 p 时，该过程的三种实现方式。$p = 0$ 时，原始环不变；随 p 增大图形无序性开始增加，到 $p = 1$ 时，所有的边都被随机地重新连接。我们得出的一个主要结论是，对于 p 处于中间值时，图为一小世界网络：像规则图一样高度集聚，又像随机图一样具有很小的特征路径长度。（见图 2）

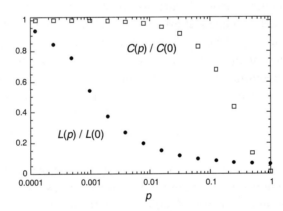

图 2. 特征路径长度 $L(p)$ 与图 1 中所描述的随机重连图形的集聚系数 $C(p)$ 的关系。L 被定义为两顶点之间最短路径上的边的数量，为所有顶点对的平均值。集聚系数 $C(p)$ 按下述定义。假设一顶点 v 有 k_v 个邻点；那么至多有 $k_v(k_v - 1)/2$ 条边连接它们（这种情况发生在 v 的每个邻点都与 v 的所有其他邻点相连）。用 C_v 来表示真正存在的边占这些所有被允许出现的边的比例。定义 C 为所有的顶点 v 对应 C_v 的平均值。对于朋友关系网络，这些统计数据有着直观的含义：L 为两个人之间最短链中朋友关系的平均数；C_v 反映 v 的朋友中有多少互相之间也是朋友；因而 C 测量朋友关系圈的小集团性。图中所示的数据为图 1 中所描述的 20 种随机重连过程的平均值，并相对规则晶格网络的 $L(0)$ 和 $C(0)$ 进行了归一化。所有的图形都有 $n = 1,000$ 个顶点，每个顶点平均有 $k = 10$ 条边。我们使用对数横坐标以解决 $L(p)$ 骤降

to the onset of the small-world phenomenon. During this drop, $C(p)$ remains almost constant at its value for the regular lattice, indicating that the transition to a small world is almost undetectable at the local level.

We quantify the structural properties of these graphs by their characteristic path length $L(p)$ and clustering coefficient $C(p)$, as defined in Fig. 2 legend. Here $L(p)$ measures the typical separation between two vertices in the graph (a global property), whereas $C(p)$ measures the cliquishness of a typical neighbourhood (a local property). The networks of interest to us have many vertices with sparse connections, but not so sparse that the graph is in danger of becoming disconnected. Specifically, we require $n \gg k \gg \ln(n) \gg 1$, where $k \gg \ln(n)$ guarantees that a random graph will be connected[16]. In this regime, we find that $L \sim n/2k \gg 1$ and $C \sim 3/4$ as $p \to 0$, while $L \approx L_{random} \sim \ln(n)/\ln(k)$ and $C \approx C_{random} \sim k/n \ll 1$ as $p \to 1$. Thus the regular lattice at $p = 0$ is a highly clustered, large world where L grows linearly with n, whereas the random network at $p = 1$ is a poorly clustered, small world where L grows only logarithmically with n. These limiting cases might lead one to suspect that large C is always associated with large L, and small C with small L.

On the contrary, Fig. 2 reveals that there is a broad interval of p over which $L(p)$ is almost as small as L_{random} yet $C(p) \gg C_{random}$. These small-world networks result from the immediate drop in $L(p)$ caused by the introduction of a few long-range edges. Such "short cuts" connect vertices that would otherwise be much farther apart than L_{random}. For small p, each short cut has a highly nonlinear effect on L, contracting the distance not just between the pair of vertices that it connects, but between their immediate neighbourhoods, neighbourhoods of neighbourhoods and so on. By contrast, an edge removed from a clustered neighbourhood to make a short cut has, at most, a linear effect on C; hence $C(p)$ remains practically unchanged for small p even though $L(p)$ drops rapidly. The important implication here is that at the local level (as reflected by $C(p)$), the transition to a small world is almost undetectable. To check the robustness of these results, we have tested many different types of initial regular graphs, as well as different algorithms for random rewiring, and all give qualitatively similar results. The only requirement is that the rewired edges must typically connect vertices that would otherwise be much farther apart than L_{random}.

The idealized construction above reveals the key role of short cuts. It suggests that the small-world phenomenon might be common in sparse networks with many vertices, as even a tiny fraction of short cuts would suffice. To test this idea, we have computed L and C for the collaboration graph of actors in feature films (generated from data available at http://us.imdb.com), the electrical power grid of the western United States, and the neural network of the nematode worm *C. elegans*[17]. All three graphs are of scientific interest. The graph of film actors is a surrogate for a social network[18], with the advantage of being much more easily specified. It is also akin to the graph of mathematical collaborations centred, traditionally, on P. Erdös (partial data available at http://www.acs.oakland.edu/~grossman/erdoshp.html). The graph of the power grid is relevant to the efficiency and robustness of power networks[19]. And *C. elegans* is the sole example of a completely mapped neural network.

的问题，对应小世界现象的开始。在这个骤降过程中，规则晶格网络的 $C(p)$ 基本保持不变，也就意味着网络结构向小世界的转变在局域上是无法探测的。

通过它们的特征路径长度 $L(p)$ 和集聚系数 $C(p)$（图 2 注给出了这两项的定义），我们将这些图形的结构属性进行量化。这里 $L(p)$ 测量图形中两顶点之间的分隔（一种全局属性），而 $C(p)$ 测量近邻顶点间的小集团性（一种局域属性）。我们感兴趣的网络有很多顶点但连接很稀疏，但是没有稀疏到存在使图断开的危险。具体地讲，我们要求 $n \gg k \gg \ln(n) \gg 1$，其中 $k \gg \ln(n)$ 确保了随机图会被连接 [16]。在这种机制下，我们发现当 $p \to 0$ 时，$L \sim n/2k \gg 1$，$C \sim 3/4$，而当 $p \to 1$ 时，$L \approx L_{随机} \sim \ln(n)/\ln(k)$，$C \approx C_{随机} \sim k/n \ll 1$。因此在 $p = 0$ 时的规则晶格网络是一高度群聚、L 随 n 线性增加的大世界，而 $p = 1$ 时的随机网络是一低群聚、L 只随 n 对数增长的小世界。这些带有限制性的情况很可能让人们认为大的 C 永远与大 L 相关，而小 C 与小 L 相关。

相反地，图 2 揭示了在 $L(p)$ 小到与 $L_{随机}$ 接近，但 $C(p) \gg C_{随机}$ 时，p 存在一个很宽的跨度。这些小世界网络源于由若干远程边的引入而引起的 $L(p)$ 的骤降。这样的"捷径"连接起了原本要相隔较 $L_{随机}$ 远很多的顶点。对于小的 p 来说，每一条捷径都产生很强的非线性效应作用于 L，不仅仅缩短了它所连接的顶点之间的距离，而且还缩短了它们与直接邻点之间的距离，以及与邻点的邻点之间的距离，等等。相比之下，从群聚的邻点中移除用于制造捷径的一条边，最多产生线性效应作用于 C；因此对于小的 p 即使 $L(p)$ 迅速减小，$C(p)$ 也仍保持不变。这里的一个很重要的启示是，就局部而言（通过 $C(p)$ 所反应的），向小世界的转变几乎无法探测。为了检验这些结果的稳健性，我们测试了多种不同类型的初始规则图，以及随机重连的不同算法，所有的测试都给出性质类似的结果。唯一的要求是重新连接的边通常必须连接原本比 $L_{随机}$ 远很多的顶点。

以上理想的构造揭示了捷径的核心作用。它意味着小世界现象在多顶点的稀疏网络中很普遍，很少的捷径就满足要求。为了验证这个想法，我们计算了故事片电影中演员的合作图（数据由 http://us.imdb.com 提供）、美国西部的电力网络以及秀丽隐杆线虫神经网络 [17] 这三个图的 L 和 C。所有这三个图都引起了科学界的兴趣。电影演员合作图是社会网络的代表 [18]，具有很容易被细化定义的优势。它也与传统上的、以保罗·埃尔德什为中心的数学合作图相似（部分数据可以在 http://www.acs.oakland.edu/~grossman/erdoshp.html 上获得）。电力网络的图与电力网络的效率和稳定性相关 [19]。而秀丽隐杆线虫是完整的神经网络的唯一例子。

Table 1 shows that all three graphs are small-world networks. These examples were not hand-picked; they were chosen because of their inherent interest and because complete wiring diagrams were available. Thus the small-world phenomenon is not merely a curiosity of social networks[13,14] nor an artefact of an idealized model—it is probably generic for many large, sparse networks found in nature.

Table 1. Empirical examples of small-world networks

	L_{actual}	L_{random}	C_{actual}	C_{random}
Film actors	3.65	2.99	0.79	0.00027
Power grid	18.7	12.4	0.080	0.005
C. elegans	2.65	2.25	0.28	0.05

Characteristic path length L and clustering coefficient C for three real networks, compared to random graphs with the same number of vertices (n) and average number of edges per vertex (k). (Actors: $n = 225,226$, $k = 61$. Power grid: $n = 4,941$, $k = 2.67$. C. elegans: $n = 282$, $k = 14$.) The graphs are defined as follows. Two actors are joined by an edge if they have acted in a film together. We restrict attention to the giant connected component[16] of this graph, which includes ~90% of all actors listed in the Internet Movie Database (available at http://us.imdb.com), as of April 1997. For the power grid, vertices represent generators, transformers and substations, and edges represent high-voltage transmission lines between them. For C. elegans, an edge joins two neurons if they are connected by either a synapse or a gap junction. We treat all edges as undirected and unweighted, and all vertices as identical, recognizing that these are crude approximations. All three networks show the small-world phenomenon: $L \gtrsim L_{random}$ but $C \gg C_{random}$.

We now investigate the functional significance of small-world connectivity for dynamical systems. Our test case is a deliberately simplified model for the spread of an infectious disease. The population structure is modelled by the family of graphs described in Fig. 1. At time $t = 0$, a single infective individual is introduced into an otherwise healthy population. Infective individuals are removed permanently (by immunity or death) after a period of sickness that lasts one unit of dimensionless time. During this time, each infective individual can infect each of its healthy neighbours with probability r. On subsequent time steps, the disease spreads along the edges of the graph until it either infects the entire population, or it dies out, having infected some fraction of the population in the process.

Two results emerge. First, the critical infectiousness r_{half}, at which the disease infects half the population, decreases rapidly for small p (Fig. 3a). Second, for a disease that is sufficiently infectious to infect the entire population regardless of its structure, the time $T(p)$ required for global infection resembles the $L(p)$ curve (Fig. 3b). Thus, infectious diseases are predicted to spread much more easily and quickly in a small world; the alarming and less obvious point is how few short cuts are needed to make the world small.

Our model differs in some significant ways from other network models of disease spreading[20-24]. All the models indicate that network structure influences the speed and extent of disease transmission, but our model illuminates the dynamics as an explicit function of structure (Fig. 3), rather than for a few particular topologies, such as random graphs, stars and chains[20-23]. In the work closest to ours, Kretschmar and Morris[24] have shown

表1显示所有三个图都是小世界网络。这些例子不是特别挑选出来的；它们之所以被选择是因为它们本身具有吸引力，而且它们的连接完备清晰。因此小世界现象不只是社会网络的奇异现象[13,14]，也不是一个理想模型的人为性质——它很可能是自然中发现的许多大的、稀疏的网络的普遍现象。

表1. 小世界网络的实际举例

	$L_{真实}$	$L_{随机}$	$C_{真实}$	$C_{随机}$
电影演员	3.65	2.99	0.79	0.00027
电力网络	18.7	12.4	0.080	0.005
秀丽隐杆线虫	2.65	2.25	0.28	0.05

三个真实网络的特征路径长度 L 和集聚系数 C，与具有相同的顶点数（n）及每顶点的平均边数（k）的随机图的比较。（电影演员：$n=225,226$，$k=61$。电力网络：$n=4,941$，$k=2.67$。秀丽隐杆线虫：$n=282$，$k=14$。）图的定义如下。两演员如果出演过同一电影，他们即被一边连接。我们仅关注其中的最大连通子集[16]，其中包含了截至1997年4月网络电影数据库（源自 http://us.imdb.com）所列出的所有演员的约90%。对于电力网络，顶点代表发电机、变压器和变电站，边代表他们之间的高压电线。对于秀丽隐杆线虫，如果两神经元通过突触或者缝隙连接相连，则用边将二者连接。我们假设所有的边无方向无加权，所有的顶点都是相同的，所以这只是一种粗糙的近似。所有三个网络显示出小世界现象：$L \gtrsim L_{随机}$ 而 $C \gg C_{随机}$。

我们现在来研究以小世界方式连接的动力学系统的功能意义。我们的测试案例是被有意简化的一传染性疾病的传播模型。人口的相互作用结构用图1所描述的一系列图给出。在时间 $t=0$ 时，一感染性个体被引入原本健康的人群。被感染的个体在患病经过一无量纲时间单位后被永久性移除（通过免疫或死亡）。在这段时间内，每个被感染的个体以概率 r 传染每个健康的邻居。在随后的时间段，疾病沿图的边传播直到把疾病传染给所有的人，或者疾病传染一部分人之后消亡。

这里出现了两个结果。首先，考查疾病传染一半人口的临界感染率 $r_{半数}$，在 p 较小时，它迅速减小（图3a）。其次，无论什么结构，对于足以传染整个人群的疾病，整体感染所需要的时间 $T(p)$ 与 $L(p)$ 曲线相似（图3b）。因此，传染性疾病可能会更容易、更迅速地在小世界中传播；需要注意但不太明显的一点是最少需要多少捷径可以构造这样的小世界。

我们的模型与其他疾病传播的网络模型[20-24]有很大程度的不同。其他所有的模型都显示网络结构影响疾病传播的速度和程度，但是我们的模型给出了动力学行为作为结构性质的一个明确函数关系（图3），而不是只考虑少数的特定拓扑结构，例如随机图、星形网络和链[20-23]。与我们最接近的研究中，克雷奇马尔和莫里斯[24]已

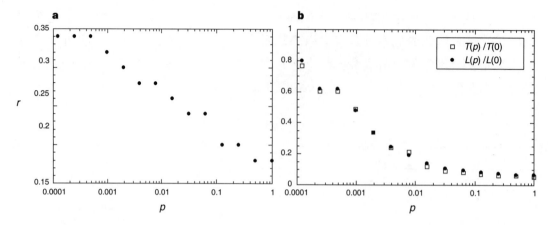

Fig. 3. Simulation results for a simple model of disease spreading. The community structure is given by one realization of the family of randomly rewired graphs used in Fig. 1. **a**, Critical infectiousness r_{half}, at which the disease infects half the population, decreases with p. **b**, The time $T(p)$ required for a maximally infectious disease ($r = 1$) to spread throughout the entire population has essentially the same functional form as the characteristic path length $L(p)$. Even if only a few per cent of the edges in the original lattice are randomly rewired, the time to global infection is nearly as short as for a random graph.

that increases in the number of concurrent partnerships can significantly accelerate the propagation of a sexually-transmitted disease that spreads along the edges of a graph. All their graphs are disconnected because they fix the average number of partners per person at $k = 1$. An increase in the number of concurrent partnerships causes faster spreading by increasing the number of vertices in the graph's largest connected component. In contrast, all our graphs are connected; hence the predicted changes in the spreading dynamics are due to more subtle structural features than changes in connectedness. Moreover, changes in the number of concurrent partners are obvious to an individual, whereas transitions leading to a smaller world are not.

We have also examined the effect of small-world connectivity on three other dynamical systems. In each case, the elements were coupled according to the family of graphs described in Fig. 1. (1) For cellular automata charged with the computational task of density classification[25], we find that a simple "majority-rule" running on a small-world graph can outperform all known human and genetic algorithm-generated rules running on a ring lattice. (2) For the iterated, multi-player "Prisoner's dilemma"[11] played on a graph, we find that as the fraction of short cuts increases, cooperation is less likely to emerge in a population of players using a generalized "tit-for-tat"[26] strategy. The likelihood of cooperative strategies evolving out of an initial cooperative/non-cooperative mix also decreases with increasing p. (3) Small-world networks of coupled phase oscillators synchronize almost as readily as in the mean-field model[2], despite having orders of magnitude fewer edges. This result may be relevant to the observed synchronization of widely separated neurons in the visual cortex[27] if, as seems plausible, the brain has a small-world architecture.

We hope that our work will stimulate further studies of small-world networks. Their

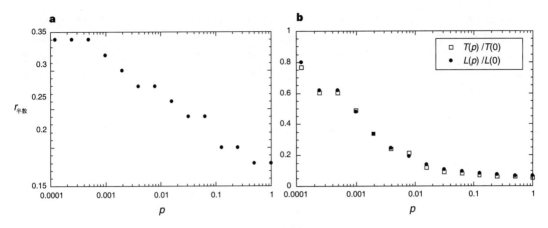

图 3. 一个简单的疾病传播模型的模拟结果。社群结构由图 1 中随机重连图族的一个实现给出。**a**，疾病感染半数人群的临界感染率 $r_{半数}$ 随 p 减小。**b**，最强感染性疾病 $(r=1)$ 传播至整个人群所需的时间 $T(p)$ 具有与特征路径长度 $L(p)$ 相同的函数形式。即使原始晶格网络中只有几个百分比的边被重连，整体感染的时间也与随机图中的一样短。

经指出并行性伴侣关系数目的增加会加速沿图边传播的性疾病的传染。他们所有的图形都是断开的，因为他们设定了每个人的伴侣平均数为 $k=1$。并行性伴侣关系数目的增加会通过增加最大连通集团中顶点的数目引起传播加速。相比而言，我们的图是相连的；因此所预测的传播动态行为上的改变是由于更多的精细结构特征而非连通性的改变。此外，并行性伴侣关系数目的改变对于个体来说影响是明显的，但对整体结构向小世界转变的影响却不明显。

我们也测试了小世界性质在其他三个动态系统中的效应。在每种情况下，元素通过图 1 所描述的系统网络发生耦合。(1)对于负责密度分类计算的元胞自动机 [25]，我们发现一个简单的在小世界图中运行的"多数规则"可以胜过所有已知在环形晶格网络中运行的人为以及遗传算法产生的规则。(2)对于一个基于网络迭代的、多人的"因徒困境" [11]，我们发现随捷径数量的增加，运用"以牙还牙" [26] 战略的参与者更不容易产生合作。由原始的合作/非合作组合发展产生的合作战略的可行性也随着 p 的增加而降低。(3)耦合相位振荡的小世界网络的同步几乎与在平均场模型 [2] 中的一样容易，尽管边的数量少了几个量级。这个结果可能与所观测到的视觉皮层中被广泛分离的神经元的同步性 [27] 相关，如此看来，大脑似乎也具有小世界体系结构。

我们希望本文可以促进小世界网络更深层次的研究。小世界网络同时具有高聚

distinctive combination of high clustering with short characteristic path length cannot be captured by traditional approximations such as those based on regular lattices or random graphs. Although small-world architecture has not received much attention, we suggest that it will probably turn out to be widespread in biological, social and man-made systems, often with important dynamical consequences.

(**393**, 440-442; 1998)

Duncan J. Watts[*] & Steven H. Strogatz

Department of Theoretical and Applied Mechanics, Kimball Hall, Cornell University, Ithaca, New York 14853, USA

[*] Present address: Paul F. Lazarsfeld Center for the Social Sciences, Columbia University, 812 SIPA Building, 420 W118 St, New York, New York 10027, USA.

Received 27 November 1997; accepted 6 April 1998.

References:

1. Winfree, A. T. *The Geometry of Biological Time* (Springer, New York, 1980).

2. Kuramoto, Y. *Chemical Oscillations, Waves, and Turbulence* (Springer, Berlin, 1984).

3. Strogatz, S. H. & Stewart, I. Coupled oscillators and biological synchronization. *Sci. Am.* **269**(6), 102-109 (1993).

4. Bressloff, P. C., Coombes, S. & De Souza, B. Dynamics of a ring of pulse-coupled oscillators: a group theoretic approach. *Phys. Rev. Lett.* **79**, 2791-2794 (1997).

5. Braiman, Y., Lindner, J. F. & Ditto, W. L. Taming spatiotemporal chaos with disorder. *Nature* **378**, 465-467 (1995).

6. Wiesenfeld, K. New results on frequency-locking dynamics of disordered Josephson arrays. *Physica B* **222**, 315-319 (1996).

7. Gerhardt, M., Schuster, H. & Tyson, J. J. A cellular automaton model of excitable media including curvature and dispersion. *Science* **247**, 1563-1566 (1990).

8. Collins, J. J., Chow, C. C. & Imhoff, T. T. Stochastic resonance without tuning. *Nature* **376**, 236-238 (1995).

9. Hopfield, J. J. & Herz, A. V. M. Rapid local synchronization of action potentials: Toward computation with coupled integrate-and-fire neurons. *Proc. Natl Acad. Sci. USA* **92**, 6655-6662 (1995).

10. Abbott, L. F. & van Vreeswijk, C. Asynchronous states in neural networks of pulse-coupled oscillators. *Phys. Rev. E* **48**(2), 1483-1490 (1993).

11. Nowak, M. A. & May, R. M. Evolutionary games and spatial chaos. *Nature* **359**, 826-829 (1992).

12. Kauffman, S. A. Metabolic stability and epigenesis in randomly constructed genetic nets. *J. Theor. Biol.* **22**, 437-467 (1969).

13. Milgram, S. The small world problem. *Psychol. Today* **2**, 60-67 (1967).

14. Kochen, M. (ed.) *The Small World* (Ablex, Norwood, NJ, 1989).

15. Guare, J. *Six Degrees of Separation: A Play* (Vintage Books, New York, 1990).

16. Bollabás, B. *Random Graphs* (Academic, London, 1985).

17. Achacoso, T. B. & Yamamoto, W. S. *AY's Neuroanatomy of C. elegans for Computation* (CRC Press, Boca Raton, FL, 1992).

18. Wasserman, S. & Faust, K. *Social Network Analysis: Methods and Applications* (Cambridge Univ. Press, 1994).

19. Phadke, A. G. & Thorp, J. S. *Computer Relaying for Power Systems* (Wiley, New York, 1988).

20. Sattenspiel, L. & Simon, C. P. The spread and persistence of infectious diseases in structured populations. *Math. Biosci.* **90**, 341-366 (1988).

21. Longini, I. M. Jr A mathematical model for predicting the geographic spread of new infectious agents. *Math. Biosci.* **90**, 367-383 (1988).

22. Hess, G. Disease in metapopulation models: implications for conservation. *Ecology* **77**, 1617-1632 (1996).

23. Blythe, S. P., Castillo-Chavez, C. & Palmer, J. S. Toward a unified theory of sexual mixing and pair formation. *Math. Biosci.* **107**, 379-405 (1991).

24. Kretschmar, M. & Morris, M. Measures of concurrency in networks and the spread of infectious disease. *Math. Biosci.* **133**, 165-195 (1996).

25. Das, R., Mitchell, M. & Crutchfield, J. P. in *Parallel Problem Solving from Nature* (eds Davido, Y., Schwefel, H.-P. & Männer, R.) 344-353 (Lecture Notes in Computer Science 866, Springer, Berlin, 1994).

26. Axelrod, R. *The Evolution of Cooperation* (Basic Books, New York, 1984).

27. Gray, C. M., König, P., Engel, A. K. & Singer, W. Oscillatory responses in cat visual cortex exhibit intercolumnar synchronization which reflects global stimulus properties. *Nature* **338**, 334-337 (1989).

Acknowledgements. We thank B. Tjaden for providing the film actor data, and J. Thorp and K. Bae for the Western States Power Grid data. This work was supported by the US National Science Foundation (Division of Mathematical Sciences).

Correspondence and requests for materials should be addressed to D.J.W. (e-mail: djw24@columbia.edu).

集度与短特征路径长度的独特特征，不能用传统的近似方法例如基于规则晶格网络或者随机图的方法来获得。尽管小世界体系结构还没有引起很大的关注，我们推测它将在生物、社会以及人工系统等广泛存在，并且通常对系统的动力学行为有重要的影响。

（崔宁 翻译；狄增如 审稿）

Deciphering the Biology of *Mycobacterium tuberculosis* from the Complete Genome Sequence

S. T. Cole *et al.*

Editor's Note

This paper reveals the complete genome sequence of the tuberculosis-causing bacterium *Mycobacterium tuberculosis*. The genome, which contains over 4 million base pairs and around 4,000 genes, provides a sequence of every potential drug target and every possible antigen that could be used in a vaccine, making it of huge importance. The data, collected by microbial scientist Stewart T. Cole and colleagues, also shed light on the bacterium's basic biology. For example, the genome contains many genes involved in lipid metabolism, thought to encode the bacterium's complex cell wall. And two new families of proteins are thought to represent a possible source of antigenic variation, perhaps explaining how the pathogen alters its surface proteins to evade the host's immune response.

Countless millions of people have died from tuberculosis, a chronic infectious disease caused by the tubercle bacillus. The complete genome sequence of the best-characterized strain of *Mycobacterium tuberculosis*, H37Rv, has been determined and analysed in order to improve our understanding of the biology of this slow-growing pathogen and to help the conception of new prophylactic and therapeutic interventions. The genome comprises 4,411,529 base pairs, contains around 4,000 genes, and has a very high guanine+cytosine content that is reflected in the biased amino-acid content of the proteins. *M. tuberculosis* differs radically from other bacteria in that a very large portion of its coding capacity is devoted to the production of enzymes involved in lipogenesis and lipolysis, and to two new families of glycine-rich proteins with a repetitive structure that may represent a source of antigenic variation.

DESPITE the availability of effective short-course chemotherapy (DOTS) and the Bacille Calmette-Guérin (BCG) vaccine, the tubercle bacillus continues to claim more lives than any other single infectious agent[1]. Recent years have seen increased incidence of tuberculosis in both developing and industrialized countries, the widespread emergence of drug-resistant strains and a deadly synergy with the human immunodeficiency virus (HIV). In 1993, the gravity of the situation led the World Health Organisation (WHO) to declare tuberculosis a global emergency in an attempt to heighten public and political awareness. Radical measures are needed now to prevent the grim predictions of the WHO becoming reality. The combination of genomics and bioinformatics has the potential to generate the information and knowledge that will

根据全基因组序列破译结核分枝杆菌生物学

科尔等

编者按

这篇文章揭示了引起结核病的细菌——结核分枝杆菌的完整基因组序列。该基因组包含超过 400 万个碱基对和大约 4,000 个基因，为每个潜在的药物靶标和每个可能用于生产疫苗的抗原提供了序列，这使得该基因组的序列尤为重要。由微生物学家斯图尔特·科尔及其同事收集的数据也揭示了该细菌的基本生物学特性。例如，该基因组包含很多参与脂类代谢的基因，这些基因被认为编码细菌复杂的细胞壁。两个新蛋白家族被认为代表了抗原变异的可能来源，这也许能解释病原体是如何改变自身表面蛋白来躲避宿主的免疫反应。

数以百万计的人死于肺结核——一种由结核杆菌引起的慢性传染病。目前研究最清楚的结核分枝杆菌菌株 H37Rv 的完整基因组序列已完成测序和分析，有助于增加我们对这种生长缓慢的病原菌的生物学理解，帮助我们研发新的预防和治疗干预措施。该基因组由 4,411,529 个碱基对组成，包含约 4,000 个基因，鸟嘌呤＋胞嘧啶含量很高，这在蛋白质的氨基酸含量偏倚上有所反映。结核分枝杆菌与其他细菌的根本不同之处在于，它很大一部分编码能力用于编码脂肪合成和分解的酶，以及两个富含甘氨酸且有重复性结构的蛋白新家族，这两个家族可能是抗原变异的来源。

尽管已有有效的短程化疗（DOTS）和卡介苗（BCG）疫苗可用，相比其他单一传染源，结核杆菌仍然夺走更多人的生命[1]。近年来，结核病发病率在发展中国家和工业化国家都呈上升趋势，耐药菌株普遍出现，并且与人类免疫缺陷病毒（HIV）共同形成致命的双重感染。1993 年，由于势态严重，世界卫生组织（WHO）宣布结核病为全球紧急事件，试图提高公众意识和政治意识。现在需要严格的措施防止世界卫生组织残酷的预言成为现实。基因组学和生物信息学的结合，有望产生新信息和新知识，构思和开发针对该空气传播疾病的新疗法及干预措施，并阐明其病原

enable the conception and development of new therapies and interventions needed to treat this airborne disease and to elucidate the unusual biology of its aetiological agent, *Mycobacterium tuberculosis*.

The characteristic features of the tubercle bacillus include its slow growth, dormancy, complex cell envelope, intracellular pathogenesis and genetic homogeneity[2]. The generation time of *M. tuberculosis*, in synthetic medium or infected animals, is typically ~24 hours. This contributes to the chronic nature of the disease, imposes lengthy treatment regimens and represents a formidable obstacle for researchers. The state of dormancy in which the bacillus remains quiescent within infected tissue may reflect metabolic shutdown resulting from the action of a cell-mediated immune response that can contain but not eradicate the infection. As immunity wanes, through ageing or immune suppression, the dormant bacteria reactivate, causing an outbreak of disease often many decades after the initial infection[3]. The molecular basis of dormancy and reactivation remains obscure but is expected to be genetically programmed and to involve intracellular signalling pathways.

The cell envelope of *M. tuberculosis*, a Gram-positive bacterium with a G+C-rich genome, contains an additional layer beyond the peptidoglycan that is exceptionally rich in unusual lipids, glycolipids and polysaccharides[4,5]. Novel biosynthetic pathways generate cell-wall components such as mycolic acids, mycocerosic acid, phenolthiocerol, lipoarabinomannan and arabinogalactan, and several of these may contribute to mycobacterial longevity, trigger inflammatory host reactions and act in pathogenesis. Little is known about the mechanisms involved in life within the macrophage, or the extent and nature of the virulence factors produced by the bacillus and their contribution to disease.

It is thought that the progenitor of the *M. tuberculosis* complex, comprising *M. tuberculosis*, *M. bovis*, *M. bovis* BCG, *M. africanum* and *M. microti*, arose from a soil bacterium and that the human bacillus may have been derived from the bovine form following the domestication of cattle. The complex lacks interstrain genetic diversity, and nucleotide changes are very rare[6]. This is important in terms of immunity and vaccine development as most of the proteins will be identical in all strains and therefore antigenic drift will be restricted. On the basis of the systematic sequence analysis of 26 loci in a large number of independent isolates[6], it was concluded that the genome of *M. tuberculosis* is either unusually inert or that the organism is relatively young in evolutionary terms.

Since its isolation in 1905, the H37Rv strain of *M. tuberculosis* has found extensive, worldwide application in biomedical research because it has retained full virulence in animal models of tuberculosis, unlike some clinical isolates; it is also susceptible to drugs and amenable to genetic manipulation. An integrated map of the 4.4 megabase (Mb) circular chromosome of this slow-growing pathogen had been established previously and ordered libraries of cosmids and bacterial artificial chromosomes (BACs) were available[7,8].

216

体——结核分枝杆菌独特的生物学机理。

结核杆菌的典型特征包括生长缓慢、休眠、细胞被膜复杂、胞内发病和遗传同质性 [2]。在合成培养基或感染的动物中，结核分枝杆菌的代时通常约为 24 小时。这导致这种病本质上是慢性的，必须采取长期的治疗方案，对研究者而言也是一个严重障碍。休眠状态下的结核杆菌在感染组织内保持静态，这可能反映了细胞介导的免疫反应导致了代谢的关闭，这些免疫反应只能抑制感染但不能消除感染。伴随着衰老或免疫抑制引起的免疫力减弱，休眠状态的细菌再次活化，导致疾病通常在最初感染的几十年后爆发 [3]。休眠和再次活化的分子基础仍然不清楚，但预计受到遗传调控并且涉及细胞内的信号通路。

结核分枝杆菌是革兰氏阳性细菌，基因组富含 G+C，它的细胞被膜在肽聚糖外还有另外一层，这一层中的罕见脂、糖脂和多糖极其丰富 [4,5]。新的生物合成途径产生细胞壁组分，如分枝菌酸、结核蜡酸、苯酚硫代醇、脂阿拉伯甘露聚糖和阿拉伯半乳糖，其中一些可能有助于结核杆菌长期存活，引发宿主炎症反应，并且在致病中发挥作用。目前对它在巨噬细胞内的生存机制了解甚少，对杆菌产生的毒力因子的水平和本质以及它们对疾病的作用也知之甚少。

结核分枝杆菌复合群包含结核分枝杆菌、牛型分枝杆菌、牛型分枝杆菌卡介苗、非洲分枝杆菌和田鼠分枝杆菌，其祖先被认为源自土壤细菌，而人型杆菌可能是在牛驯化后来源于牛型分枝杆菌。复合群菌株缺乏菌株间的遗传多样性，核苷酸变化非常少 [6]。就免疫和疫苗研发来说这一点很重要，因为大多数蛋白质在所有菌株中完全相同，所以抗原漂移会受限。基于对大量独立分离株的 26 个位点进行系统序列分析 [6] 得出的结论是结核分枝杆菌的基因组要么异常不活跃，要么该生物在进化上相对年轻。

自 1905 年分离到结核分枝杆菌的 H37Rv 菌株，该菌株已在世界范围的生物医学研究中得到了深入的应用，因为它在肺结核动物模型中保留了全部的毒力，这与有些临床菌株不同；并且它还对药物敏感，易于遗传操作。我们已经建立了该生长缓慢的病原菌的 4.4 Mb 环状染色体的整合图谱，并且规则排序的黏粒和细菌人工染色体 (BACs) 文库也已建好 [7,8]。

Organization and Sequence of the Genome

Sequence analysis. To obtain the contiguous genome sequence, a combined approach was used that involved the systematic sequence analysis of selected large-insert clones (cosmids and BACs) as well as random small-insert clones from a whole-genome shotgun library. This culminated in a composite sequence of 4,411,529 base pairs (bp) (Figs 1, 2; Editorial note: Figure 2, showing a linear map of the chromosome of *M. tuberculosis* H37Rv, was originally included as a fold-out insert. For details, see http://www.sanger. ac.uk/resources/downloads/bacteria/mycobacterium.html), with a G+C content of 65.6%. This represents the second-largest bacterial genome sequence currently available (after that of *Escherichia coli*)[9]. The initiation codon for the *dnaA* gene, a hallmark for the origin of replication, *oriC*, was chosen as the start point for numbering. The genome is rich in repetitive DNA, particularly insertion sequences, and in new multigene families and duplicated housekeeping genes. The G+C content is relatively constant throughout the genome (Fig. 1) indicating that horizontally transferred pathogenicity islands of atypical base composition are probably absent. Several regions showing higher than average G+C content (Fig. 1) were detected; these correspond to sequences belonging to a large gene family that includes the polymorphic G+C-rich sequences (PGRSs).

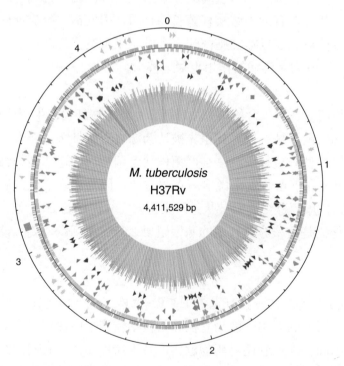

Fig. 1. Circular map of the chromosome of *M. tuberculosis* H37Rv. The outer circle shows the scale in Mb, with 0 representing the origin of replication. The first ring from the exterior denotes the positions of stable RNA genes (tRNAs are blue, others are pink) and the direct repeat region (pink cube); the second ring inwards shows the coding sequence by strand (clockwise, dark green; anticlockwise, light green); the third ring depicts repetitive DNA (insertion sequences, orange; 13E12 REP family, dark pink; prophage, blue);

基因组结构和序列

序列分析 为得到连续的基因组序列，我们采取了一个组合方法，这个方法包括对选定的大插入片段克隆（黏粒和 BACs）和一个全基因组鸟枪文库的随机小插入片段克隆进行系统的序列分析。最终得到 4,411,529 个碱基对（bp）的组合序列（图 1 和图 2；编者注：图 2 显示了结核分枝杆菌 H37Rv 染色体的线性图谱，在原文中是一个折叠插入图。更多的细节请看 http://www.sanger.ac.uk/resources/downloads/bacteria/mycobacterium.html），其中 G+C 含量为 65.6%。这是目前已有的第二大细菌基因组序列（在大肠杆菌之后）[9]。dnaA 基因的起始密码子，复制起始位点的标志 oriC 被选定为编码的起点。该基因组富含重复 NDA 序列，特别是插入序列，以及新的多基因家族和重复的管家基因。G+C 含量在整个基因组中相对恒定（图 1），这表明可能没有横向转移来的具有非典型碱基组成的致病岛。我们检测到几个 G+C 含量高于平均水平的区域（图 1）；它们对应属于同一个大基因家族的序列，这些序列富含 G+C 且具有多态性（PGRS）。

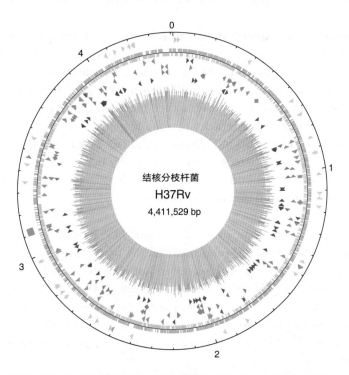

图 1. 结核分枝杆菌 H37Rv 染色体的环形图。外环单位的刻度为 Mb，0 代表复制起点。从外侧数第一环表示稳定 RNA 基因的位置（tRNA 是蓝色的，其他是粉红色）和正向重复区域（粉红条）；向内第二环按链表示编码序列（顺时针，暗绿色；逆时针，浅绿色）；第三环表示重复 DNA（插入序列，橙色；13E12 REP 家族，深粉色；原噬菌体，蓝色）；第四环表示 PPE 家族成员的位置（绿色）；第五环表示

the fourth ring shows the positions of the PPE family members (green); the fifth ring shows the PE family members (purple, excluding PGRS); and the sixth ring shows the positions of the PGRS sequences (dark red). The histogram (centre) represents G+C content, with < 65% G+C in yellow, and > 65% G+C in red. The figure was generated with software from DNASTAR.

Genes for stable RNA. Fifty genes coding for functional RNA molecules were found. These molecules were the three species produced by the unique ribosomal RNA operon, the 10Sa RNA involved in degradation of proteins encoded by abnormal messenger RNA, the RNA component of RNase P, and 45 transfer RNAs. No 4.5S RNA could be detected. The *rrn* operon is situated unusually as it occurs about 1,500 kilobases (kb) from the putative *oriC*; most eubacteria have one or more *rrn* operons near to *oriC* to exploit the gene-dosage effect obtained during replication[10]. This arrangement may be related to the slow growth of *M. tuberculosis*. The genes encoding tRNAs that recognize 43 of the 61 possible sense codons were distributed throughout the genome and, with one exception, none of these uses A in the first position of the anticodon, indicating that extensive wobble occurs during translation. This is consistent with the high G+C content of the genome and the consequent bias in codon usage. Three genes encoding tRNAs for methionine were found; one of these genes (*metV*) is situated in a region that may correspond to the terminus of replication (Figs 1, 2). As *metV* is linked to defective genes for integrase and excisionase, perhaps it was once part of a phage or similar mobile genetic element.

Insertion sequences and prophages. Sixteen copies of the promiscuous insertion sequence IS*6110* and six copies of the more stable element IS*1081* reside within the genome of H37Rv[8]. One copy of IS*1081* is truncated. Scrutiny of the genomic sequence led to the identification of a further 32 different insertion sequence elements, most of which have not been described previously, and of the 13E12 family of repetitive sequences which exhibit some of the characteristics of mobile genetic elements (Fig. 1). The newly discovered insertion sequences belong mainly to the IS*3* and IS*256* families, although six of them define a new group. There is extensive similarity between IS*1561* and IS*1552* with insertion sequence elements found in *Nocardia* and *Rhodococcus* spp., suggesting that they may be widely disseminated among the actinomycetes.

Most of the insertion sequences in *M. tuberculosis* H37Rv appear to have inserted in intergenic or non-coding regions, often near tRNA genes (Fig. 1). Many are clustered, suggesting the existence of insertional hot-spots that prevent genes from being inactivated, as has been described for *Rhizobium*[11]. The chromosomal distribution of the insertion sequences is informative as there appears to have been a selection against insertions in the quadrant encompassing *oriC* and an overrepresentation in the direct repeat region that contains the prototype IS*6110*. This bias was also observed experimentally in a transposon mutagenesis study[12].

At least two prophages have been detected in the genome sequence and their presence may explain why *M. tuberculosis* shows persistent low-level lysis in culture. Prophages phiRv1 and phiRv2 are both ~10 kb in length and are similarly organized, and some of their

PE 家族成员（紫色，除 PGRS 外）；第六环表示 PGRS 序列的位置（暗红色）。柱状图（中心）代表 G+C 含量，＜65% 的 G+C 含量为黄色，＞65% 的 G+C 含量为红色。此图使用软件 DNASTAR 生成。

编码稳定 RNA 的基因　我们发现了 50 个编码功能性 RNA 分子的基因。这些分子分别是：由独特的核糖体 RNA 操纵子产生的三类分子、参与降解异常信使 RNA 所编码的蛋白质的 10Sa RNA、核糖核酸酶 P（RNase P）的 RNA 组分和 45 个转运 RNA 分子。没有检测到 4.5S RNA。*rrn* 操纵子的位置异常，因为它距假定的 *oriC* 1,500 kb；大多数真细菌在 *oriC* 附近有一个或多个 *rrn* 操纵子，从而利用复制期间得到的基因剂量效应[10]。这种排列可能与结核分枝杆菌生长缓慢有关。编码 tRNA 的基因分布在基因组中，能识别分散在基因组中 61 个可能的有义密码子中的 43 个，但有一个例外，所有反密码子的第一位都不用 A，表明翻译过程中摆动广泛存在。这与基因组中 G+C 含量高及其导致的密码子偏好相符。我们发现了三个甲硫氨酸 tRNA 编码基因；其中一个基因（*metV*）位于可能对应复制终点的区域（图 1 和图 2）。*metV* 与整合酶和切除酶的缺陷基因相关，或许它曾经是噬菌体或相似的可移动遗传元件的一部分。

插入序列和原噬菌体　H37Rv 基因组中有 16 个拷贝的混杂的 IS*6110* 插入序列和 6 个拷贝的较稳定的 IS*1081* 元件[8]。其中一个 IS*1081* 拷贝是截短的。仔细查看基因组序列又进一步鉴定了 32 个不同的插入序列元件，其中大部分以前没有描述过，另外还鉴定了 13E12 家族重复序列，该序列具有可移动遗传元件的某些特征（图 1）。新发现的插入序列大多属于 IS*3* 和 IS*256* 家族，但其中 6 个定义了一个新的种类。IS*1561* 和 IS*1552* 与诺卡氏菌属和红球菌属中发现的插入序列元件有广泛的相似性，这表明它们可能在放线菌中广泛传播。

结核分枝杆菌 H37Rv 中的大多数插入序列似乎插到了基因间区或非编码区，通常靠近 tRNA 基因（图 1）。许多插入序列聚集成簇，意味着存在插入热点以防止基因被灭活，这与根瘤菌中所描述的类似[11]。插入序列在染色体上的分布可以提供有用的信息，因为似乎存在某种选择避免插入到 *oriC* 周围四分之一的区域，并且插入序列似乎在包含 IS*6110* 原型的正向重复区域中出现次数过多。这种倾向在一个研究转座子突变的实验中发现过[12]。

在基因组序列中至少发现了两个原噬菌体，它们的存在可能解释了为什么结核分枝杆菌在培养中存在持续低水平裂解。原噬菌体 phiRv1 和 phiRv2 长度都约为 10 kb，结构相似，并且它们的一些基因产物与链霉菌和腐生分枝杆菌编码的产物有显著的

gene products show marked similarity to those encoded by certain bacteriophages from *Streptomyces* and saprophytic mycobacteria. The site of insertion of phiRv1 is intriguing as it corresponds to part of a repetitive sequence of the 13E12 family that itself appears to have integrated into the biotin operon. Some strains of *M. tuberculosis* have been described as requiring biotin as a growth supplement, indicating either that phiRv1 has a polar effect on expression of the distal *bio* genes or that aberrant excision, leading to mutation, may occur. During the serial attenuation of *M. bovis* that led to the vaccine strain *M. bovis* BCG, the phiRv1 prophage was lost[13]. In a systematic study of the genomic diversity of prophages and insertion sequences (S.V.G. *et al.*, manuscript in preparation), only IS*1532* exhibited significant variability, indicating that most of the prophages and insertion sequences are currently stable. However, from these combined observations, one can conclude that horizontal transfer of genetic material into the free-living ancestor of the *M. tuberculosis* complex probably occurred in nature before the tubercle bacillus adopted its specialized intracellular niche.

Genes encoding proteins. 3,924 open reading frames were identified in the genome (see Methods), accounting for ~91% of the potential coding capacity (Figs 1, 2). A few of these genes appear to have in-frame stop codons or frameshift mutations (irrespective of the source of the DNA sequenced) and may either use frameshifting during translation or correspond to pseudogenes. Consistent with the high $G+C$ content of the genome, GTG initiation codons (35%) are used more frequently than in *Bacillus subtilis* (9%) and *E. coli* (14%), although ATG (61%) is the most common translational start. There are a few examples of atypical initiation codons, the most notable being the ATC used by *infC*, which begins with ATT in both *B. subtilis* and *E. coli* [9,14]. There is a slight bias in the orientation of the genes (Fig. 1) with respect to the direction of replication as ~59% are transcribed with the same polarity as replication, compared with 75% in *B. subtilis*. In other bacteria, genes transcribed in the same direction as the replication forks are believed to be expressed more efficiently[9,14]. Again, the more even distribution in gene polarity seen in *M. tuberculosis* may reflect the slow growth and infrequent replication cycles. Three genes (*dnaB*, *recA* and Rv1461) have been invaded by sequences encoding inteins (protein introns) and in all three cases their counterparts in *M. leprae* also contain inteins, but at different sites[15] (S.T.C. *et al.*, unpublished observations).

Protein function, composition and duplication. By using various database comparisons, we attributed precise functions to ~40% of the predicted proteins and found some information or similarity for another 44%. The remaining 16% resembled no known proteins and may account for specific mycobacterial functions. Examination of the amino-acid composition of the *M. tuberculosis* proteome by correspondence analysis[16], and comparison with that of other microorganisms whose genome sequences are available, revealed a statistically significant preference for the amino acids Ala, Gly, Pro, Arg and Trp, which are all encoded by $G+C$-rich codons, and a comparative reduction in the use of amino acids encoded by $A+T$-rich codons such as Asn, Ile, Lys, Phe and Tyr (Fig. 3). This approach also identified two groups of proteins rich in Asn or Gly that belong to new families, PE and PPE (see below). The fraction of the proteome that has arisen through gene

相似性。phiRv1 的插入位点很有趣，因为它与 13E12 家族的一段重复序列对应，这段序列本身似乎已经整合到生物素操纵子中。有一些结核分枝杆菌菌株被描述为需要生物素作为生长补充，这说明要么 phiRv1 对远端 bio 基因的表达有极性效应，要么可能发生了异常切除，导致基因突变。在牛型分枝杆菌连续衰减产生牛型分枝杆菌卡介苗疫苗株的过程中，phiRv1 原噬菌体丢失[13]。对原噬菌体基因组多样性和插入序列的系统研究（戈登等，稿件准备中）显示，只有 IS1532 显示出显著变化，这表明大多数原噬菌体和插入序列目前处于稳定状态。但是，综合这些观察结果我们可以得出结论，遗传物质向营自由生活的结核分枝杆菌复合群祖先发生水平转移，实际上很可能发生在结核菌采用它独特的细胞内生活方式之前。

基因编码蛋白　从基因组中鉴定到 3,924 个开放读码框（见方法），约占潜在编码能力的 91%（图 1 和图 2）。其中少数基因似乎有读码框内终止密码子或移码突变（不考虑测序 DNA 的来源），这些基因可能在翻译过程中有移码，或是假基因的转录产物。与基因组高 G+C 含量一致，GTG 起始密码子的使用频率（35%）高于枯草芽孢杆菌（9%）和大肠杆菌（14%），尽管 ATG 在翻译起始最常见（61%）。这里有少数非典型起始密码子的例子，最值得注意的是 infC 使用 ATC，而在枯草芽孢杆菌和大肠杆菌中则都使用 ATT[9,14]。相对于复制方向，基因定位有轻微偏好（图 1），约 59% 的转录方向与复制方向相同，而枯草芽孢杆菌中为 75%。在其他细菌中，基因转录与复制又方向相同并可以更高效地表达[9,14]。同样，结核分枝杆菌基因极性分布更均衡，这可以反映出其生长缓慢和复制周期稀少。三个基因（dnaB、recA 和 Rv1461）都有内含肽（蛋白内含子）编码序列插入，并且这三个基因在麻风分枝杆菌中的同源基因也都包含内含肽，只是插入位点不同[15]（科尔等，未发表的观察结果）。

蛋白质功能、组成和复制　通过使用各种数据库进行比较，我们对约 40% 的预测蛋白的功能有了精确的注释，另外也发现了其他 44% 的蛋白质的一些信息或相似性。其余的 16% 与已知蛋白质没有相似性，可能负责特定的结核杆菌功能。通过对应分析检查结核分枝杆菌的蛋白质组的氨基酸组成[16]，并与其他基因组序列可用的微生物进行比较，统计上显示结核分枝杆菌显著偏好丙氨酸（Ala）、甘氨酸（Gly）、脯氨酸（Pro）、精氨酸（Arg）和色氨酸（Trp）等，这些氨基酸均由富含 G+C 的密码子编码，相对少使用富含 A+T 的密码子编码的氨基酸，如天冬酰胺（Asn）、异亮氨酸（Ile）、赖氨酸（Lys）、苯丙氨酸（Phe）和酪氨酸（Tyr）（见图 3）。这种方法还发现了两组富含天冬酰胺（Asn）或甘氨酸（Gly）的蛋白质新家族，PE 和 PPE（见

duplication is similar to that seen in *E. coli* or *B. subtilis* (~51%; refs 9, 14), except that the level of sequence conservation is considerably higher, indicating that there may be extensive redundancy or differential production of the corresponding polypeptides. The apparent lack of divergence following gene duplication is consistent with the hypothesis that *M. tuberculosis* is of recent descent[6].

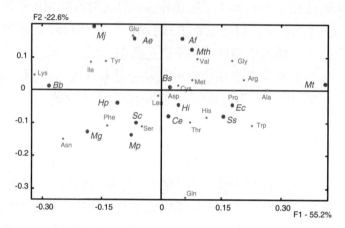

Fig. 3. Correspondence analysis of the proteomes from extensively sequenced organisms as a function of amino-acid composition. Note the extreme position of *M. tuberculosis* and the shift in amino-acid preference reflecting increasing G+C content from left to right. Abbreviations used: *Ae, Aquifex aeolicus*; *Af, Archaeoglobus fulgidis*; *Bb, Borrelia burgdorfei*; *Bs, B. subtilis*; *Ce, Caenorhabditis elegans*; *Ec, E. coli*; *Hi, Haemophilus influenzae*; *Hp, Helicobacter pylori*; *Mg, Mycoplasma genitalium*; *Mj, Methanococcus jannaschi*; *Mp, Mycoplasma pneumoniae*; *Mt, M. tuberculosis*; *Mth, Methanobacterium thermoautotrophicum*; *Sc, Saccharomyces cerevisiae*; *Ss, Synechocystis* sp. strain PCC6803. F1 and F2, first and second factorial axes[16].

General Metabolism, Regulation and Drug Resistance

Metabolic pathways. From the genome sequence, it is clear that the tubercle bacillus has the potential to synthesize all the essential amino acids, vitamins and enzyme co-factors, although some of the pathways involved may differ from those found in other bacteria. *M. tuberculosis* can metabolize a variety of carbohydrates, hydrocarbons, alcohols, ketones and carboxylic acids[2,17]. It is apparent from genome inspection that, in addition to many functions involved in lipid metabolism, the enzymes necessary for glycolysis, the pentose phosphate pathway, and the tricarboxylic acid and glyoxylate cycles are all present. A large number (~200) of oxidoreductases, oxygenases and dehydrogenases is predicted, as well as many oxygenases containing cytochrome P450, that are similar to fungal proteins involved in sterol degradation. Under aerobic growth conditions, ATP will be generated by oxidative phosphorylation from electron transport chains involving a ubiquinone cytochrome *b* reductase complex and cytochrome *c* oxidase. Components of several anaerobic phosphorylative electron transport chains are also present, including genes for nitrate reductase (*narGHJI*), fumarate reductase (*frdABCD*) and possibly nitrite reductase (*nirBD*), as well as a new reductase (*narX*) that results from a rearrangement of a homologue of the *narGHJI* operon. Two genes encoding haemoglobin-like proteins,

下文)。通过基因复制蛋白质组的比例升高,与大肠杆菌或枯草杆菌相似(大约51%；参考文献9和14),但序列的保守水平要高很多,这表明相应的多肽可能存在大量冗余或差异化表达。基因复制后明显缺少分化,这与结核分枝杆菌在进化上属于近代物种的假说是一致的[6]。

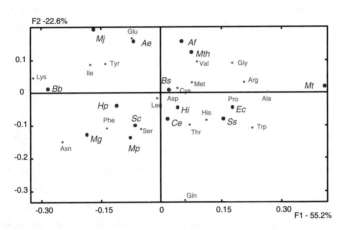

图 3. 大量测序的有机蛋白质组作为氨基酸组成的函数的对应分析。注意结核分枝杆菌的极端位置和氨基酸偏好的改变,该改变反映了 G+C 含量由左至右增加。缩写的使用:*Ae*,风产液菌；*Af*,闪亮古生球菌；*Bb*,伯氏疏螺旋体；*Bs*,枯草芽孢杆菌；*Ce*,秀丽隐杆线虫；*Ec*,大肠杆菌；*Hi*,流感嗜血杆菌；*Hp*,幽门螺杆菌；*Mg*,生殖支原体；*Mj*,詹氏甲烷球菌；*Mp*,肺炎支原体；*Mt*,结核分枝杆菌；*Mth*,嗜热自养甲烷杆菌；*Sc*,酿酒酵母；*Ss*,集胞藻 PCC6803 菌株。F1 和 F2,第一和第二因子轴[16]。

基本代谢、调控及耐药性

代谢途径 从基因组序列可以清楚地看到,尽管结核杆菌一些途径可能不同于在其他细菌中发现的途径,但它有能力合成所有必需氨基酸、维生素和辅酶。结核分枝杆菌可以代谢多种碳水化合物、烃、醇、酮和羧酸[2,17]。基因组检测清楚地显示,除了脂代谢的许多功能,糖酵解、戊糖磷酸途径、三羧酸和乙醛酸循环等所必需的酶也都存在。预测有大量(约 200 种)的氧化还原酶、氧化酶和脱氢酶,还有许多含有细胞色素 P450 的氧化酶,它们和参与固醇降解的真菌蛋白类似。有氧生长条件下,通过电子传递链包括泛醌-细胞色素 *b* 还原酶复合体和细胞色素 *c* 氧化酶,氧化磷酸化产生腺苷三磷酸(ATP)。还存在几个厌氧磷酸化电子传递链组分,这包括硝酸还原酶基因(*narGHJI*)、富马酸还原酶基因(*frdABCD*),可能还有亚硝酸盐还原酶基因(*nirBD*),另外还有一个新的还原酶基因(*narX*),它可能由一个与 *narGHJI* 操纵子同源的基因重排而来。我们还发现两个血红蛋白类似蛋白的编码基因,可能防

which may protect against oxidative stress or be involved in oxygen capture, were found. The ability of the bacillus to adapt its metabolism to environmental change is significant as it not only has to compete with the lung for oxygen but must also adapt to the microaerophilic/anaerobic environment at the heart of the burgeoning granuloma.

Regulation and signal transduction. Given the complexity of the environmental and metabolic choices facing *M. tuberculosis*, an extensive regulatory repertoire was expected. Thirteen putative sigma factors govern gene expression at the level of transcription initiation, and more than 100 regulatory proteins are predicted (Table 1)(Editorial note: this large table, giving a functional classification of all protein-coding genes in *M. tuberculosis*, has been omitted in this edited version. Note that the original version published in *Nature* contained errors, which were corrected in an erratum in *Nature* **396**, 190–198 (1998)). Unlike *B. subtilis* and *E. coli*, in which there are > 30 copies of different two-component regulatory systems[14], *M. tuberculosis* has only 11 complete pairs of sensor histidine kinases and response regulators, and a few isolated kinase and regulatory genes. This relative paucity in environmental signal transduction pathways is probably offset by the presence of a family of eukaryotic-like serine/threonine protein kinases (STPKs), which function as part of a phosphorelay system[18]. The STPKs probably have two domains: the well-conserved kinase domain at the amino terminus is predicted to be connected by a transmembrane segment to the carboxy-terminal region that may respond to specific stimuli. Several of the predicted envelope lipoproteins, such as that encoded by *lppR* (Rv2403), show extensive similarity to this putative receptor domain of STPKs, suggesting possible interplay. The STPKs probably function in signal transduction pathways and may govern important cellular decisions such as dormancy and cell division, and although their partners are unknown, candidate genes for phosphoprotein phosphatases have been identified.

Drug resistance. *M. tuberculosis* is naturally resistant to many antibiotics, making treatment difficult[19]. This resistance is due mainly to the highly hydrophobic cell envelope acting as a permeability barrier[4], but many potential resistance determinants are also encoded in the genome. These include hydrolytic or drug-modifying enzymes such as β-lactamases and aminoglycoside acetyl transferases, and many potential drug–efflux systems, such as 14 members of the major facilitator family and numerous ABC transporters. Knowledge of these putative resistance mechanisms will promote better use of existing drugs and facilitate the conception of new therapies.

Lipid Metabolism

Very few organisms produce such a diverse array of lipophilic molecules as *M. tuberculosis*. These molecules range from simple fatty acids such as palmitate and tuberculostearate, through isoprenoids, to very-long-chain, highly complex molecules such as mycolic acids and the phenolphthiocerol alcohols that esterify with mycocerosic acid to form the scaffold for attachment of the mycosides. Mycobacteria contain examples of every known lipid and

御氧化应激或参与氧捕获。结核分枝杆菌的代谢适应环境变化的能力很有意义，因为它不仅要和肺竞争氧气，还必须适应迅速生长的肉芽瘤中心的微需氧/厌氧环境。

调控和信号传导　鉴于结核分枝杆菌面临的复杂环境和代谢选择，预计会有一套广泛的调控基因集。13 个假定的 σ 因子在转录起始水平控制基因表达，并且预测有 100 多个调节蛋白（表 1）。（编者注：给出结核分枝杆菌所有蛋白编码基因的大表在此版中被删除了。注意发表在《自然》中的版本含有错误，在《自然》1998 年第 396 卷 190 ~ 198 页的勘误中得以校正。）与枯草芽孢杆菌和大肠杆菌中有 > 30 个拷贝的不同的双组分调控系统[14]不同，结核分枝杆菌只有完整的 11 对传感器组氨酸激酶和反应调节子，以及少量单独的激酶和调控基因。环境信号传导通路的相对不足可能通过一个真核样丝氨酸/苏氨酸蛋白激酶（STPK）家族来抵消，这个家族是磷酸传递系统的一部分[18]。STPK 可能有两个结构域：氨基端为高度保守的激酶结构域，羧基端则可能对一些特定的刺激产生反应，据推测，这两个区域通过一个跨膜片段相连。一些预测的细胞被膜脂蛋白，如 lppR（Rv2403）编码的蛋白，与 STPK 中假定的受体结构域有广泛的相似性，这暗示可能存在相互作用。STPK 可能在信号传导通路中起作用，并可能控制重要的细胞决策，如休眠和细胞分裂，尽管它们的搭档未知，但已经发现磷蛋白磷酸酶的候选基因。

耐药性　结核分枝杆菌自身能抵抗多种抗生素，这使治疗变得困难[19]。这种抗性的主要原因是细胞被膜高度疏水成为渗透屏障[4]，但许多潜在的耐药性决定成分也是由基因组编码的。这些成分包括水解酶或药物修饰酶，如 β-内酰胺酶和氨基糖苷乙酰转移酶，以及许多潜在的药物外排系统，如主要协助转运蛋白家族的 14 个成员和大量 ABC 转运体。关于这些假设的耐药机制的知识将促进对现有药物的更好利用并有助于设计新的治疗方案。

脂 质 代 谢

极少数生物体像结核分枝杆菌一样产生一系列不同的亲脂性分子。这些分子包括简单脂肪酸，如棕榈酸酯和结核硬脂酸，以及类异戊二烯和极长链、高度复杂的分子，如分枝菌酸和苯酚结核菌醇，后者酯化结核蜡酸形成供海藻糖苷附着的支架。结核分枝杆菌包含已知的各类脂质和聚酮生物合成系统，包括通常在哺乳动物和植

polyketide biosynthetic system, including enzymes usually found in mammals and plants as well as the common bacterial systems. The biosynthetic capacity is overshadowed by the even more remarkable radiation of degradative, fatty acid oxidation systems and, in total, there are ~250 distinct enzymes involved in fatty acid metabolism in *M. tuberculosis* compared with only 50 in *E. coli*[20].

Fatty acid degradation. *In vivo*-grown mycobacteria have been suggested to be largely lipolytic, rather than lipogenic, because of the variety and quantity of lipids available within mammalian cells and the tubercle[2] (Fig. 4a). The abundance of genes encoding components of fatty acid oxidation systems found by our genomic approach supports this proposition, as there are 36 acyl-CoA synthases and a family of 36 related enzymes that could catalyse the first step in fatty acid degradation. There are 21 homologous enzymes belonging to the enoyl-CoA hydratase/isomerase superfamily of enzymes, which rehydrate the nascent product of the acyl-CoA dehydrogenase. The four enzymes that convert the 3-hydroxy fatty acid into a 3-keto fatty acid appear less numerous, mainly because they are difficult to distinguish from other members of the short-chain alcohol dehydrogenase family on the basis of primary sequence. The five enzymes that complete the cycle by thiolysis of the β-ketoester, the acetyl-CoA C-acetyltransferases, do indeed appear to be a more limited family. In addition to this extensive set of dissociated degradative enzymes, the genome also encodes the canonical FadA/FadB β-oxidation complex (Rv0859 and Rv0860). Accessory activities are present for the metabolism of odd-chain and multiply unsaturated fatty acids.

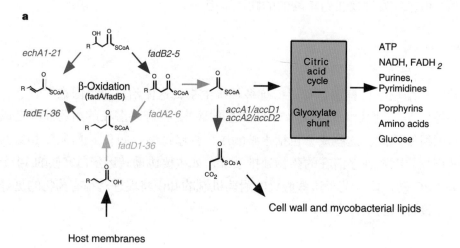

物以及一般细菌系统中发现的酶。与辐射分布更加显著的脂肪酸氧化降解系统相比，生物合成能力甚至相形见绌，总的来说在结核分枝杆菌中共有约250种不同的酶参与脂肪酸代谢，相比之下大肠杆菌中只有50种[20]。

脂肪酸的降解　研究发现体内生长的分枝杆菌主要参与脂肪降解，而不是脂肪合成，因为哺乳动物细胞和结节中有大量各种可用的脂类[2]（图4a）。我们用基因组学方法发现大量脂肪酸氧化系统组分的编码基因支持这一观点，有36个酰基辅酶A合酶和一个含有36种酶，能催化脂肪酸降解第一步的酶家族。有21个属于烯脂酰辅酶A水合酶/异构酶超家族的同源酶，它们对酯酰辅酶A脱氢酶的新生产物进行再水合。有四个酶负责将3-羟基脂肪酸转化为3-酮脂肪酸，这个数目似乎少了很多，主要是因为在一维序列水平上难以将它们和短链醇脱氢酶家族的其他成员区分开。乙酰辅酶A C-乙酰转移酶通过硫解β-酮酸酯完成循环，该家族确实较小，只有五个成员。除了这一系列数目众多的游离降解酶，基因组也编码典型的FadA/FadB β-氧化复合体（Rv0859和Rv0860），还包括代谢奇数链脂肪酸和多不饱和脂肪酸等辅助活性。

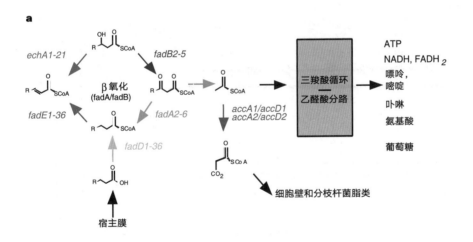

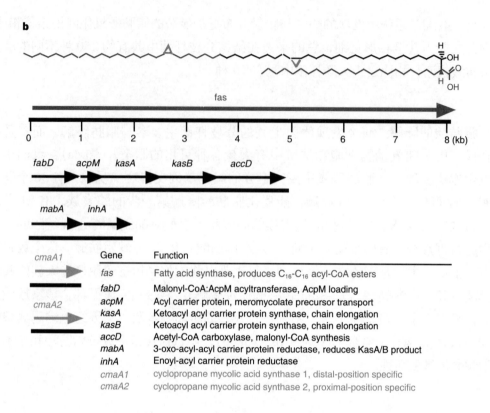

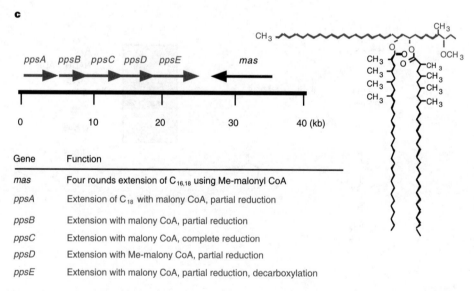

Fig. 4. Lipid metabolism. **a**, Degradation of host-cell lipids is vital in the intracellular life of *M. tuberculosis*. Host-cell membranes provide precursors for many metabolic processes, as well as potential precursors of mycobacterial cell-wall constituents, through the actions of a broad family of β-oxidative enzymes encoded by multiple copies in the genome. These enzymes produce acetyl CoA, which can be converted into many different metabolites and fuel for the bacteria through the actions of the enzymes of the citric acid cycle and the glyoxylate shunt of this cycle. **b**, The genes that synthesize mycolic acids, the dominant lipid component of the mycobacterial cell wall, include the type I fatty acid synthase (*fas*) and

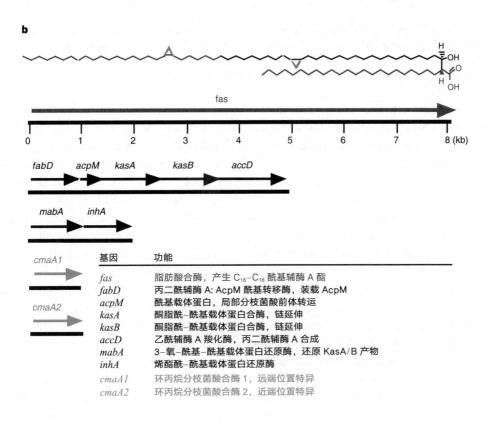

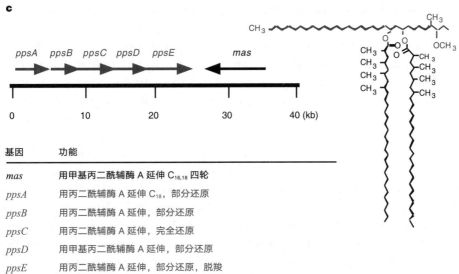

图 4. 脂质代谢。**a**, 宿主细胞脂质的降解对于结核分枝杆菌的胞内生活是至关重要的。在基因组中多拷贝基因编码的 β 氧化酶大家族的作用下, 宿主细胞膜为许多代谢过程提供前体, 也为分枝杆菌细胞壁成分提供潜在的前体。这些酶产生乙酰辅酶 A, 在柠檬酸循环及乙醛酸旁路循环酶的作用下, 乙酰辅酶 A 转换成许多不同的代谢物和供细菌生长用的原料。**b**, 合成分枝菌酸(分枝杆菌细胞壁的主要脂分)的基因, 包括 I 型脂肪酸合酶(*fas*)和独特的 II 型系统, II 型系统依靠扩展一个结合在酰基载体蛋白的前体

a unique type II system which relies on extension of a precursor bound to an acyl carrier protein to form full-length (~80-carbon) mycolic acids. The *cma* genes are responsible for cyclopropanation. **c**, The genes that produce phthiocerol dimycocerosate form a large operon and represent type I (*mas*) and type II (the *pps* operon) polyketide synthase systems. Functions are colour coordinated.

Fatty acid biosynthesis. At least two discrete types of enzyme system, fatty acid synthase (FAS) I and FAS II, are involved in fatty acid biosynthesis in mycobacteria (Fig. 4b). FAS I (Rv2524, *fas*) is a single polypeptide with multiple catalytic activities that generates several shorter CoA esters from acetyl-CoA primers[5] and probably creates precursors for elongation by all of the other fatty acid and polyketide systems. FAS II consists of dissociable enzyme components which act on a substrate bound to an acyl-carrier protein (ACP). FAS II is incapable of *de novo* fatty acid synthesis but instead elongates palmitoyl-ACP to fatty acids ranging from 24 to 56 carbons in length[17,21]. Several different components of FAS II may be targets for the important tuberculosis drug isoniazid, including the enoyl-ACP reductase InhA[22], the ketoacyl-ACP synthase KasA and the ACP AcpM[21]. Analysis of the genome shows that there are only three potential ketoacyl synthases: KasA and KasB are highly related, and their genes cluster with *acpM*, whereas KasC is a more distant homologue of a ketoacyl synthase III system. The number of ketoacyl synthase and ACP genes indicates that there is a single FAS II system. Its genetic organization, with two clustered ketoacyl synthases, resembles that of type II aromatic polyketide biosynthetic gene clusters, such as those for actinorhodin, tetracycline and tetracenomycin in *Streptomyces* species[23]. InhA seems to be the sole enoyl-ACP reductase and its gene is co-transcribed with a *fabG* homologue, which encodes 3-oxoacyl-ACP reductase. Both of these proteins are probably important in the biosynthesis of mycolic acids.

Fatty acids are synthesized from malonyl-CoA and precursors are generated by the enzymatic carboxylation of acetyl (or propionyl)-CoA by a biotin-dependent carboxylase (Fig. 4b). From study of the genome we predict that there are three complete carboxylase systems, each consisting of an α- and a β-subunit, as well as three β-subunits without an α-counterpart. As a group, all of the carboxylases seem to be more related to the mammalian homologues than to the corresponding bacterial enzymes. Two of these carboxylase systems (*accA1*, *accD1* and *accA2*, *accD2*) are probably involved in degradation of odd-numbered fatty acids, as they are adjacent to genes for other known degradative enzymes. They may convert propionyl-CoA to succinyl-CoA, which can then be incorporated into the tricarboxylic acid cycle. The synthetic carboxylases (*accA3*, *accD3*, *accD4*, *accD5* and *accD6*) are more difficult to understand. The three extra β-subunits might direct carboxylation to the appropriate precursor or may simply increase the total amount of carboxylated precursor available if this step were rate-limiting.

Synthesis of the paraffinic backbone of fatty and mycolic acids in the cell is followed by extensive postsynthetic modifications and unsaturations, particularly in the case of the mycolic acids[24,25]. Unsaturation is catalysed either by a FabA-like β-hydroxyacyl-ACP dehydrase, acting with a specific ketoacyl synthase, or by an aerobic terminal mixed function desaturase that uses both molecular oxygen and NADPH. Inspection of the genome revealed no obvious candidates for the FabA-like activity. However, three potential aerobic

232

形成全长（约 80 碳）分枝菌酸。*cma* 基因负责形成环丙烷。c，产生结核菌醇双结核蜡酸酯的基因形成一个大操纵子，代表 I 型（*mas*）和 II 型（*pps* 操纵子）聚酮合酶系统。功能和颜色相对应。

脂肪酸生物合成　至少有两个不同类型的酶系统，脂肪酸合酶（FAS）I 和 FAS II，参与结核分枝杆菌的脂肪酸生物合成（图 4b）。FAS I（Rv2524，*fas*）是一个具有多种催化活性的单一多肽，它能利用乙酰辅酶 A 引物产生一些较短的辅酶 A 酯[5]，并且可能通过所有其他脂肪酸和聚酮系统产生延伸所需的前体。FAS II 由多个可解离的酶组分组成，这些酶组分都作用于一个结合在酰基载体蛋白（ACP）上的底物。FAS II 不能从头合成脂肪酸，但可以将软脂酰-ACP 延伸为长度为 24 到 56 个碳的脂肪酸[17,21]。FAS II 的几个不同组分可能是重要的抗结核病药物异烟肼的靶标，包括烯脂酰-ACP 还原酶 InhA[22]、酮脂酰-ACP 合酶 KasA 和 ACP AcpM[21]。基因组分析表明只有三个潜在酮脂酰合酶：KasA 和 KasB 高度相关，它们的基因与 *acpM* 聚集成簇，而 KasC 则是酮脂酰合酶 III 系统的一个更远源的同源蛋白。酮脂酰合酶和 ACP 基因的数量表明有单一的 FAS II 系统。该系统的遗传结构有两个聚集成簇的酮脂酰酶，类似于 II 型芳香族聚酮生物合成基因簇，如链霉菌中的放线紫红素、四环素、特曲霉素合成基因簇[23]。InhA 似乎是唯一的烯脂酰-ACP 还原酶，其基因与编码 3-氧酰基-ACP 还原酶的 *fabG* 同源基因进行共转录。这两种蛋白质很可能在分枝菌酸合成中起重要作用。

脂肪酸是由丙二酰辅酶 A 合成的，前体由依赖生物素的羧化酶酶促羧化乙酰（或丙酰）-辅酶 A（图 4b）生成。通过对基因组的研究，我们预测出三个完整的羧化酶系统，每一个都包括一个 α 亚基和一个 β 亚基，另外还预测出三个 β 亚基，没有 α 亚基与之配对。整体而言，所有羧化酶和哺乳动物来源的同源蛋白的亲缘关系似乎高于细菌来源的同源酶。羧化酶系统（*accA1*、*accD1* 和 *accA2*、*accD2*）中的两个很可能参与奇数碳原子脂肪酸的降解，因为在基因组上它们与其他已知降解酶基因邻近。他们可能将丙酰辅酶 A 转换为琥珀酰辅酶 A，然后琥珀酰辅酶 A 被纳入三羧酸循环。具有合成作用的羧化酶（*accA3*、*accD3*、*accD4*、*accD5* 和 *accD6*）更难理解。如果这步是限速步骤的话，这三个额外的 β 亚基可能介导适当的前体羧化或只是简单地增加可用的羧化前体总数。

在细胞内合成脂肪酸和分枝菌酸的石蜡骨架后接着进行大量的合成后修饰和不饱和化，特别是在分枝菌酸的合成中[24,25]。不饱和化由一个 FabA 样的 β-羟脂酰-ACP 脱水酶催化，该酶与特定的酮酯酰合酶共同起作用，或者由一个使用分子氧和 NADPH 的需氧末端混合功能去饱和酶催化。检测基因组没有发现明显具有 FabA 样活性的候选基因。然而，很显然，三个潜在的需氧去饱和酶（由 *desA1*、

desaturases (encoded by *desA1*, *desA2* and *desA3*) were evident that show little similarity to related vertebrate or yeast enzymes (which act on CoA esters) but instead resemble plant desaturases (which use ACP esters). Consequently, the genomic data indicate that unsaturation of the meromycolate chain may occur while the acyl group is bound to AcpM.

Much of the subsequent structural diversity in mycolic acids is generated by a family of *S*-adenosyl-L-methionine-dependent enzymes, which use the unsaturated meromycolic acid as a substrate to generate *cis* and *trans* cyclopropanes and other mycolates. Six members of this family have been identified and characterized[25] and two clustered, convergently transcribed new genes are evident in the genome (*umaA1* and *umaA2*). From the functions of the known family members and the structures of mycolic acids in *M. tuberculosis*, it is tempting to speculate that these new enzymes may introduce the *trans* cyclopropanes into the meromycolate precursor. In addition to these two methyltransferases, there are two other unrelated lipid methyltransferases (Ufa1 and Ufa2) that share homology with cyclopropane fatty acid synthase of *E. coli*[25]. Although cyclopropanation seems to be a relatively common modification of mycolic acids, cyclopropanation of plasma-membrane constituents has not been described in mycobacteria. Tuberculostearic acid is produced by methylation of oleic acid, and may be synthesized by one of these two enzymes.

Condensation of the fully functionalized and preformed meromycolate chain with a 26-carbon α-branch generates full-length mycolic acids that must be transported to their final location for attachment to the cell-wall arabinogalactan. The transfer and subsequent transesterification is mediated by three well-known immunogenic proteins of the antigen 85 complex[26]. The genome encodes a fourth member of this complex, antigen 85C′ (*fbpC2*, Rv0129), which is highly related to antigen 85C. Further studies are needed to show whether the protein possesses mycolyltransferase activity and to clarify the reason behind the apparent redundancy.

Polyketide synthesis. Mycobacteria synthesize polyketides by several different mechanisms. A modular type I system, similar to that involved in erythromycin biosynthesis[23], is encoded by a very large operon, *ppsABCDE*, and functions in the production of phenolphthiocerol[5]. The absence of a second type I polyketide synthase suggests that the related lipids phthiocerol A and B, phthiodiolone A and phthiotriol may all be synthesized by the same system, either from alternative primers or by differential postsynthetic modification. It is physiologically significant that the *pps* gene cluster occurs immediately upstream of *mas*, which encodes the multifunctional enzyme mycocerosic acid synthase (MAS), as their products phthiocerol and mycocerosic acid esterify to form the very abundant cell-wall-associated molecule phthiocerol dimycocerosate (Fig. 4c).

Members of another large group of polyketide synthase enzymes are similar to MAS, which also generates the multiply methyl-branched fatty acid components of mycosides and phthiocerol dimycocerosate, abundant cell-wall-associated molecules[5]. Although some of these polyketide synthases may extend type I FAS CoA primers to produce other long-chain methyl-branched fatty acids such as mycolipenic, mycolipodienic and mycolipanolic acids or the phthioceranic and hydroxyphthioceranic acids, or may even show functional overlap[5],

desA2 和 *desA3* 编码）与脊椎动物或酵母中的相关酶（作用于辅酶 A 酯）几乎没有相似性，而与植物中的去饱和酶（使用 ACP 酯）相似。因此，基因组数据表明，当酰基结合到 AcpM 上可能使得局部分枝菌酸酯链去饱和。

分枝菌酸许多后来的结构多样性是由 S–腺苷–L–甲硫氨酸依赖的酶家族产生的，它以不饱和的局部分枝菌酸为底物生成顺式–和反式–环丙烷以及其他分枝菌酸。现在已经鉴定和描述了这个家族的六个成员[25]，两个基因簇集中转录的新基因在基因组中很明显（*umaA1* 和 *umaA2*）。从已知家族成员的功能和结核分枝杆菌分枝菌酸的结构推测，这些新酶可能将反式环丙烷引入到局部分枝菌酸前体中。除了这两个甲基转移酶，还有另外两个无关的脂质甲基转移酶（Ufa1 和 Ufa2）与大肠杆菌的环丙烷脂肪酸合酶有同源性[25]。虽然环丙烷化似乎是分枝菌酸的一种相对常见的修饰，但质膜组分的环丙烷化在分枝杆菌中还未见描述。结核硬脂酸通过油酸的甲基化产生，该合成过程可能由这两个酶中的一个完成。

完全功能的预先形成的局部分枝菌酸链和 26 碳 α 分枝缩合产生全长分枝菌酸，分枝菌酸必须转移到最终的位置以结合细胞壁阿拉伯半乳糖。转移和随后的酯交换反应由抗原 85 复合体中三个众所周知的免疫原性蛋白介导[26]。基因组编码这个复合体的第四个成员，抗原 85C′（*fbpC2*，Rv0129）与抗原 85C 高度相关。这个蛋白是否具有分枝酸转移酶活性，以及其明显冗余背后的原因的阐明，都需要进一步的研究。

聚酮合成　分枝杆菌通过几种不同机制合成聚酮。模块化的 I 型系统，类似于参与红霉素生物合成的系统[23]，由一个非常大的操纵子 *ppsABCDE* 编码，并在苯酚结核菌醇的产生中起作用[5]。基因组中不存在第二个 I 型聚酮合酶说明相关的脂质结核菌醇 A 和 B、结核菌二醇 A 和结核菌三醇都可能由同一系统合成，这要么使用了不同的引物，要么合成后进行了不同的修饰。*pps* 基因簇紧邻 *mas* 上游，*mas* 编码多功能酶分枝杆菌结核蜡酸合酶（MAS），这在生理上很重要，因为它们的产物结核菌醇和结核蜡酸酯化形成了非常丰富的细胞壁关联分子结核菌醇双结核蜡酸酯（图 4c）。

另一个聚酮合酶大家族的成员与 MAS 相似，它也产生丰富的细胞壁关联分子——海藻糖苷和结核菌醇双结核蜡酸酯的多种甲基化分枝脂肪酸组分[5]。虽然其中一些聚酮合酶可能延长 I 型 FAS 辅酶 A 引物，产生其他长链甲基化分枝脂肪酸，如霉脂酸、霉脂二烯酸和霉脂羟基酸或分枝菌蜡酸和羟基分枝菌蜡酸，甚至出现功能重叠[5]，但是这些酶的数目比已知代谢物多得多。因此，可能存在只在特定条件

there are many more of these enzymes than there are known metabolites. Thus there may be new lipid and polyketide metabolites that are expressed only under certain conditions, such as during infection and disease.

A fourth class of polyketide synthases is related to the plant enzyme superfamily that includes chalcone and stilbene synthase[23]. These polyketide synthases are phylogenetically divergent from all other polyketide and fatty acid synthases and generate unreduced polyketides that are typically associated with anthocyanin pigments and flavonoids. The function of these systems, which are often linked to apparent type I modules, is unknown. An example is the gene cluster spanning *pks10*, *pks7*, *pks8* and *pks9*, which includes two of the chalcone-synthase-like enzymes and two modules of an apparent type I system. The unknown metabolites produced by these enzymes are interesting because of the potent biological activities of some polyketides such as the immunosuppressor rapamycin.

Siderophores. Peptides that are not ribosomally synthesized are made by a process that is mechanistically analogous to polyketide synthesis[23,27]. These peptides include the structurally related iron-scavenging siderophores, the mycobactins and the exochelins[2,28], which are derived from salicylate by the addition of serine (or threonine), two lysines and various fatty acids and possible polyketide segments. The *mbt* operon, encoding one apparent salicylate-activating protein, three amino-acid ligases, and a single module of a type I polyketide synthase, may be responsible for the biosynthesis of the mycobacterial siderophores. The presence of only one non-ribosomal peptide-synthesis system indicates that this pathway may generate both siderophores and that subsequent modification of a single ε-amino group of one lysine residue may account for the different physical properties and function of the siderophores[28].

Immunological Aspects and Pathogenicity

Given the scale of the global tuberculosis burden, vaccination is not only a priority but remains the only realistic public health intervention that is likely to affect both the incidence and the prevalence of the disease[29]. Several areas of vaccine development are promising, including DNA vaccination, use of secreted or surface-exposed proteins as immunogens, recombinant forms of BCG and rational attenuation of *M. tuberculosis*[29]. All of these avenues of research will benefit from the genome sequence as its availability will stimulate more focused approaches. Genes encoding ~90 lipoproteins were identified, some of which are enzymes or components of transport systems, and a similar number of genes encoding preproteins (with type I signal peptides) that are probably exported by the Sec-dependent pathway. *M. tuberculosis* seems to have two copies of *secA*. The potent T-cell antigen Esat-6 (ref. 30), which is probably secreted in a Sec-independent manner, is encoded by a member of a multigene family. Examination of the genetic context reveals several similarly organized operons that include genes encoding large ATP-hydrolysing membrane proteins that might act as transporters. One of the surprises of the genome project was the discovery of two extensive families of novel glycine-rich proteins, which may be of immunological significance as they are predicted to be abundant and potentially polymorphic antigens.

下，如在感染和疾病时表达的新的脂和聚酮代谢产物。

第四类聚酮合酶与包含查尔酮和二苯乙烯合酶的植物酶超家族[23]有关。这些聚酮合酶在系统发生上与其他所有聚酮和脂肪酸合酶不同，它们产生通常与花青素和黄酮类化合物相关联的非还原聚酮。这些系统的功能未知，经常与一些明显的Ⅰ型模块相连接。一个例子是涵盖 *pks10*、*pks7*、*pks8* 和 *pks9* 的基因簇，它包括两个查尔酮合酶的类似酶和明显的Ⅰ型系统的两个模块。因为一些聚酮如免疫抑制剂雷帕霉素具有强大的生物活性，所以这些酶产生的未知代谢产物很有趣。

铁载体　非核糖体合成的肽链是通过与聚酮合成相似的过程产生的[23,27]。这些肽包括结构相关的清除铁的铁载体，分枝杆菌素和胞外螯合素[2,28]，它们是通过向水杨酸盐添加丝氨酸（或苏氨酸）、两个赖氨酸和各种脂肪酸和可能的聚酮片段得到的。*mbt* 操纵子可能负责分枝杆菌铁载体的合成，它编码一个明显的水杨酸盐活化蛋白、三个氨基酸连接酶和Ⅰ型聚酮合酶的一个单一模块。基因组中只有一个非核糖体肽合成系统表明该途径可能产生这两种铁载体，并且一个赖氨酸残基的 ε-氨基的后续修饰可能是铁载体具有不同物理性质和功能的原因[28]。

免疫学特性和致病性

鉴于全球结核病的发生规模，接种疫苗不只是一个优先选择，也可能是对结核病的发病率和流行都发挥作用的唯一实际的公共健康干预措施[29]。疫苗开发的几个领域是有前途的，这包括使用分泌蛋白或表面暴露蛋白作为免疫原的 DNA 疫苗、重组形式的卡介苗和合理减毒的结核分枝杆菌[29]。所有这些研究途径将从基因组序列中受益，基因组序列的获得将促进产生更加聚焦的解决方法。我们鉴定了约 90 个脂蛋白编码基因，其中有些是酶或转运系统组分，还有数量相当的基因编码蛋白质前体（含Ⅰ型信号肽），它们很可能通过 Sec 依赖途径向胞外转运。结核分枝杆菌似乎有两个拷贝的 *secA*。有效的 T 细胞抗原 Esat-6（参考文献 30）由多基因家族的一个成员编码，很可能以不依赖 Sec 的方式分泌。对基因组邻近区域的检查发现了几个结构相似的操纵子，其中包括编码具有 ATP 水解功能的高分子量膜蛋白基因，而这些膜蛋白可能是转运体。本基因组项目的惊喜之一是发现了新的富含甘氨酸的蛋白质，它们分别属于两个大家族，可能有免疫学意义，因为它们被预测是丰富且潜在的多态性抗原。

The PE and PPE multigene families. About 10% of the coding capacity of the genome is devoted to two large unrelated families of acidic, glycine-rich proteins, the PE and PPE families, whose genes are clustered (Figs 1, 2) and are often based on multiple copies of the polymorphic repetitive sequences referred to as PGRSs, and major polymorphic tandem repeats (MPTRs), respectively[31,32]. The names PE and PPE derive from the motifs Pro–Glu (PE) and Pro–Pro–Glu (PPE) found near the N terminus in most cases[33]. The 99 members of the PE protein family all have a highly conserved N-terminal domain of ~110 amino-acid residues that is predicted to have a globular structure, followed by a C-terminal segment that varies in size, sequence and repeat copy number (Fig. 5). Phylogenetic analysis separated the PE family into several subfamilies. The largest of these is the highly repetitive PGRS class, which contains 61 members; members of the other subfamilies, share very limited sequence similarity in their C-terminal domains (Fig. 5). The predicted molecular weights of the PE proteins vary considerably as a few members contain only the N-terminal domain, whereas most have C-terminal extensions ranging in size from 100 to 1,400 residues. The PGRS proteins have a high glycine content (up to 50%), which is the result of multiple tandem repetitions of Gly–Gly–Ala or Gly–Gly–Asn motifs, or variations thereof.

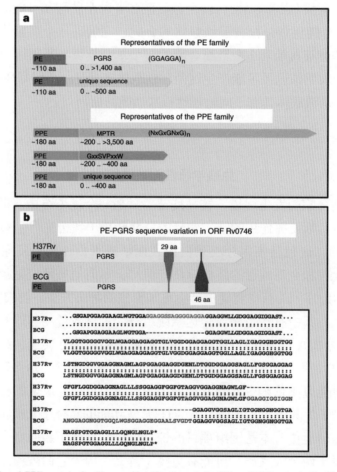

Fig. 5. The PE and PPE protein families. **a**, Classification of the PE and PPE protein families. **b**, Sequence

PE 和 PPE 多基因家族 基因组约 10% 的编码能力用于编码酸性富含甘氨酸但相互之间不相关的两大蛋白家族，PE 和 PPE 家族，它们的基因聚集成簇（图 1 和图 2），且通常分别基于多个拷贝的被称为 PGRS 的多态性重复序列和主要多态性串联重复序列（MPTR）[31,32]。命名为 PE 和 PPE 是因为在大多数情况下靠近 N 端有脯氨酸–谷氨酸（PE）和脯氨酸–脯氨酸–谷氨酸（PPE）基序 [33]。PE 蛋白家族的 99 个成员都有一个高度保守的 N 末端结构域，该结构域含有约 110 个氨基酸残基，预测为球状结构，其后是大小、序列和重复拷贝数都不同的 C 端片段（图 5）。系统发育分析将 PE 家族分为几个亚家族。其中最大的是高度重复的 PGRS 类，包含 61 个成员；其他亚家族成员 C 端结构域的序列相似性很有限（图 5）。PE 蛋白的预测分子量差别很大，因为少数成员只包含 N 端结构域，而大多数成员具有大小从 100 到 1,400 个残基不等的 C 端延伸。PGRS 蛋白甘氨酸含量高（达 50%），这是甘氨酸–甘氨酸–丙氨酸或甘氨酸–甘氨酸–天冬酰胺基序或其他变异基序多次串联重复的结果，或者是由于突变导致的。

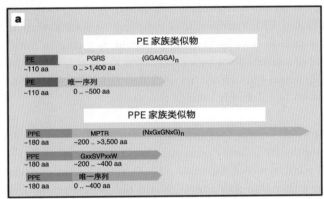

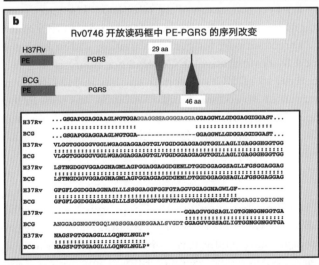

图 5. PE 和 PPE 蛋白家族。**a**，PE 和 PPE 蛋白家族的分类。**b**，结核分枝杆菌 H37Rv 和牛分枝杆菌卡介

variation between *M. tuberculosis* H37Rv and *M. bovis* BCG-Pasteur in the PE-PGRS encoded by open reading frame (ORF) Rv0746.

The 68 members of the PPE protein family (Fig. 5) also have a conserved N-terminal domain that comprises ~180 amino-acid residues, followed by C-terminal segments that vary markedly in sequence and length. These proteins fall into at least three groups, one of which constitutes the MPTR class characterized by the presence of multiple, tandem copies of the motif Asn–X–Gly–X–Gly–Asn–X–Gly. The second subgroup contains a characteristic, well-conserved motif around position 350, whereas the third contains proteins that are unrelated except for the presence of the common 180-residue PPE domain.

The subcellular location of the PE and PPE proteins is unknown and in only one case, that of a lipase (Rv3097), has a function been demonstrated. On examination of the protein database from the extensively sequenced *M. leprae*[15], no PGRS- or MPTR-related polypeptides were detected but a few proteins belonging to the non-MPTR subgroup of the PPE family were found. These proteins include one of the major antigens recognized by leprosy patients, the serine-rich antigen[34]. Although it is too early to attribute biological functions to the PE and PPE families, it is tempting to speculate that they could be of immunological importance. Two interesting possibilities spring to mind. First, they could represent the principal source of antigenic variation in what is otherwise a genetically and antigenically homogeneous bacterium. Second, these glycine-rich proteins might interfere with immune responses by inhibiting antigen processing.

Several observations and results support the possibility of antigenic variation associated with both the PE and the PPE family proteins. The PGRS member Rv1759 is a fibronectin-binding protein of relative molecular mass 55,000 (ref. 35) that elicits a variable antibody response, indicating either that individuals mount different immune responses or that this PGRS protein may vary between strains of *M. tuberculosis*. The latter possibility is supported by restriction fragment length polymorphisms for various PGRS and MPTR sequences in clinical isolates[33]. Direct support for genetic variation within both the PE and the PPE families was obtained by comparative DNA sequence analysis (Fig. 5). The gene for the PE–PGRS protein Rv0746 of BCG differs from that in H37Rv by the deletion of 29 codons and the insertion of 46 codons. Similar variation was seen in the gene for the PPE protein Rv0442 (data not shown). As these differences were all associated with repetitive sequences they could have resulted from intergenic or intragenic recombinational events or, more probably, from strand slippage during replication[32]. These mechanisms are known to generate antigenic variability in other bacterial pathogens[36].

There are several parallels between the PGRS proteins and the Epstein–Barr virus nuclear antigens (EBNAs). Members of both polypeptide families are glycine-rich, contain extensive Gly–Ala repeats, and exhibit variation in the length of the repeat region between different isolates. The Gly–Ala repeat region of EBNA1 functions as a *cis*-acting inhibitor of the ubiquitin/proteasome antigen-processing pathway that generates peptides presented in

苗-巴斯德开放阅读框(ORF)Rv0746编码的PE–PGRS的序列变异。

PPE蛋白家族的68个成员(图5)也有保守的N端结构域,由180个氨基酸残基组成,其后是序列和长度都有很大差异的C端片段。这些蛋白质至少可分为三组,其中一组组成了MPTR类,其特点是含有多个天冬酰胺–X–甘氨酸–X–甘氨酸–天冬酰胺–X–甘氨酸基序的串联拷贝。第二组在位置350左右包含一个独特的高保守基序,而第三组包含的蛋白,除了都含有长180个氨基酸残基的PPE结构域外没有其他相关性。

PE和PPE蛋白的亚细胞定位未知,仅在一个脂酶(Rv3097)的例子中证明是有功能的。对全面测序的麻风分枝杆菌[15]蛋白数据库进行检测,没有检测到PGRS或MPTR相关多肽,但发现了少量非MPTR亚组的PPE蛋白家族。这些蛋白质包括麻风病患者识别的一个主要抗原——富含丝氨酸的抗原[34]。尽管认定PE和PPE家族有生物学功能还为时过早,但对它们可能有免疫学功能的推测是很吸引人的。我们想到两个有趣的可能性。第一,它们可能代表结核分枝杆菌抗原变异的主要来源,否则结核分枝杆菌将是遗传和抗原均一的细菌。第二,这些富含甘氨酸的蛋白质可能会通过抑制抗原加工干扰免疫应答。

少量观察数据和结果支持抗原变异与PE和PPE家族蛋白存在相关的可能性。PGRS成员Rv1759是纤连蛋白结合蛋白,相对分子量为55,000(参考文献35),诱导不同的抗体反应,这表明要么不同个体产生了不同的免疫应答,要么结核分枝杆菌不同菌株的PGRS蛋白是不同的。后者的可能性得到临床分离株中不同的PGRS和MPTR序列的限制性片段长度多态性的支持[33]。PE和PPE家族内的遗传变异得到了比较DNA序列分析的直接支持(图5)。BCG的PE–PGRS蛋白Rv0746的基因与H37Rv中的不同,前者含29个密码子的缺失和46个密码子的插入。类似的变化在PPE蛋白Rv0442的基因中也可以看到(数据未显示)。由于这些差异都与重复序列相关,它们可能来自基因间或基因内的重组事件,或者更可能是复制过程中链的滑动引起的[32]。已知其他病原菌可利用这些机制产生抗原变异[36]。

PGRS蛋白和EB病毒核抗原(EBNAs)之间有一些相似之处。两个多肽家族成员都富含甘氨酸,含有丰富的甘氨酸–丙氨酸重复序列,不同分离株重复区域的长度不同。EBNA1的甘氨酸–丙氨酸重复区域是泛素/蛋白酶体抗原加工途径的顺式作用抑制剂,这一途径产生的肽出现在主要组织相容性复合物(MHC)I类分子存在的

the context of major histocompatibility complex (MHC) class I molecules[37,38]. MHC class I knockout mice are very susceptible to *M. tuberculosis*, underlining the importance of a cytotoxic T-cell response in protection against disease[3,39]. Given the many potential effects of the PPE and PE proteins, it is important that further studies are performed to understand their activity. If extensive antigenic variability or reduced antigen presentation were indeed found, this would be significant for vaccine design and for understanding protective immunity in tuberculosis, and might even explain the varied responses seen in different BCG vaccination programmes[40].

Pathogenicity. Despite intensive research efforts, there is little information about the molecular basis of mycobacterial virulence[41]. However, this situation should now change as the genome sequence will accelerate the study of pathogenesis as never before, because other bacterial factors that may contribute to virulence are becoming apparent. Before the completion of the genome sequence, only three virulence factors had been described[41]: catalase-peroxidase, which protects against reactive oxygen species produced by the phagocyte; *mce*, which encodes macrophage-colonizing factor[42]; and a sigma factor gene, *sigA* (aka *rpoV*), mutations in which can lead to attenuation[41]. In addition to these single-gene virulence factors, the mycobacterial cell wall[4] is also important in pathology, but the complex nature of its biosynthesis makes it difficult to identify critical genes whose inactivation would lead to attenuation.

On inspection of the genome sequence, it was apparent that four copies of *mce* were present and that these were all situated in operons, comprising eight genes, organized in exactly the same manner. In each case, the genes preceding *mce* code for integral membrane proteins, whereas *mce* and the following five genes are all predicted to encode proteins with signal sequences or hydrophobic stretches at the N terminus. These sets of proteins, about which little is known, may well be secreted or surface-exposed; this is consistent with the proposed role of Mce in invasion of host cells[42]. Furthermore, a homologue of *smpB*, which has been implicated in intracellular survival of *Salmonella typhimurium*, has also been identified[43]. Among the other secreted proteins identified from the genome sequence that could act as virulence factors are a series of phospholipases C, lipases and esterases, which might attack cellular or vacuolar membranes, as well as several proteases. One of these phospholipases acts as a contact-dependent haemolysin (N. Stoker, personal communication). The presence of storage proteins in the bacillus, such as the haemoglobin-like oxygen captors described above, points to its ability to stockpile essential growth factors, allowing it to persist in the nutrient-limited environment of the phagosome. In this regard, the ferritin-like proteins, encoded by *bfrA* and *bfrB*, may be important in intracellular survival as the capacity to acquire enough iron in the vacuole is very limited.

Methods

Sequence analysis. Initially, ~3.2 Mb of sequence was generated from cosmids[8] and the remainder was obtained from selected BAC clones[7] and 45,000 whole-genome shotgun clones. Sheared

条件下 [37,38]。MHC I 类敲除小鼠非常容易感染结核分枝杆菌，这强调了细胞毒性 T 细胞反应在抗病保护中的重要性 [3,39]。鉴于 PPE 和 PE 蛋白质的多种潜在作用，进一步研究了解它们的活性有很重要的意义。如果确实发现大量抗原变异或抗原呈递减少，将对设计疫苗和理解肺结核中保护性免疫有重大意义，甚至可能解释在不同的卡介苗接种方案中看到的不同反应 [40]。

致病性 尽管进行了大量研究，但关于分枝杆菌毒性的分子基础仍知之甚少 [41]。然而，这种情况现在应该会有所改变，因为基因组序列将前所未有地加快致病机制的研究，其他可能与毒力有关的细菌因子正在变得更加清楚。在完成基因组测序之前，只有三个毒力因子被描述过 [41]：过氧化氢酶–过氧化物酶，可以保护其免受巨噬细胞产生的活性氧的伤害；*mce*，编码巨噬细胞集落因子 [42]；一个 σ 因子基因，*sigA*（又名 *rpoV*），该基因的突变可导致减毒 [41]。除了这些单基因毒力因子，结核菌的细胞壁 [4] 在病理学上也很重要，但因其生物合成的复杂性致使鉴定失活导致减毒的关键基因很困难。

检查基因组序列发现，很显然存在 *mec* 的四个拷贝，而且都位于操纵子中，这些操纵子包括 8 个基因，组织形式完全相同。在每个操纵子中，*mec* 前面的基因编码整合膜蛋白，而 *mec* 及其后面的五个基因都被预测编码 N 端有信号肽序列或疏水序列的蛋白质。我们对这些蛋白质知之甚少，它们很可能是分泌或表面暴露的蛋白，这与有人已经提出的 Mce 在侵入宿主细胞中的功能一致 [42]。此外，我们还鉴定了 *smpB* 的一个同源基因，该基因与鼠伤寒沙门菌在细胞内的生存密切相关 [43]。根据基因组序列鉴定的其他分泌蛋白可能作为毒力因子的蛋白包括一系列磷脂酶 C、脂肪酶和酯酶，它们可能攻击细胞膜或者液泡膜，还包括几个蛋白酶。其中一个磷脂酶发挥接触依赖性溶血素的作用（斯托克，个人交流）。芽孢杆菌中存在储存蛋白，例如上述血红蛋白样氧捕捉蛋白，表明其具备储存必需生长因子的能力，使之能够在吞噬小体养分有限的环境中生存。在这点上，*bfrA* 和 *bfrB* 编码的铁蛋白样蛋白，可能对细胞内生存很重要，因为在液泡中获得足够铁的能力是非常有限的。

方　法

序列分析 起初，黏粒测序获得了约 3.2 Mb 的序列 [8]，其余序列是从选定的 BAC 克隆 [7] 和 45,000 个全基因组鸟枪克隆中得到的。将黏粒和 BAC 的打断片段（1.4~2.0 kb）克隆

fragments (1.4–2.0 kb) from cosmids and BACs were cloned into M13 vectors, whereas genomic DNA was cloned in pUC18 to obtain both forward and reverse reads. The PGRS genes were grossly underrepresented in pUC18 but better covered in the BAC and cosmid M13 libraries. We used small-insert libraries[44] to sequence regions prone to compression or deletion and, in some cases, obtained sequences from products of the polymerase chain reaction or directly from BACs[7]. All shotgun sequencing was performed with standard dye terminators to minimize compression problems, whereas finishing reactions used dRhodamine or BigDye terminators (http://www.sanger.ac.uk). Problem areas were verified by using dye primers. Thirty differences were found between the genomic shotgun sequences and the cosmids; twenty of which were due to sequencing errors and ten to mutations in cosmids (1 error per 320 kb). Less than 0.1% of the sequence was from areas of single-clone coverage, and ~0.2% was from one strand with only one sequencing chemistry.

Informatics. Sequence assembly involved PHRAP, GAP4 (ref. 45) and a customized perl script that merges sequences from different libraries and generates segments that can be processed by several finishers simultaneously. Sequence analysis and annotation was managed by DIANA (B.G.B. *et al.*, unpublished). Genes encoding proteins were identified by TB-parse[46] using a hidden Markov model trained on known *M. tuberculosis* coding and non-coding regions and translation-initiation signals, with corroboration by positional base preference. Interrogation of the EMBL, TREMBL, SwissProt, PROSITE[47] and in-house databases involved BLASTN, BLASTX[48], DOTTER (http://www.sanger. ac.uk) and FASTA[49]. tRNA genes were located and identified using tRNAscan and tRNAscan-SE[50]. The complete sequence, a list of annotated cosmids and linking regions can be found on our website (http:// www. sanger.ac.uk) and in MycDB (http://www.pasteur.fr/mycdb/).

(**393**, 537-544; 1998)

S. T. Cole*, R. Brosch*, J. Parkhill, T. Garnier*, C. Churcher, D. Harris, S. V. Gordon*, K. Eiglmeier*, S. Gas*, C. E. Barry III†, F. Tekaia‡, K. Badcock, D. Basham, D. Brown, T. Chillingworth, R. Connor, R. Davies, K. Devlin, T. Feltwell, S. Gentles, N. Hamlin, S. Holroyd, T. Hornsby, K. Jagels, A. Krogh§, J. McLean, S. Moule, L. Murphy, K. Oliver, J. Osborne, M. A. Quail, M.-A. Rajandream, J. Rogers, S. Rutter, K. Seeger, J. Skelton, R. Squares, S. Squares, J. E. Sulston, K. Taylor, S. Whitehead & B. G. Barrell

Sanger Centre, Wellcome Trust Genome Campus, Hinxton CB10 1SA, UK

* Unité de Génétique Moléculaire Bactérienne, and ‡ Unité de Génétique Moléculaire des Levures, Institut Pasteur, 28 rue du Docteur Roux, 75724 Paris Cedex 15, France

† Tuberculosis Research Unit, Laboratory of Intracellular Parasites, Rocky Mountain Laboratories, National Institute of Allergy and Infectious Diseases, National Institutes of Health, Hamilton, Montana 59840, USA

§ Center for Biological Sequence Analysis, Technical University of Denmark, Lyngby, Denmark

Received 15 April; accepted 8 May 1998.

References:

1. Snider, D. E. Jr, Raviglione, M. & Kochi, A. in *Tuberculosis: Pathogenesis, Protection, and Control* (ed. Bloom, B. R.) 2-11 (Am. Soc. Microbiol., Washington DC, 1994).

2. Wheeler, P. R. & Ratledge, C. in *Tuberculosis: Pathogenesis, Protection, and Control* (ed. Bloom, B. R.) 353-385 (Am. Soc. Microbiol., Washington DC, 1994).

3. Chan, J. & Kaufmann, S. H. E. in *Tuberculosis: Pathogenesis, Protection, and Control* (ed. Bloom, B. R.) 271-284 (Am. Soc. Microbiol., Washington DC, 1994).

4. Brennan, P. J. & Draper, P. in *Tuberculosis: Pathogenesis, Protection, and Control* (ed. Bloom, B. R.) 271-284 (Am. Soc. Microbiol., Washington DC, 1994).

5. Kolattukudy, P. E., Fernandes, N. D., Azad, A. K., Fitzmaurice, A. M. & Sirakova, T. D. Biochemistry and molecular genetics of cell-wall lipid biosynthesis in mycobacteria. *Mol. Microbiol.* **24**, 263-270 (1997).

到 M13 载体上，将基因组 DNA 克隆到质粒 pUC18 上获得正向和反向可读片段。PGRS 基因在 pUC18 中出现的频率严重低于正常水平，但在 BAC 和黏粒 M13 文库中覆盖得很好。我们使用小插入片段文库[44]对易压缩或缺失的区域测序，在某些情况下，从 PCR 反应产物中或直接从 BACs 中获得序列[7]。所有鸟枪测序采用标准染料终止剂，以尽量减少压缩问题，终止反应用 dRhodamine 或 BigDye 终止剂（http://www.sanger.ac.uk）。存在问题的区域用染料标记引物验证。在基因组鸟枪测序和黏粒测序的序列之间发现了 30 个差异；其中 20 个是由测序错误引起的，10 个是由黏粒中的突变（1 个错误/320 kb）引起的。小于 0.1% 的序列来自单个克隆覆盖的区域，大约 0.2% 来自只有一个测序化学反应的一条链。

信息学 序列组装使用了 PHRAP、GAP4（参考文献 45）和定制的 perl 脚本，该脚本用于合并不同文库中的序列，并生成可以同时被几个基因组完成工具处理的片段。序列分析和注释由 DIANA（巴雷尔等，未发表）完成。使用 TB-parse[46]鉴定蛋白质编码基因，使用马尔可夫模型分析已知的结核分枝杆菌编码和非编码区域及翻译起始信号，通过位置碱基偏好进行验证。综合使用 EMBL、TREMBL、SwissProt、PROSITE[47]数据库和内部数据库 BLASTN、BLASTX[48]、DOTTER（http://www.sanger.ac.uk）和 FASTA[49]等。使用 tRNAscan 和 tRNAscan-SE[50]定位和鉴定 tRNA 基因。在我们的网站（http://www.sanger.ac.uk）上和 MycDB（http://www.pasteur.fr/mycdb/）中可找到完整的基因组序列、一系列已注释的黏粒和相连接区域。

（李梅 翻译；解彬彬 审稿）

6. Sreevatsan, S. *et al.* Restricted structural gene polymorphism in the *Mycobacterium tuberculosis* complex indicates evolutionarily recent global dissemination. *Proc. Natl Acad. Sci. USA* **94,** 9869-9874 (1997).

7. Brosch, R. *et al.* Use of a *Mycobacterium tuberculosis* H37Rv bacterial artificial chromosome library for genome mapping, sequencing and comparative genomics. *Infect. Immun.* **66,** 2221-2229 (1998).

8. Philipp, W. J. *et al.* An integrated map of the genome of the tubercle bacillus, *Mycobacterium tuberculosis* H37Rv, and comparison with *Mycobacterium leprae*. *Proc. Natl Acad. Sci. USA* **93,** 3132-3137 (1996).

9. Blattner, F. R. *et al.* The complete genome sequence of *Escherichia coli* K-12. *Science* **277,** 1453-1462 (1997).

10. Cole, S. T. & Saint-Girons, I. Bacterial genomics. *FEMS Microbiol. Rev.* **14,** 139-160 (1994).

11. Freiberg, C. *et al.* Molecular basis of symbiosis between *Rhizobium* and legumes. *Nature* **387,** 394-401 (1997).

12. Bardarov, S. *et al.* Conditionally replicating mycobacteriophages: a system for transposon delivery to *Mycobacterium tuberculosis*. *Proc. Natl Acad. Sci. USA* **94,** 10961-10966 (1997).

13. Mahairas, G. G., Sabo, P. J., Hickey, M. J., Singh, D. C. & Stover, C. K. Molecular analysis of genetic differences between *Mycobacterium bovis* BCG and virulent *M. bovis*. *J. Bacteriol.* **178,** 1274-1282 (1996).

14. Kunst, F. *et al.* The complete genome sequence of the gram-positive bacterium *Bacillus subtilis*. *Nature* **390,** 249-256 (1997).

15. Smith, D. R. *et al.* Multiplex sequencing of 1.5 Mb of the *Mycobacterium leprae* genome. *Genome Res.* **7,** 802-819 (1997).

16. Greenacre, M. *Theory and Application of Correspondence Analysis* (Academic, London, 1984).

17. Ratledge, C. R. in *The Biology of the Mycobacteria* (eds Ratledge, C. & Stanford, J.) 53-94 (Academic, San Diego, 1982).

18. Av-Gay, Y. & Davies, J. Components of eukaryotic-like protein signaling pathways in *Mycobacterium tuberculosis*. *Microb. Comp. Genomics* **2,** 63-73 (1997).

19. Cole, S. T. & Telenti, A. Drug resistance in *Mycobacterium tuberculosis*. *Eur. Resp. Rev.* **8,** 701S-713S (1995).

20. Riley, M. & Labedan, B. in Escherichia coli *and* Salmonella (ed. Neidhardt, F. C.) 2118-2202 (ASM, Washington, 1996).

21. Mdluli, K. *et al.* Inhibition of a *Mycobacterium tuberculosis* β-ketoacyl ACP synthase by isoniazid. *Science* **280,** 1607-1610 (1998).

22. Banerjee, A. *et al. inhA*, a gene encoding a target for isoniazid and ethionamide in *Mycobacterium tuberculosis*. *Science* **263,** 227-230 (1994).

23. Hopwood, D. A. Genetic contributions to understanding polyketide synthases. *Chem. Rev.* **97,** 2465-2497 (1997).

24. Minnikin, D. E. in *The Biology of the Mycobacteria* (eds Ratledge, C. & Stanford, J.) 95-184 (Academic, London, 1982).

25. Barry, C. E. III *et al.* Mycolic acids: structure, biosynthesis, and physiological functions. *Prog. Lipid Res.* (in the press).

26. Belisle, J. T. *et al.* Role of the major antigen of *Mycobacterium tuberculosis* in cell wall biogenesis. *Science* **276,** 1420-1422 (1997).

27. Marahiel, M. A., Stachelhaus, T. & Mootz, H. D. Modular peptide synthetases involved in nonribosomal peptide synthesis. *Chem. Rev.* **97,** 2651-2673 (1997).

28. Gobin, J. *et al.* Iron acquisition by *Mycobacterium tuberculosis*: isolation and characterization of a family of iron-binding exochelins. *Proc. Natl Acad. Sci. USA* **92,** 5189-5193 (1995).

29. Young, D. B. & Fruth, U. in *New Generation Vaccines* (eds Levine, M., Woodrow, G., Kaper, J. & Cobon, G. S.) 631-645 (Marcel Dekker, New York, 1997).

30. Sorensen, A. L., Nagai, S., Houen, G., Andersen, P. & Anderson, A. B. Purification and characterization of a low-molecular-mass T-cell antigen secreted by *Mycobacterium tuberculosis*. *Infect. Immun.* **63,** 1710-1717 (1995).

31. Hermans, P. W. M., van Soolingen, D. & van Embden, J. D. A. Characterization of a major polymorphic tandem repeat in *Mycobacterium tuberculosis* and its potential use in the epidemiology of *Mycobacterium kansasii* and *Mycobacterium gordonae*. *J. Bacteriol.* **174,** 4157-4165 (1992).

32. Poulet, S. & Cole, S. T. Characterisation of the polymorphic GC-rich repetitive sequence (PGRS) present in *Mycobacterium tuberculosis*. *Arch. Microbiol.* **163,** 87-95 (1995).

33. Cole, S. T. & Barrell, B. G. in *Genetics and Tuberculosis* (eds Chadwick, D. J. & Cardew, G., *Novartis Foundation Symp. 217*) 160-172 (Wiley, Chichester, 1998).

34. Vega-Lopez, F. *et al.* Sequence and immunological characterization of a serine-rich antigen from *Mycobacterium leprae*. *Infect. Immun.* **61,** 2145-2153 (1993).

35. Abou-Zeid, C. *et al.* Genetic and immunological analysis of *Mycobacterium tuberculosis* fibronectin-binding proteins. *Infect. Immun.* **59,** 2712-2718 (1991).

36. Robertson, B. D. & Meyer, T. F. Genetic variation in pathogenic bacteria. *Trends Genet.* **8,** 422-427 (1992).

37. Levitskaya, J. *et al.* Inhibition of antigen processing by the internal repeat region of the Epstein-Barr virus nuclear antigen-1. *Nature* **375,** 685-688 (1995).

38. Levitskaya, J., Sharipo, A., Leonchiks, A., Ciechanover, A. & Masucci, M. G. Inhibition of ubiquitin/ proteasome-dependent protein degradation by the Gly-Ala repeat domain of the Epstein-Barr virus nuclear antigen 1. *Proc. Natl Acad. Sci. USA* **94,** 12616-12621 (1997).

39. Flynn, J. L., Goldstein, M. A., Treibold, K. J., Koller, B. & Bloom, B. R. Major histocompatibility complex class-I restricted T cells are required for resistance to *Mycobacterium tuberculosis* infection. *Proc. Natl Acad. Sci. USA* **89,** 12013-12017 (1992).

40. Bloom, B. R. & Fine, P. E. M. in *Tuberculosis: Pathogenesis, Protection, and Control* (ed. Bloom, B. R.) 531-557 (Am. Soc. Microbiol., Washington DC, 1994).

41. Collins, D. M. In search of tuberculosis virulence genes. *Trends Microbiol.* **4,** 426-430 (1996).

42. Arruda, S., Bomfim, G., Knights, R., Huima-Byron, T. & Riley, L. W. Cloning of an *M. tuberculosis* DNA fragment associated with entry and survival inside cells. *Science* **261,** 1454-1457 (1993).

43. Baumler, A. J., Kusters, J. G., Stojikovic, I. & Heffron, F. *Salmonella typhimurium* loci involved in survival within macrophages. *Infect. Immun.* **62,** 1623-1630 (1994).

44. McMurray, A. A., Sulston, J. E. & Quail, M. A. Short-insert libraries as a method of problem solving in genome sequencing. *Genome Res.* **8,** 562-566 (1998).

45. Bonfield, J. K., Smith, K. F. & Staden, R. A new DNA sequence assembly program. *Nucleic Acids Res.* **24,** 4992-4999 (1995).

46. Krogh, A., Mian, I. S. & Haussler, D. A hidden Markov model that finds genes in *E. coli* DNA. *Nucleic Acids Res.* **22,** 4768-4778 (1994).

47. Bairoch, A., Bucher, P. & Hofmann, K. The PROSITE database, its status in 1997. *Nucleic Acids Res.* **25,** 217-221 (1997).

48. Altschul, S., Gish, W., Miller, W., Myers, E. & Lipman, D. A basic local alignment search tool. *J. Mol. Biol.* **215,** 403-410 (1990).

49. Pearson, W. & Lipman, D. Improved tools for biological sequence comparisons. *Proc. Natl Acad. USA* **85,** 2444-2448 (1988).

50. Lowe, T. M. & Eddy, S. R. tRNAscan-SE: a program for improved detection of transfer RNA genes in genomic DNA. *Nucleic Acids Res.* **25,** 955-964 (1997).

Acknowledgements. We thank Y. Av-Gay, F.-C. Bange, A. Danchin, B. Dujon, W. R. Jacobs Jr, L. Jones, M. McNeil, I. Moszer, P. Rice and J. Stephenson for advice, reagents and support. This work was supported by the Wellcome Trust. Additional funding was provided by the Association Francaise Raoul Follereau, the World Health Organisation and the Institut Pasteur. S.V.G. received a Wellcome Trust travelling research fellowship.

Correspondence and requests for materials should be addressed to B.G.B. (barrell@sanger.ac.uk) or S.T.C. (stcole@pasteur.fr). The complete sequence has been deposited in EMBL/GenBank/DDBJ as MTBH37RV, accession number AL123456.

Two Feathered Dinosaurs from Northeastern China

Ji Qiang *et al.*

Editor's Note

Sinosauropteryx was soon followed by even more spectacular creatures. This report from Ji and colleagues describes two: *Protarchaeopteryx* and *Caudipteryx* which, like *Sinosauropteryx*, came from Liaoning Province in northeastern PRC. Phylogenetic analysis placed these long-legged, ground-living runners close to the origin of birds, although they were both more primitive than *Archaeopteryx* and presumably incapable of flight. Unlike *Sinosauropteryx*, however, there was no doubting that these creatures had feathers. *Protarchaeopteryx* had a switch of feathers on the end of its tail, and *Caudipteryx* appeared to have feathers fringing its forelimbs. These creatures provided a graphic demonstration that feathers appeared in evolution long before dinosaurs became capable of flight.

Current controversy over the origin and early evolution of birds centres on whether or not they are derived from coelurosaurian theropod dinosaurs. Here we describe two theropods from the Upper Jurassic/Lower Cretaceous Chaomidianzi Formation of Liaoning Province, China. Although both theropods have feathers, it is likely that neither was able to fly. Phylogenetic analysis indicates that they are both more primitive than the earliest known avialan (bird), *Archaeopteryx*. These new fossils represent stages in the evolution of birds from feathered, ground-living, bipedal dinosaurs.

Dinosauria Owen 1842
Theropoda Marsh 1881
Maniraptora Gauthier 1986
Unnamed clade
Protarchaeopteryx robusta Ji & Ji 1997

Holotype. National Geological Museum of China, NGMC 2125 (Figs 1, 2 and 3).

Locality and horizon. Sihetun area near Beipiao City, Liaoning, China. Jiulongsong Member of Chaomidianzi Formation, Jehol Group[1]. This underlies the Yixian Formation, the age of which has been determined to be Late Jurassic to Early Cretaceous[2-4].

Diagnosis. Large straight premaxillary teeth, and short, bulbous maxillary and dentary teeth, all of which are primitively serrated. Rectrices form a fan at the end of the tail.

Description. The skull of *Protarchaeopteryx* is shorter than the femur (Table 1). There are

中国东北地区两类长羽毛的恐龙

季强等

编者按

继中华龙鸟发现之后更多引人注目的生物相继被发现。这篇季强和同事们发表的文章描述了其中两种：原始祖鸟和尾羽龙，它们跟中华龙鸟一样来自中国东北地区的辽宁省。系统发育分析显示这些长腿、陆地生活的擅跑者与鸟类的起源关系密切，尽管它们都比始祖鸟原始，并且可能不会飞行。不过，与中华龙鸟不同的是，它们毫无疑问确实存在羽毛。原始祖鸟尾巴的末端发育一簇羽毛，尾羽龙的前肢边缘也覆盖羽毛。这些生物为羽毛的出现远远早于恐龙变得能够飞行提供了形象的说明。

最近关于鸟类起源及其早期演化的争论集中在鸟类是否起源于兽脚类恐龙的虚骨龙类。本文描述了在中国辽宁省上侏罗统/下白垩统炒米甸子组的两类兽脚类恐龙。虽然这两类都有羽毛，但很可能都不会飞行。系统发育分析表明两者都比已知最早的鸟类——始祖鸟更加原始。这些新的化石代表了从长有羽毛、地面生活和两足行走的恐龙向鸟类演化的阶段。

恐龙总目 Dinosauria Owen 1842

兽脚亚目 Theropoda Marsh 1881

手盗龙类 Maniraptora Gauthier 1986

未命名分类单元 Unnamed clade

粗壮原始祖鸟 *Protarchaeopteryx robusta* Ji & Ji 1997

正模标本　中国地质博物馆，标本编号 NGMC 2125（图 1、2 和 3）。

产地与层位　中国辽宁省北票市四合屯，热河群炒米甸子组九龙松段 [1]。位于义县组之下，义县组时代为晚侏罗世至早白垩世 [2-4]。

鉴定特征　大且直的前颌骨齿，短而呈球状的上颌骨齿和齿骨齿，所有的牙齿都是原始的锯齿状。尾羽在尾部末端形成扇形。

描述　原始祖鸟的头骨比股骨短（表 1）。有四颗带锯齿的前颌骨齿（图 1c），齿

four serrated premaxillary teeth (Fig. 1c), with crown heights of up to 12 mm. Premaxillary teeth of coelophysids[5], compsognathids[6,7] and early birds lack serrations, but premaxillary denticles are present in most other theropods. Six maxillary and seven dentary teeth are preserved (Fig. 1), all of which are less than a quarter the height of the premaxillary teeth. They most closely resemble those of *Archaeopteryx*[8] in shape (Figs 1b, c and 2b, c), but have anterior and posterior serrations (7–10 serrations per mm).

Table 1. Lengths of elements in *Protarchaeopteryx* and *Caudipteryx*

Element	NGMC 2125	NGMC 97-4-A	NGMC 97-9-A
Body length	690	890	725
Skull	70	76	79
Sternal plates	25	36	–
Humerus	88	69	70
Arm (humerus to end of phalange II-2)	297	214	220
Ilium	95	101	–
Ischium	–	77	–
Leg (femur to end of phalange III-4)	450	550	540
Femur	122	147	149
Tibia	160	188	182
Metatarsal III	85	115	117

Length measurements are given in millimetres. NGMC 2125, *Protarchaeopteryx*; NGMC 97-4-A and NGMC 97-9-A, *Caudipteryx*.

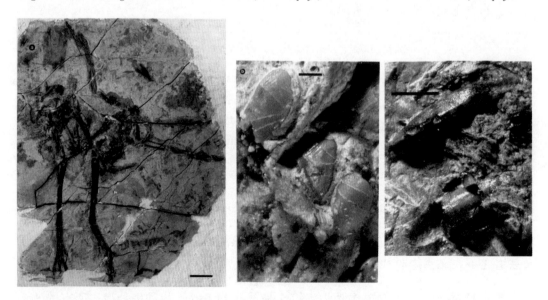

Fig. 1. *Protarchaeopteryx robusta*. **a**, NGMC 2125, holotype. Scale bar, 5 cm. **b**, Fourth to sixth left dentary teeth. Scale bar, 1 mm. **c**, Premaxillary teeth showing small serrations. Scale bar, 5 mm.

冠高达 12 毫米。腔骨龙类 [5]、美颌龙类 [6,7] 和早期鸟类的前颌骨齿缺少锯齿，但前颌骨齿的锯齿出现在大多数其他的兽脚类恐龙中。标本保存了六颗上颌骨齿和七颗齿骨齿（图 1），这些牙齿要短于前颌骨齿高度的四分之一。它们在形态上非常接近始祖鸟的牙齿 [8]（图 1b、1c 和图 2b、2c），但其前缘和后缘具有锯齿（每毫米 7～10 个锯齿）。

表 1. 原始祖鸟和尾羽龙骨骼成分的长度

骨骼成分	NGMC 2125	NGMC 97-4-A	NGMC 97-9-A
身长	690	890	725
头骨	70	76	79
胸骨板	25	36	–
肱骨	88	69	70
前臂（肱骨到第 II 指第 2 指骨末端）	297	214	220
髂骨	95	101	–
坐骨	–	77	–
后肢（股骨到第 III 趾第 4 趾骨末端）	450	550	540
股骨	122	147	149
胫骨	160	188	182
第 III 跖骨	85	115	117

长度测量单位是毫米。NGMC 2125 为原始祖鸟，NGMC 97-4-A 和 NGMC 97-9-A 为尾羽龙。

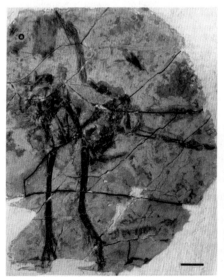

图 1. 粗壮原始祖鸟。**a**，NGMC 2125，正模标本，比例尺为 5 厘米。**b**，左侧第 4 到第 6 齿骨齿，比例尺为 1 毫米。**c**，前颌骨齿的小锯齿，比例尺为 5 毫米。

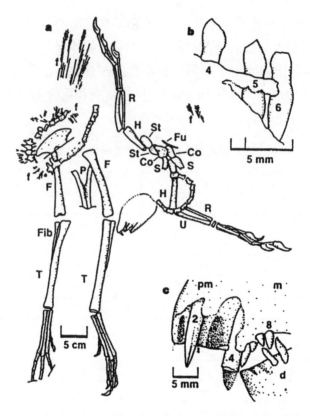

Fig. 2. *Protarchaeopteryx robusta*. **a**, Outline of the specimen shown in Fig. 1a. **b**, Outline of the left dentary teeth shown in Fig. 1b. **c**, Drawing of the front of the jaws, showing the large size of the premaxillary teeth compared with maxillary and dentary ones. Abbreviations: Co, coracoid; d, dentary; F, femur; f, feathers; Fib, fibula; Fu, furcula; H, humerus; m, maxilla; P, pubis; pm, premaxilla; R, radius; S, scapula; St, sternal plate; T, tibia; U, ulna. Numbers represent tooth positions from front to back.

The amphicoelous posterior cervicals are the same length as the posterior dorsals, which have large pleurocoels. If the lengths of missing segments of the tail are accounted for, there were fewer than 28 caudals. Vertebrae increase in length from proximal to mid-caudals, as in most non-avian coelurosaurs.

There are two thin, flat, featureless sternal plates. The clavicles are fused into a broad, U-shaped furcula (interclavicular angle is about 60°) as in *Archaeopteryx*, *Confuciusornis* and many non-avian theropods. The forelimb is shorter than the hindlimb. The arm is shorter (compared to the femur) than it is in birds, but is longer than those of long-armed non-avian coelurosaurs such as dromaeosaurids and oviraptorids (Table 2). The better preserved right wrist of NGMC 2125 has a single semilunate carpal capping the first two metacarpals. The hand has the normal theropod phalangeal formula of 2-3-4-x-x. The manus is longer than either the humerus or radius. Compared to femur length, the hand is more elongate than those of any theropods other than *Archaeopteryx*[9] and *Confuciusornis* (Table 2). More advanced birds such as *Cathayornis* have shorter hands[10]. Phalanges III-1 and III-2 in the hand of *Protarchaeopteryx* are almost the same size, and are about half the length of III-3. The unguals are long and sharp, and keratinous sheaths are preserved on two of them.

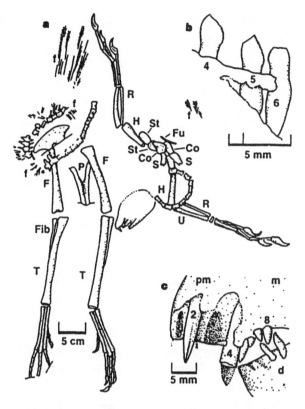

图 2. 粗壮原始祖鸟。**a**，根据图 1a 绘制的标本轮廓。**b**，根据图 1b 绘制的左侧齿骨齿轮廓。**c**，颌部前部素描图，显示与上颌骨齿和齿骨齿相比较大的前颌骨齿。缩写：Co，乌喙骨；d，齿骨；F，股骨；f，羽毛；Fib，腓骨；Fu，叉骨；H，肱骨；m，上颌骨；P，耻骨；pm，前颌骨；R，桡骨；S，肩胛骨；St，胸骨板；T，胫骨；U，尺骨。数字代表牙齿位置的前后顺序。

双凹型的后部颈椎与后部背椎等长，背椎具有大的侧凹。如果算上尾巴缺失的片段，尾椎应该少于 28 节。尾椎的椎体从近端到中部逐渐加长，这与大多数非鸟虚骨龙类一样。

有两块薄而扁平且没有明显特征的胸骨板。锁骨愈合成宽阔的 U 形叉骨（锁骨间夹角为 60 度），类似于始祖鸟、孔子鸟和很多非鸟兽脚类恐龙。前肢比后肢短。前臂（与股骨相比较）要比鸟类的对应部位短，但要比那些非鸟虚骨龙类如驰龙类、窃蛋龙类等较长的前臂长（表 2）。NGMC 2125 的右腕保存很好，单一的半月形腕骨覆盖前两个掌骨。手指指式是兽脚类恐龙通常的 2-3-4-x-x。手部比肱骨和桡骨都长。和股骨长度相比较，手指要比任何兽脚类（除了始祖鸟[9]和孔子鸟）的都要长（表 2）。较进步鸟类诸如华夏鸟的手指比较短[10]。原始祖鸟的第Ⅲ指第 1、2 指骨几乎等长，大约为第 3 指骨长的一半。指爪长而锋利，其中两枚指爪上都保存有角质鞘。

Table 2. Relative proportions of elements in relevant avian and non-avian theropods

Element	Drom	Ov	Tro	Cx	Px	Ax	Con
Arm/F	1.8–2.6	1.5–1.8	1.8	1.5	2.4	3.7	3.9
S/H	0.8	1.0–1.2	–	1.1	–	0.6	0.8
R/H	0.7–0.8	0.8–0.9	0.6–0.7	0.9	0.8	0.9	0.8
Manus/H	0.9–1.2	1.2–1.4	1.3	1.2	1.6	1.2	1.3
Manus/F	1.0	0.7–1.0	0.8	0.6	1.2	1.5	1.6
McI/McII	0.4–0.5	0.4–0.6	0.3	0.4	0.4	0.3	0.4
T/F	1.1–1.4	1.2	1.1–1.2	1.2	1.3	1.4	1.1
Leg/F	3.6	3.3	3.8	3.7	3.7	3.8	3.3
Leg/arm	1.4	1.7	2.1	2.5	1.5	1.1	0.8

All data were collected from original specimens by P.J.C. Ax, *Archaeopteryx*; Con, *Confuciusornis*; Cx, *Caudipteryx*; Drom, dromaeosaurids; F, femur; H, humerus; Mc, metacarpal; Ov, oviraptorids; Px, *Protarchaeopteryx*; R, radius; S, scapula; T, tibia; Tro, troodontids.

The preacetabular blade of the ilium is about the same length as the postacetabular blade. The pubic boot expands posteriorly. Anteriorly, the pubis is not exposed.

The tibia is longer than the femur, as it is in most advanced theropods and early birds. It is not known if the fibula extended to the tarsus.

The metatarsals are separate from each other and the distal tarsals. Metatarsal I is centred halfway up the posteromedial edge of the second metatarsal. In perching birds such as *Sinornis*[9], metatarsal I is positioned near the end of metatarsal II and is retroverted. Its condition in *Archaeopteryx* is intermediate. Pedal unguals are smaller than manual unguals.

A clump of at least six plumulaceous feathers is preserved anterior to the chest, with some showing well-developed vanes (Fig. 3a). Evenly distributed plumulaceous feathers up to 27 mm long are associated with ten proximal caudal vertebrae. Twenty-millimetre plumulaceous feathers are preserved along the lateral side of the right femur and the proximal end of the left femur.

Parts of more than twelve rectrices are preserved[11] attached to the distal caudals. One of the symmetrical tail feathers (Fig. 3b) extends 132 mm from the closest tail vertebra, and has a long tapering rachis with a basal diameter of 1.5 mm. The well-formed pennaceous vanes of *Protarchaeopteryx* show that barbules were present. The vane is 5.3 mm wide on either side of the rachis. At midshaft, five barbs come off the rachis every 5 mm (compared with six in *Archaeopteryx*), and individual barbs are 15 mm long. As in modern rectrices, the barbs at the base of the feather are plumulaceous.

表 2. 相关的鸟类和非鸟兽脚类恐龙的骨骼成分比例

骨骼成分	Drom	Ov	Tro	Cx	Px	Ax	Con
Arm/F	1.8～2.6	1.5～1.8	1.8	1.5	2.4	3.7	3.9
S/H	0.8	1.0～1.2	–	1.1	–	0.6	0.8
R/H	0.7～0.8	0.8～0.9	0.6～0.7	0.9	0.8	0.9	0.8
Manus/H	0.9～1.2	1.2～1.4	1.3	1.2	1.6	1.2	1.3
Manus/F	1.0	0.7～1.0	0.8	0.6	1.2	1.5	1.6
McI/McII	0.4～0.5	0.4～0.6	0.3	0.4	0.4	0.3	0.4
T/F	1.1～1.4	1.2	1.1～1.2	1.2	1.3	1.4	1.1
Leg/F	3.6	3.3	3.8	3.7	3.7	3.8	3.3
Leg/arm	1.4	1.7	2.1	2.5	1.5	1.1	0.8

所有数据由菲利普·柯里从原始标本收集。Arm，前臂；Ax，始祖鸟；Con，孔子鸟；Cx，尾羽龙；Drom，驰龙类；F，股骨；H，肱骨；Leg，后肢；Manus，手部；Mc，掌骨；Ov，窃蛋龙类；Px，原始祖鸟；R，桡骨；S，肩胛骨；T，胫骨；Tro，伤齿龙类。

髂骨的前、后髋臼区域大约等长。耻骨的靴状突向后扩展。耻骨前部没有暴露出来。

胫骨比股骨长，这与大多数进步的兽脚类和早期鸟类一样。但不太清楚腓骨是否延伸到跗骨。

跖骨彼此分离，并与跗骨远端分离。第Ⅰ跖骨位于第Ⅱ跖骨后内侧缘的一半位置。在栖禽类如中国鸟中[9]，第Ⅰ跖骨的位置接近第Ⅱ跖骨末端并向后倾。始祖鸟的情形介于中间。趾爪比指爪小得多。

一丛至少包括 6 枚绒羽的羽毛保存在胸部前方，有一些显示发育很好的羽片（图 3a）。均匀分布的长达 27 毫米的绒羽，与近端 10 节尾椎相关联。20 毫米长的绒羽沿着右股骨侧面及左股骨近端保存。

多于 12 根的部分尾羽保存下来，附着在远端尾椎上[11]。其中一根对称的尾羽（图 3b）从最近的尾椎延伸 132 毫米，羽轴长而向末端逐渐变尖，羽轴基部直径为 1.5 毫米。原始祖鸟已经成型的羽片保存有羽小枝。在羽轴任一侧的羽片宽 5.3 毫米。在羽轴中部，每隔 5 毫米有 5 个羽枝从羽轴伸出（始祖鸟是 6 个），单个羽枝长 15 毫米。而现生鸟类的尾羽中，羽毛基部的羽枝为似绒羽状。

Fig. 3. *Protarchaeopteryx robusta*, NGMC 2125. **a**, Contour and plumulaceous feathers. Scale bar, 10 mm. **b**, Rectrices. Scale bar, 5 mm.

Maniraptora Gauthier 1986
Unnamed clade

Diagnosis. The derived presence of a short tail (less than 23 caudal vertebrae) and arms with remiges attached to the second digit.

Caudipteryx zoui gen. et sp. nov.

Etymology. "*Caudipteryx*" means "tail feather"; "*zoui*" refers to Zou Jiahua, vice-premier of China and an avid supporter of the scientific work in Liaoning.

Holotype. NGMC 97-4-A (Figs 4 and 5b).

256

图 3. 粗壮原始祖鸟，NGMC 2125。**a**，廓羽及绒羽，比例尺为 10 毫米。**b**, 尾羽，比例尺为 5 毫米。

手盗龙类 Maniraptora Gauthier 1986

未命名分类单元 Unnamed clade

鉴定特征 具有的进步特征为短尾（尾椎少于 23 节）以及前臂飞羽着生于第 Ⅱ 指上。

邹氏尾羽龙（新属新种） *Caudipteryx zoui* gen. et sp. nov.

词源 "*Caudipteryx*"意为"尾羽"；"*zoui*"指的是时任中国国务院副总理的邹家华，他大力支持辽宁的化石科学研究。

正模标本 NGMC 97-4-A（图 4 和图 5b）。

Fig. 4. *Caudipteryx zoui*, holotype, NGMC 97-4-A. Scale bar, 5 cm.

Paratype. NGMC 97-9-A (Fig. 5d).

Locality and horizon. Sihetun area, Liaoning. Jiulongsong Member of the Chaomidianzi Formation.

Diagnosis. Elongate, hooked premaxillary teeth with broad roots; maxilla and dentary edentulous. Tail short (one-quarter of the length of the body). Arm is long for a non-avian theropod; short manual claws. Leg-to-arm ratio, 2.5.

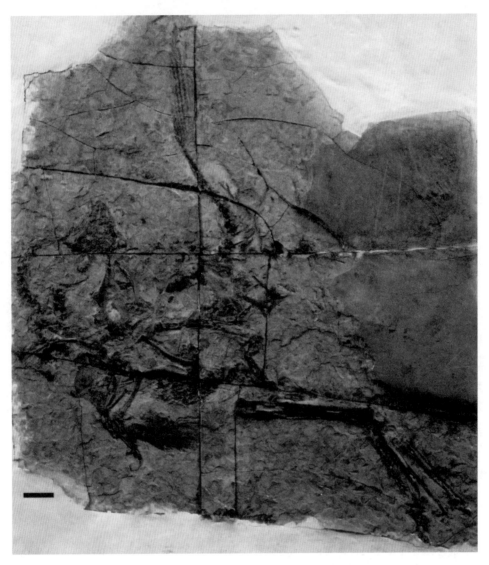

图 4. 邹氏尾羽龙，正模标本，NGMC 97-4-A。比例尺为 5 厘米。

副模标本 NGMC 97-9-A（图 5d）。

产地与层位 辽宁省四合屯，炒米甸子组九龙松段。

鉴定特征 长的钩状前颌骨齿，齿根比较宽；上颌骨和齿骨缺少牙齿。尾巴短（约为身体长度的四分之一）。作为非鸟兽脚类恐龙，其前臂较长；指爪短小。后肢与前臂长度之比为 2.5。

Fig. 5. *Caudipteryx zoui*. **a**, Haemal spines from the fourth, sixth, eighth, eleventh and thirteenth caudal vertebrae (from left to right) of NGMC 97-4-A in left lateral view. **b**, Drawing of the specimen shown in Fig. 4. **c**, Wrist of NGMC 97-4-A. **d**, Drawing of NGMC 97-9-A. **e**, Proximal tarsals of NGMC 97-9-A. Abbreviations: a, astragalus; c, calcaneum; Co, coracoid; F, femur; g, gastroliths; H, humerus; I, ilium; Is, ischium; P, pubis; R, radius; S, scapula; St, sternal plate; T, tibia; U, ulna; ?, possibly fragment of gastralia. Roman numerals represent digit numbers.

Description. The skulls of both specimens of *Caudipteryx* are shorter than the corresponding femora because of a reduction in the length of the antorbital region. The relatively large premaxilla (Figs 6 and 7) borders most of the large external naris. The maxilla and nasal are short, but the frontals and jugals are long. The lacrimal of NGMC 97-4-A is an inverted L-shaped, pneumatic bone. Scleral plates are preserved in the 20-mm-diameter orbits of both specimens. The tall quadratojugal seems to have contacted the squamosal and abutted the lateral surface of the quadrate. The single-headed quadrate is vertical in orientation. The ectopterygoid has a normal theropod hooklike jugal process. There is a broad, beak-like margin at the symphysis of the dentaries. Posteriorly, the dentary bifurcates around a large external mandibular fenestra as in oviraptorids. A well-developed, sliding intramandibular joint is present between dentary and surangular.

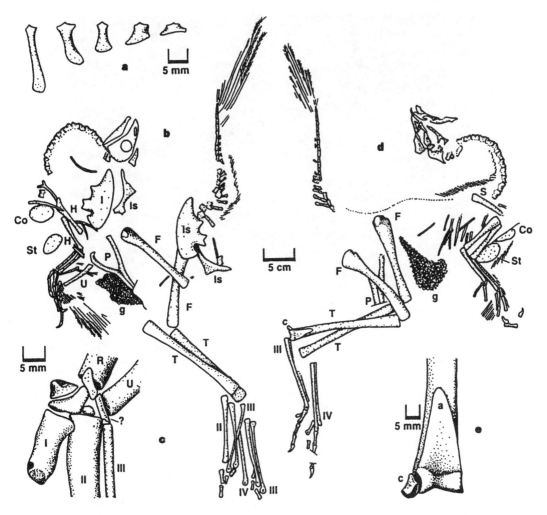

图 5. 邹氏尾羽龙。**a**，NGMC 97-4-A 的第 4、6、8、11、13 尾椎相应的脉弧左侧视（从左至右）。**b**，图 4 中标本的素描图。**c**，NGMC 97-4-A 的腕部。**d**，NGMC 97-9-A 的素描图。**e**，NGMC 97-9-A 的近端跗骨。缩写：a，距骨；c，跟骨；Co，乌喙骨；F，股骨；g，胃石；H，肱骨；I，髂骨；Is，坐骨；P，耻骨；R，桡骨；S，肩胛骨；St，胸骨板；T，胫骨；U，尺骨；？，可能的腹膜肋片段。罗马数字表示指/趾的顺序。

描述 两件尾羽龙标本的头骨长度要比各自的股骨短，这是由眶前区缩短造成的。相对较大的前颌骨（图 6 和图 7）构成较大的外鼻孔的大部分边缘。上颌骨和鼻骨较短，而额骨和轭骨较长。NGMC 97-4-A 标本的泪骨呈倒转的"L"形，为含气骨骼。两件标本的巩膜板均保存在直径 20 毫米的眼眶中。高的方轭骨似乎与鳞骨连接且与方骨的侧面相邻。单头的方骨近于直立。外翼骨具有兽脚类恐龙通常具有的钩状轭骨突。齿骨联合的边缘表面宽阔且呈鸟喙状。像窃蛋龙类一样，齿骨后部围绕着大的外下颌孔分叉。非常发育的、可滑动的颌内关节存在于齿骨与上隅骨之间。

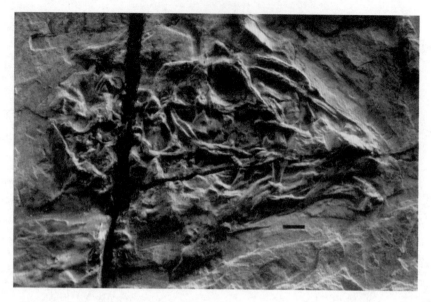

Fig. 6. *Caudipteryx zoui*, skull of NGMC 97-9-A in right lateral view. Scale bar, 1 cm.

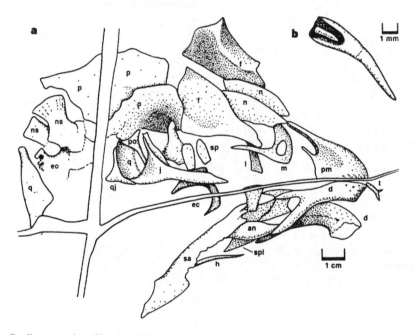

Fig. 7. *Caudipteryx zoui*. **a**, Sketch of skull shown in Fig. 6. **b**, Premaxillary tooth of NGMC 97-4-A, showing resorption pit and germ tooth. Abbreviations: an, angular; d, dentary; ec, ectopterygoid; eo, exoccipital; f, frontal; h, hyoid; j, jugal; l, lacrymal; m, maxilla; n, nasal; ns, neural spine; p, parietal; pm, premaxilla; po, postorbital; q, quadrate; qj, quadratojugal; sa, surangular; sp, scleral plate; spl, splenial; t, premaxillary teeth.

There are four teeth in each premaxilla. They have elongate, needlelike crowns, and the roots are five times wider than the crowns (Fig. 7b). The lingual wall of the root of the third right tooth has been resorbed for the crown of a replacement tooth. The teeth seem

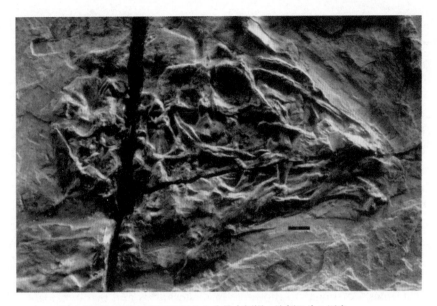

图 6. 邹氏尾羽龙，NGMC 97-9-A 头骨右侧视。比例尺为 1 厘米。

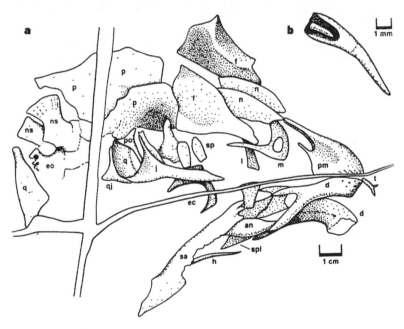

图 7. 邹氏尾羽龙。**a**，图 6 的头骨素描图。**b**，NGMC 97-4-A 的前颌骨齿，显示出再吸收窝和胚齿。缩写：an，隔骨；d，齿骨；ec，外翼骨；eo，外枕骨；f，额骨；h，舌骨；j，轭骨；l，泪骨；m，上颌骨；n，鼻骨；ns，神经棘；p，顶骨；pm，前颌骨；po，眶后骨；q，方骨；qj，方轭骨；sa，上隔骨；sp，巩膜板；spl，夹板骨；t，前颌骨齿。

每侧前颌骨上各有 4 颗牙齿。它们具有拉长的、像针一样的齿冠，根部是冠部的 5 倍宽（图 7b）。右侧第 3 颗牙齿齿根的舌面被一颗替换齿的齿冠再吸收。牙齿似

263

to have been procumbent, with an inflection at the gumline. *Caudipteryx* had no maxillary or dentary teeth.

There are ten amphicoelous cervical vertebrae and five sacrals as in most non-avian theropods and *Archaeopteryx*[8,12]. The tail of NGMC 97-4-A is articulated and well-preserved, and includes 22 vertebrae, as in *Archaeopteryx*. It is shorter than the 30-segment tails of oviraptorids. Most other non-avian theropods have much longer tails. Caudals do not become longer posteriorly, as they do in most non-avian theropods and *Archaeopteryx*. Almost two-thirds of the tail of NGMC 97-4-A is preserved as a straight rod, but the vertebrae are not fused. The first six haemal spines are elongate, rodlike structures. More posterior haemal spines decrease in height, but expand anteriorly and posteriorly (Fig. 5a).

Each segment of gastralia is formed by two pairs of slender, tapering rods, as in all non-avian theropods, *Protarchaeopteryx* and early birds[8,9,13,14].

The paired sternals are similar to those of dromaeosaurids and oviraptorids. *Confuciusornis* had a relatively larger, unkeeled sternum. Some short bones with slight expansions at each end are found near the sternal plates of NGMC 97-9-A, and may be sternal ribs.

The scapula is longer than the humerus, whereas the scapula-to-humerus ratio is less than 1.0 in flying birds (Table 2) because of humerus elongation. The clavicles are fused into a broad, U-shaped furcula in NGMC 97-9-A as in *Archaeopteryx*, *Confuciusornis* and many non-avian theropods.

Compared to the humerus, forearm length is similar to that in oviraptorosaurs (Table 2), *Archaeopteryx* and *Protarchaeopteryx*. In more advanced birds[15,16], the radius is longer than the humerus. The external surface of the ulna, as in *Archaeopteryx*[8], lacks any evidence of quill nodes.

There are three carpals preserved in NGMC 97-4-A, including a large semi-lunate one that caps metacarpals I and II as in dromaeosaurids, oviraptorids, troodontids, *Archaeopteryx*, *Confuciusornis* and other birds. Four carpals have been recognized in *Archaeopteryx*[12]. A large triangular radiale sits between the semi-lunate and the radius. A small carpal articulates with the third metacarpal. A thin wedge of bone at the end of the ulna is probably a fragment of gastralia.

The unfused metacarpals and digits of both specimens are well preserved. The third metacarpal is almost as long as the second, but is more slender. The hand has the normal theropod phalangeal formula of 2-3-4-x-x. The manus is longer than either the humerus or the radius, which is a primitive characteristic shared with most non-avian coelurosaurs, *Archaeopteryx*[9] and *Confuciusornis*. In contrast with *Archaeopteryx*, *Confuciusornis*, *Protarchaeopteryx* and many non-avian theropods (ornithomimids, troodontids, dromaeosaurids and oviraptorids), the manus is relatively short compared with the femur.

平呈平伏状，齿龈有些弯曲。尾羽龙没有上颌骨齿和齿骨齿。

与大多数非鸟兽脚类恐龙和始祖鸟一样 [8,12] 有 10 个双凹型的颈椎和 5 个荐椎。NGMC 97-4-A 尾部骨骼相关节且保存完好，同始祖鸟一样共包括 22 节尾椎。这比窃蛋龙类的 30 节要短。大多数其他非鸟兽脚类恐龙都具有长得多的尾巴。尾椎没有像大多非鸟兽脚类恐龙和始祖鸟一样向后变长。NGMC 97-4-A 几乎三分之二的尾巴呈杆状，但尾椎没有愈合。前 6 个脉弧是伸长的棒状结构。靠后的脉弧高度减小，但前后向都扩展 (图 5a)。

每节腹膜肋由两对细长且两端变尖的棒状小骨组成，这与所有非鸟兽脚类恐龙、原始祖鸟以及早期鸟类一样 [8,9,13,14]。

成对的胸骨与驰龙类和窃蛋龙类的相似。孔子鸟具有一个相对较大的、无龙骨突的胸骨。NGMC 97-9-A 的每个胸骨板附近都有一些短小、两端微弱膨大的骨骼，可能是胸肋。

肩胛骨比肱骨长，但在飞行鸟类中，肱骨的加长使其肩胛骨与肱骨的比率小于1.0(表 2)。NGMC 97-9-A 的锁骨愈合成宽阔的 U 形叉骨，就像始祖鸟、孔子鸟和许多非鸟兽脚类恐龙一样。

以肱骨比较，尾羽龙前臂的长度与窃蛋龙类 (表 2)、始祖鸟和原始祖鸟的是较为相似的。在较进步的鸟类中 [15,16]，桡骨比肱骨长。尺骨的外表面比较像始祖鸟 [8]，缺乏羽茎节点存在的证据。

NGMC 97-4-A 有三个腕骨保存，包括一块大的覆盖第 Ⅰ、Ⅱ 掌骨的半月形腕骨，这与驰龙类、窃蛋龙类、伤齿龙类、始祖鸟、孔子鸟和其他鸟类一样。始祖鸟保存了四个腕骨 [12]。一块大的三角形桡腕骨位于半月形腕骨和桡骨之间，小的腕骨和第Ⅲ掌骨相关节。尺骨末端有块细小的楔形骨，大概是腹膜肋的碎块。

两件标本中未愈合的掌骨和指骨保存很完整。第Ⅲ掌骨和第Ⅱ掌骨差不多一样长，但较纤细。手指指式是兽脚类恐龙通常的 2-3-4-x-x。手部比肱骨和桡骨都长，这是与大多数非鸟虚骨龙类、始祖鸟 [9] 和孔子鸟共有的原始特征。同始祖鸟、孔子鸟和原始祖鸟及许多非鸟兽脚类恐龙(似鸟龙类、伤齿龙类、驰龙类和窃蛋龙类)相反的是，手部相对股骨而言要短。

The curved second manual ungual is about two-thirds the size of the same element in *Protarchaeopteryx*, and is less than 70% the length of the penultimate phalanx.

Pelvic elements are unfused, as they are in all non-avian theropods (except some ceratosaurs) and the most primitive birds[16]. The acetabulum is large, comprising almost a quarter of the length of the ilium (the ratio of acetabulum-to-ilium length is less than 0.11 in birds[17]). It has a deeper, shorter, more squared-off preacetabular region than that of *Protarchaeopteryx*, and closely resembles the ilium of dromaeosaurids[18]. The tapering postacetabular region is lower and longer than the preacetabular. The pubic peduncle is anteroposteriorly elongated, and has a notch (Figs 4 and 5b) in the ventral margin that divides the suture into two surfaces. This notch and the deep pubic peduncle of the ischium are characteristic of opisthopubic pelves. The ischium has no dorsal process such as that found in *Archaeopteryx* and *Confuciusornis*, and the shaft curves down and back. A well-developed ventromedial flange is present, perhaps indicating contact between elements. In general appearance, the ischium most closely resembles those of non-avian coelurosaurs.

The ratio of hindlimb-to-forelimb length is higher than in other coelurosaurs (Table 2) except alvarezsaurids[19], which had exceptionally short arms. The greater trochanter is separated from the lesser trochanter of the femur by a shallow notch, and forms a raised, semi-lunate rim that is similar to the trochanter femoris of birds, troodontids and avimimids.

None of the fibulae is complete, but NGMC 97-9-A has a socket for the distal end of the fibula formed by the calcaneum, astragalus and tibia. The astragalus is not fused to the tibia. The ascending process of NGMC 97-9-A (Fig. 5e) extends 22% of the distance up the front surface of the tibia, compared with 12% in *Archaeopteryx*[12]. As in *Archaeopteryx*[8], *Confuciusornis*[10] and most non-avian theropods, the calcaneum is retained as a separate, disk-like element. Two distal tarsals are positioned over the third and fourth metatarsals, as in *Archaeopteryx*, *Boluochia*[20] and all non-avian theropods that lack fused tarsometatarsals.

The metatarsals of *Caudipteryx* are not fused; this is the plesiomorphic condition expressed in most non-avian theropods. Metatarsal I is centred about a quarter of the way up the posteromedial corner of the second metatarsal. The third is the longest of the metatarsals, and in anterior view completely separates the second and fourth metatarsals, unlike in the arctometatarsalian condition of many theropods[21]. Nevertheless, at midshaft the third metatarsal is thin anteroposteriorly and is triangular in cross-section. The pedal unguals are triangular in cross-section and are about the same size as the manual unguals.

At least fourteen remiges are attached to the second metacarpal, phalanx II-1, and the base of phalanx II-2 of NGMC 97-4-A (Fig. 8a). Each remex has a well-preserved rachis and vane. The most distal remex is less than 30 mm long. The second most distal remex is 63.5 mm long, is symmetrical, and has 6.5-mm-long barbs on either side of the rachis. The fourth most distal primary remex is 95 mm long and is longer than the humerus. Unfortunately, the distal ends of the remaining remiges are not preserved. In flying birds

弯曲的第 Ⅱ 指爪长度是原始祖鸟同一骨骼成分的三分之二，也比倒数第 2 指骨长度的 70% 短。

腰带各骨骼未愈合，类似于所有的非鸟兽脚类恐龙（一些角鼻龙类除外）和大多数原始的鸟类 [16]。髋臼很大，几乎占髂骨长度的四分之一（在鸟类中，髋臼与髂骨的比率小于 0.11[17]）。它的前髋臼区域比原始祖鸟的深而短，更接近方形，十分接近驰龙类的髂骨 [18]。向后端变窄的后髋臼区域要比前髋臼区域低而长。耻骨柄前后延长，在靠近腹侧边缘发育的凹槽（图 4 和图 5b）将缝合线分开为两面。这个凹槽和深的坐骨的耻骨柄是后伸型耻骨具有的特征。坐骨没有如同始祖鸟和孔子鸟一样的背突，坐骨柄向下向后弯曲。存在发育很好的腹中缘，或许表明两骨骼之间的接触。大体上看，坐骨表现出同非鸟虚骨龙类十分相似的特征。

后肢和前肢长度的比率要比其他的虚骨龙类（除了阿瓦拉慈龙类 [19]，其前臂异常短小）的高（表 2）。股骨的大转子与小转子被一个浅的凹槽分开，并形成了一个突起的、半月形边缘，形态和鸟类、伤齿龙类以及拟鸟龙类的股骨转子相似。

腓骨都保存不完整，但是 NGMC 97-9-A 的跟骨、距骨和胫骨形成一个腓骨末端的窝。距骨没有和胫骨愈合。NGMC 97-9-A 的上升突（图 5e）延伸了胫骨的前表面之上距离的 22%，而始祖鸟的则是 12%[12]。类似于始祖鸟 [8]、孔子鸟 [10] 和大多数非鸟兽脚类恐龙，跟骨仍保留为一分离的圆盘状骨骼。两个远端跗骨的位置超过第 Ⅲ、Ⅳ 跖骨，像始祖鸟、波罗赤鸟 [20] 以及所有缺少愈合的跗跖骨的非鸟兽脚类恐龙一样。

尾羽龙的跖骨没有愈合，这是与大多非鸟兽脚类恐龙相似的近祖性状。第 Ⅰ 跖骨位于第 Ⅱ 跖骨后内侧角向上四分之一处。第 Ⅲ 跖骨是最长的跖骨，从前视上看，完全地将第 Ⅱ、Ⅳ 跖骨分开，这与很多兽脚类恐龙 [21] 的窄跖型情形不同。然而，第 Ⅲ 跖骨的骨干前后向窄细，横截面呈三角形。趾爪的横截面是三角形，大约与指爪的尺寸相同。

NGMC 97-4-A 标本中，至少有 14 根飞羽附着在第 Ⅱ 掌骨和第 Ⅱ 指的第 1 指骨上，以及第 Ⅱ 指第 2 指骨的基部（图 8a）。每根飞羽上很好地保留着羽轴和羽片。最远端的飞羽长不超过 30 毫米。第二远端的飞羽长 63.5 毫米，左右对称，羽轴的两侧有 6.5 毫米长的羽枝。第四远端的初级飞羽长 95 毫米，比肱骨都长。遗憾的是，

(even *Archaeopteryx*[12]), each remex is longer than they are in *Caudipteryx*, and the most distal remiges are the longest. For example, the remiges of *Archaeopteryx*[22] are more than double the length of the femur. The barbs on either side of the rachis are symmetrical, contrasting with *Archaeopteryx* and modern flying birds[23].

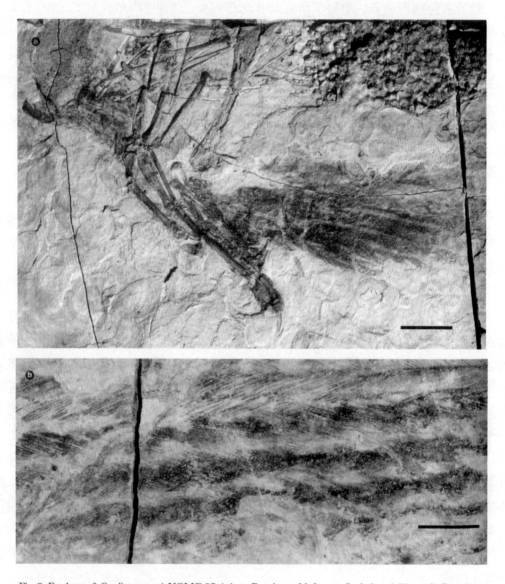

Fig. 8. Feathers of *Caudipteryx zoui*, NGMC 97-4-A. **a**, Remiges of left arm. Scale bar, 1.75 cm. **b**, Rectrices, showing colour banding. Scale bar, 1 cm.

The holotype preserves ten complete and two partial rectrices. Eleven are attached to the left side of the tail, and were probably paired with another eleven feathers on the right side (only the terminal feather is preserved). Two rectrices are attached to each side of the last five or six caudal vertebrae, but not to more anterior ones. NGMC 97-9-A preserves most of nine rectrices. In *Archaeopteryx*, rectrices are associated with all but the first five or

其他飞羽的远端未能保存下来。在飞行鸟类（甚至始祖鸟[12]）中，每个飞羽都要比尾羽龙的飞羽长，且最远端的飞羽是最长的。例如，始祖鸟[22]的飞羽要长于股骨的两倍。与始祖鸟和现生飞行鸟类[23]不同，尾羽龙羽轴两侧的羽枝是左右对称的。

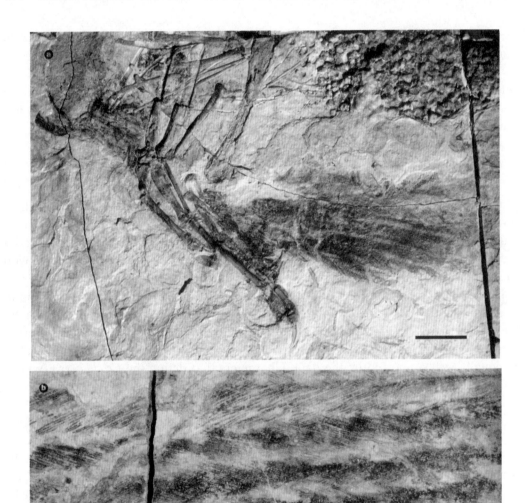

图 8. 邹氏尾羽龙的羽毛，NGMC 97-4-A。**a**，左臂上的飞羽，比例尺为 1.75 厘米。**b**，尾羽，显示出色带，比例尺为 1 厘米。

正模标本保存了 10 根完整的和 2 根局部的尾羽。其中 11 根羽毛附着在尾巴的左侧，可能右侧存在另外 11 根与之成对的羽毛（只有末端的羽毛保留着）。2 根尾羽附着在最后 5 或 6 节尾椎的两侧，但没有更靠前一些。NGMC 97-9-A 保留着 9 根尾羽的大部分。在始祖鸟中，尾羽和所有尾椎（除了前 5 或 6 节）相互关联[12,22]。每个

six caudals[12,22]. Each rachis has a basal diameter of 0.74 mm and tapers distally. All the feathers appear to be symmetrical (Fig. 8b), although in most cases the tips of the barbs of adjacent feathers overlap. The vane of the sixth feather is 6 mm wide on either side of the rachis.

The body of NGMC 97-4-A, especially the hips and the base of the tail, is covered by small, plumulaceous feathers of up to 14 mm long.

Both specimens have concentrations of small polished and rounded pebbles in the stomach region. These gastroliths are up to 4.5 mm in diameter, although most are considerably less than 4 mm wide.

Phylogenetic analysis

We examined the systematic positions of *Protarchaeopteryx* and *Caudipteryx* by coding these specimens for the 90 characters used in an analysis of avialan phylogeny[24] (for a matrix of these characters, see Supplementary Information). Characters were unordered, and a tree was produced using the branch-and-bound option of PAUP[25]. We rooted the tree with Velociraptorinae[26,27]. A single tree resulted with a length of 110 steps, a retention index of 0.849 and a consistency index of 0.855. Analysis shows *Caudipteryx* to be the sister group to the Avialae, and *Protarchaeopteryx* to be unresolved from the Velociraptorinae root (Fig. 9). The placement of *Protarchaeopteryx* as the sister group to *Caudipteryx* + Avialae, as the sister group to Velociraptorinae, or as the sister group to Velociraptorinae + (*Caudipteryx* + Avialae) are equally well supported by the data. Characters that define the *Caudipteryx* + Avialae clade in the shortest tree include unambiguous (uninfluenced by missing data or optimization) characters 2 and 12 and several more ambiguous ones (characters 4, 5, 10, 11, 15, 19, 24, 37, 85 and 86). *Caudipteryx* is separated from the Avialae by three unambiguous characters (7, 8 and 71) and additional ambiguous ones (characters 5, 6, 9, 10, 11, 18, 24, 39, 40, 56 and 69). The important characteristic of this phylogeny is that the Avialae (not including *Protarchaeopteryx* and *Caudipteryx*) is monophyletic; this placement is supported by the unequivocal presence of a quadratojugal that is joined to the quadrate by a ligament[17] (character 7), the absence of a quadratojugal squamosal contact (character 8) and a reduced or absent process of the ischium (character 71).

羽轴的基部直径是 0.74 毫米，并且末端越来越细。所有的羽毛都是对称的（图 8b），尽管在大多数情况下相邻羽毛的羽枝末梢是相互交叠的。第 6 根羽毛羽轴两侧的羽片均为 6 毫米宽。

在 NGMC 97-4-A 的身体上，特别是臀部和尾巴基部，覆盖着小的绒羽，长度可达 14 毫米。

两件标本在胃部区域具有密集的磨光的圆形小卵石。这些胃石的直径最大可达 4.5 毫米，不过大部分胃石的宽度远小于 4 毫米。

系统发育分析

我们将标本编码到初鸟类包含 90 个特征的系统发育分析中，检验了原始祖鸟和尾羽龙的系统发育位置 [24]（这些特征的矩阵可见补充信息。编者注：本书未收录补充信息）。特征是无序的，系统树是通过 PAUP 程序 [25] 的分支界定计算的。我们以疾走龙类作为系统树的根 [26,27]。运行 110 步得出单一的系统树，保留指数是 0.849，稠度指数是 0.855。分析显示尾羽龙和初鸟类是姊妹群，而原始祖鸟在疾走龙类分支上的位置并未能解决（图 9）。原始祖鸟与尾羽龙 + 初鸟类构成姊妹群，或与疾走龙类构成姊妹群，或与疾走龙类 +（尾羽龙 + 初鸟类）构成姊妹群，这些结果同等程度地被数据所支持。在该最短的树中，定义尾羽龙 + 初鸟类这一分支的特征包括确定的（未受缺失数据或最优化影响）特征 2 和 12，以及另外一些不甚明确的特征（特征 4、5、10、11、15、19、24、37、85 和 86）。尾羽龙可从初鸟类中划分开来，这被三个确定的特征（7、8 和 71）和其他不确定的特征（特征 5、6、9、10、11、18、24、39、40、56 和 69）所支持。这一系统关系树的重要特征是初鸟类（不包括原始祖鸟和尾羽龙）是单系的，它的定位被以下特征所支持：确切存在的方轭骨以及其与方骨以韧带连接 [17]（特征 7），方轭骨和鳞骨不接触（特征 8）以及减弱或缺失的坐骨突（特征 71）。

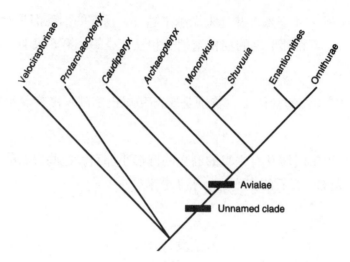

Fig. 9. Cladogram of proposed relationships of *Protoarchaeopteryx* and *Caudipteryx*. This tree is based on 90 characters and has a length of 110 steps.

As characters dealing with feathers cannot be scored relative to outgroup conditions, they were not used in the phylogenetic analysis. However, our analysis indicates that feathers can no longer be used in the diagnosis of the Avialae.

Discussion

The three *Protarchaeopteryx* and *Caudipteryx* individuals were close to maturity at the time of death. The neural spines seem to be fused to cervical and dorsal centra in *Protarchaeopteryx*. Sternal plates ossify late in the ontogeny of non-avian theropods, and are present in both *Caudipteryx* and *Protarchaeopteryx*. Well-ossified sternal ribs, wrist bones and ankle bones in *Caudipteryx* also indicate the maturity of the specimens.

The remiges of *Caudipteryx* and the rectrices of both *Protarchaeopteryx* and *Caudipteryx* have symmetrical veins, whereas even those of *Archaeopteryx* are asymmetrical. Birds with asymmetrical feathers are generally considered to be capable of flight[23], but it is possible that an animal with symmetrical feathers could also fly. Relative arm length of *Protarchaeopteryx* is shorter than that of *Archaeopteryx*, but is longer than in non-avian coelurosaurs. The arms of *Caudipteryx*, in contrast, are shorter than those of most non-avian coelurosaurs; the remiges are only slightly longer than the humerus; and the distal remiges are shorter than more proximal ones. It seems unlikely that this animal was capable of active flight. The relatively long legs of *Protarchaeopteryx* and *Caudipteryx*, both of which have the hallux positioned high and orientated anteromedially, indicate that they were ground-dwelling runners.

Paired rectrices of *Protarchaeopteryx* and *Caudipteryx* are restricted to the end of the tail, whereas in *Archaeopteryx* they extend over more than two-thirds the length of the tail[12].

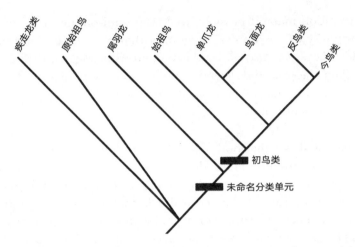

图 9. 原始祖鸟和尾羽龙系统关系的进化分支图。该系统树基于 90 个特征和 110 步得出。

由于羽毛这一特征不能与外类群中的情形作出量化联系，因此羽毛特征没有用于系统发育分析当中。然而，我们的分析指出羽毛不能再作为鉴定初鸟类的特征指标。

讨　论

这三件原始祖鸟和尾羽龙的标本个体在死亡时都比较接近成熟。原始祖鸟的颈椎与背椎椎体似乎与对应的神经棘已经愈合。在非鸟兽脚类恐龙个体发育过程中，胸骨板的骨化相对较晚，但这些在尾羽龙和原始祖鸟中都是存在的。骨化较好的胸肋、腕骨和踝骨同样显示尾羽龙是成熟个体。

尾羽龙的飞羽、原始祖鸟和尾羽龙的尾羽都具有对称的纹理，但始祖鸟的这些羽毛是不对称的。具有不对称羽毛的鸟类一般被认为能够飞行 [23]，但具有对称羽毛的动物也可能会飞。原始祖鸟前臂的相对长度要比始祖鸟的短一些，但还是要比非鸟虚骨龙类的长。而尾羽龙的前臂则相反，要比大多数非鸟虚骨龙类的短；飞羽仅比肱骨长一点点；远端飞羽短于近端飞羽。这些特征都表明这种动物具备主动飞行能力是不太可能的。原始祖鸟和尾羽龙都具相对较长的腿，它们的拇趾位置比较高且指向前内侧，表明它们是陆地生活的奔跑者。

原始祖鸟和尾羽龙成对的尾羽仅局限于尾的末端，而在始祖鸟中尾羽延伸的长度可超过尾部长度的三分之二 [12]。不管是在什么情况保存下来的，我们发现半羽和

273

Wherever preservation made it possible, we found semi-plumes and down-like feathers around the periphery of the bodies, suggesting that most of the bodies were feather-covered, possibly like *Archaeopteryx*[28]. Feathers found with *Otogornis*[29] were also apparently plumulaceous. Plumulaceous and downy feathers cover the bodies of *Protarchaeopteryx* and *Caudipteryx*, and possibly that of *Sinosauropteryx* as well[7]. This suggests that the original function of feathers was insulation.

Phylogenetic analysis shows that both *Caudipteryx* and *Protarchaeopteryx* lie outside Avialae and are non-avian coelurosaurs. This indicates that feathers are irrelevant in the diagnosis of birds. It can no longer be certain that isolated down and semi-plume feathers[30-33] discovered in Mesozoic rocks belonged to birds rather than to non-avian dinosaurs. Furthermore, the presence of feathers on flightless theropods suggests that the hypothesis that feathers and flight evolved together is incorrect. Finally, the presence of remiges, rectrices and plumulaceous feathers on non-avian theropods provides unambiguous evidence supporting the theory that birds are the direct descendants of theropod dinosaurs.

(**393**, 753-761; 1998)

Ji Qiang*, Philip J. Currie†, Mark A. Norell‡ & Ji Shu-An*
* National Geological Museum of China, Yangrou Hutong 15, Xisi, 100034 Beijing, People's Republic of China
† Royal Tyrrell Museum of Palaeontology, Box 7500, Drumheller, Alberta T0J 0Y0, Canada
‡ American Museum of Natural History, Central Park West at 79th Street, New York, New York 10024-5192, USA

Received 19 January; accepted 27 May 1998.

References:

1. Ji, Q. *et al.* On the sequence and age of the protobird bearing deposits in the Sihetun-Jianshangou area, Beipiao, western Liaoning. *Prof. Pap. Strat. Paleo.* (in the press).

2. Hou, L.-H., Zhou, Z.-H., Martin, L. D. & Feduccia, A. A beaked bird from the Jurassic of China. *Nature* **377**, 616-618 (1995).

3. Smith, P. E. *et al.* Dates and rates in ancient lakes: ⁴⁰Ar-³⁹Ar evidence for an Early Cretaceous age for the Jehol Group, northeast China. *Can. J. Earth Sci.* **32**, 1426-1431 (1995).

4. Smith, J. B., Hailu, Y. & Dodson, P. in *The Dinofest Symposium, Abstracts* (eds Wolberg, D. L. *et al.*) 55 (Academy of Natural Sciences, Philadelphia, 1998).

5. Colbert, E. H. The Triassic dinosaur *Coelophysis*. *Bull. Mus. N. Arizona* **57**, 1-160 (1989).

6. Ostrom, J. H. The osteology of *Compsognathus longipipes* Wagner. *Zitteliana* **4**, 73-118 (1978).

7. Chen, P.-j., Dong, Z.-m. & Zhen, S.-n. An exceptionally well-preserved theropod dinosaur from the Yixian Formation of China. *Nature* **391**, 147-152 (1998).

8. Wellnhofer, P. A new specimen of *Archaeopteryx* from the Solnhofen Limestone. *Nat. Hist. Mus. Los Angeles County Sci. Ser.* **36**, 3-23 (1992).

9. Sereno, P. C. & Rao, C. G. Early evolution of avian flight and perching: new evidence from the Lower Cretaceous of China. *Science* **255**, 845-848 (1992).

10. Zhou, Z. H. in *Sixth Symposium on Mesozoic Terrestrial Ecosystems and Biota, Short Papers* (eds Sun, A. & Wang, Y.) 209-214 (China Ocean, Beijing, 1995).

11. Ji, Q. & Ji, S. A. Protarchaeopterygid bird (*Protarchaeopteryx* gen. nov.)—fossil remains of archaeopterygids from China. *Chinese Geol.* **238**, 38-41 (1997).

12. Wellnhofer, P. Das fünfte skelettexemplar von *Archaeopteryx*. *Palaeontogr.* A **147**, 169-216 (1974).

13. Wellnhofer, P. Das siebte Examplar von *Archaeopteryx* aus den Solnhofener Schichten. *Archaeopteryx* **11**, 1-48 (1993).

14. Hou, L. H. A carinate bird from the Upper Jurassic of western Liaoning, China. *Chinese Sci. Bull.* **42**, 413-416 (1997).

15. Dong, Z. M. A lower Cretaceous enantiornithine bird from the Ordos Basin of Inner Mongolia, People's Republic of China. *Can. J. Earth Sci.* **30**, 2177-2179 (1993).

16. Forster, C. A., Sampson, S. D., Chiappe, L. M. & Krause, D. W. The theropod ancestry of birds: new evidence from the Late Cretaceous of Madagascar. *Science* **279**, 1915-1919 (1998).

17. Chiappe, L. M., Norell, M. A. & Clark, J. M. The skull of a relative of the stem-group bird *Mononykus*. *Nature* **392**, 275-278 (1998).

18. Norell, M. A. & Makovicky, P. Important features of the dromaeosaur skeleton: information from a new specimen. *Am. Mus. Novit.* **3215**, 1-28 (1997).

19. Perle, A., Chiappe, L. M., Barsbold, R., Clark, J. M. & Norell, M. Skeletal morphology of *Mononykus olecranus* (Theropoda: Avialae) from the Late Cretaceous of Mongolia. *Am. Mus. Novit.* **3105**, 1-29 (1994).

20. Zhou, Z. H. The discovery of Early Cretaceous birds in China. *Courier Forschungsinstitut Senckenberg* **181**, 9-22 (1995).

绒羽状的羽毛围绕在身体周围，初步判断身体表面大部分披有羽毛，这些特点大概有些类似于始祖鸟 [28]。在鄂托克鸟 [29] 上发现的羽毛同样很显然是绒羽。绒羽和绒羽状羽毛覆盖原始祖鸟和尾羽龙的身体，也许中华龙鸟也是这样 [7]。这表明羽毛的最初功能是用来保温的。

系统发育分析表明了尾羽龙和原始祖鸟位于初鸟类之外，而属于非鸟虚骨龙类恐龙。这显示羽毛对于定义鸟类是一个毫不相关的特征。发现于中生代岩石中的孤立存在的半羽和绒羽状羽毛 [30-33] 将不再确切地属于鸟类，而可能为非鸟兽脚类恐龙的羽毛。而且，在不会飞的兽脚类恐龙身上存在羽毛表明羽毛和飞行同步演化的假说是不正确的。最后，非鸟兽脚类恐龙身上的飞羽、尾羽和绒羽的存在，为鸟类是兽脚类恐龙直接后裔的理论提供了确凿的证据。

（张玉光 翻译；姬书安 审稿）

21. Holtz, T. R. Jr The phylogenetic position of the Tyrannosauridae: implications for theropod systematics. *J. Paleontol.* **68**, 1100-1117 (1994).

22. deBeer, G. *Archaeopteryx lithographica. Br. Mus. Nat. Hist.* **244**, 1-68 (1954).

23. Feduccia, A. & Tordoff, H. B. Feathers of *Archaeopteryx*: asymmetric vanes indicate aerodynamic function. *Science* **203**, 1021-1022 (1979).

24. Chiappe, L. M. in *The Encyclopedia of Dinosaurs* (eds Currie, P. J. & Padian, K.) 32-38 (Academic, San Diego, 1997).

25. Swofford, D. & Begle, D. P. *Phylogenetic Analysis Using Parsimony. Version 3.1.1.* (Smithsonian Institution, Washington DC, 1993).

26. Gauthier, J. in *The Origin of Birds and the Evolution of Flight* (ed. Padian, K.) 1-55 (Calif. Acad. Sci., San Francisco, 1986).

27. Holtz, T. R. Jr Phylogenetic taxonomy of the Coelurosauria (Dinosauria: Theropoda). *J. Paleontol.* **70**, 536-538 (1996).

28. Owen, R. On the *Archaeopteryx* of von Meyer, with a description of the fossil remains of a long-tailed species, from the Lithographic Stone of Solenhofen. *Phil. Trans., Lond.* **153**, 33-47 (1863).

29. Hou, L. H. A late Mesozoic bird from Inner Mongolia. *Vert. PalAsiatica* **32**, 258-266 (1994).

30. Kurochkin, E. N. A true carinate bird from Lower Cretaceous deposits in Mongolia and other evidence of Early Cretaceous birds in Asia. *Cretaceous Res.* **6**, 271-278 (1985).

31. Sanz, J. L., Bonapart, J. F. & Lacasa, A. Unusual Early Cretaceous birds from Spain. *Nature* **331**, 433- 435 (1988).

32. Kellner, A. W. A., Maisey, J. G. & Campos, D. A. Fossil down feather from the Lower Cretaceous of Brazil. *Palaeontol.* **37**, 489-492 (1994).

33. Grimaldi, D. & Case, G. R. A feather in amber from the Upper Cretaceous of New Jersey. *Am. Mus. Novit.* **3126**, 1-6 (1995).

Supplementary information is available on *Nature*'s World-Wide Web site (http://www.nature.com) or as paper copy from the London editorial office of *Nature*.

Acknowledgements. We thank A. Brush, B. Creisler, M. Ellison, W.-D. Heinrich, N. Jacobsen, E. and R. Koppelhus, P. Makovicky, A. Milner, G. Olshevsky, J. Ostrom and H.-P. Schultze for advice, access to collections and logistic support; and the National Geographic Society, National Science Foundation (USA), the American Museum of Natural History, National Natural Science Foundation of China and the Ministry of Geology for support. Photographs were taken by O. L. Mazzatenta and K. Aulenback; the latter was also responsible for preliminary preparation of the *Caudipteryx* specimens. Line drawings are by P. J. C.

Correspondence and requests for materials should be addressed to P.J.C. (e-mail: pcurrie@mcd.gov.ab.ca).

An Electrophoretic Ink for All-printed Reflective Electronic Displays

B. Comiskey *et al.*

Editor's Note

The printed word retains some advantages over electronic displays. The subliminal flickering of screens tires the eye, and they are fragile, expensive, energy-hungry, and lose contrast in bright ambient light. This has stimulated a quest for "electronic paper" that combines the benefits of printed paper with the convenience and versatility of electronic information systems. Here researchers at MIT's Media Lab report one of the first candidate materials: a layer of transparent microcapsules on a flexible plastic sheet. Each capsule contains white and black pigment particles that can be drawn to the top surface by an electric field. This foundational paper led to a spinoff company, E-Ink, which, in partnership with microelectronics companies, now sells a range of portable display devices.

It has for many years been an ambition of researchers in display media to create a flexible low-cost system that is the electronic analogue of paper. In this context, microparticle-based displays[1-5] have long intrigued researchers. Switchable contrast in such displays is achieved by the electromigration of highly scattering or absorbing microparticles (in the size range 0.1–5 μm), quite distinct from the molecular-scale properties that govern the behaviour of the more familiar liquid-crystal displays[6]. Microparticle-based displays possess intrinsic bistability, exhibit extremely low power d.c. field addressing and have demonstrated high contrast and reflectivity. These features, combined with a near-lambertian viewing characteristic, result in an "ink on paper" look[7]. But such displays have to date suffered from short lifetimes and difficulty in manufacture. Here we report the synthesis of an electrophoretic ink based on the microencapsulation of an electrophoretic dispersion[8]. The use of a microencapsulated electrophoretic medium solves the lifetime issues and permits the fabrication of a bistable electronic display solely by means of printing. This system may satisfy the practical requirements of electronic paper.

PREVIOUS approaches to fabricating particle-based displays have been based on rotating bichromal spheres in glass cavities[1] or elastomeric slabs[2,3], and electrophoresis in glass cavities[4,5]. The advantageous optical and electronic characteristics of microparticle displays, as compared to liquid-crystal displays, result from the ability to electrostatically migrate highly scattering pigments such as titanium dioxide ($n = 2.7$), yielding a difference in optical index of refraction with the surrounding dielectric liquid of $\Delta n = 1.3$ and black pigments with very high absorption and hiding power. This results

一种可用于全印刷反射式电子显示器的电泳墨水

科米斯基等

编者按

与电子显示相比，印刷的文字仍然具有一定的优势。电子显示器的轻微闪烁容易使眼睛疲劳，另外，它们易碎、昂贵、耗能，在明亮的环境光线下对比度下降。因而刺激了能够兼顾纸质印刷物的优点和电子信息系统的便利性与广泛性的"电子纸张"的需求。麻省理工学院媒体实验室的研究人员在本文中报道了首选材料之一：置于软塑料片上的透明微胶囊层。每个胶囊中都含有在电场中可移向顶面的黑、白两种颜色的微粒。这篇基础性的文章导致一个派生公司——E-Ink 的问世，该公司与微电子公司合作后，一系列便携式显示设备已进入市场。

长期以来，显示媒介领域的研究人员一直有一种追求，即发明一种柔韧的低成本系统，也就是实现纸张的电子模拟。在这种情况下，基于微粒的显示器 [1-5] 早已引起了研究人员们的兴趣。这类显示器中可切换的对比度是通过强散射或强吸收微粒（尺寸范围为 0.1~5 μm）的电迁移来实现的，完全不同于人们熟知的通过分子的性质来调控的液晶显示器 [6]。基于微粒的显示器拥有本征双稳态，表现出极低的直流电场编址，并且具有很高的对比度和反射率。这些特征，再加之近朗伯显示特性，就能产生"纸上墨迹"的视觉效果 [7]。但是这类显示器却存在着寿命短，制作难度大的缺点。本文中我们报道了一种基于电泳分散微胶囊化 [8] 的电泳型墨水的合成。使用微胶囊化的电泳介质解决了寿命问题，并且实现了仅用印刷技术便可制造双稳态电子显示器的要求。该系统可以达到电子纸张应用于实际的要求。

以往生产基于微粒的显示器的方法是以在玻璃腔 [1] 或弹性板 [2,3] 中的双色拧转球，和它们在玻璃腔中的电泳 [4,5] 为基础的。与液晶显示器相比，微粒显示器具有优越的光学和电学特性，基于高散射颜料例如二氧化钛（$n = 2.7$）的静电迁移能力，使其和周围介电液体及高吸收和遮盖力黑色颜料之间的光学折射率差异较大（$\Delta n = 1.3$）。从而导致散射长度很短（约 1 μm）而相应的反射率和对比度很高，效果

in a very short scattering length (~1 µm) and a correspondingly high reflectivity and contrast, similar to that of ink on paper. For comparison, typical index differences in a scattering-mode liquid crystal are less than $\Delta n = 0.25$ (ref. 9).

Despite these favourable attributes, microparticle displays at present suffer from a number of shortcomings. These include, in the case of the bichromal sphere system, difficulty in obtaining complete rotation (and thus complete contrast) due to a fall-off in the dipole force close to normal angles and difficulty of manufacture. In the case of electrophoretic systems, the shortcomings include reduced lifetime due to colloidal instability[10]. Finally, bistable display media of all types, to date, are not capable of being manufactured with simple processes such as printing. (Although recent results have been obtained in ink-jetting of electroluminescent doped polymer films for organic light-emitting structures[11], such emissive non-bistable displays do not meet a key criterion of "electronic paper", namely the persistent display of information with zero power consumption.)

To realize a printable bistable display system, we have synthesized an electrophoretic ink by microencapsulating droplets of an electrophoretic dispersion in individual microcapsules with diameters in the range of 30–300 µm. Figure 1a indicates schematically the operation of a system of microencapsulated differently coloured and charged microparticles, in which one or the other species of particles may be migrated towards the viewer by means of an externally applied electric field. Figure 1b shows cross-sectional photomicrographs of a single microcapsule addressed with positive and negative fields.

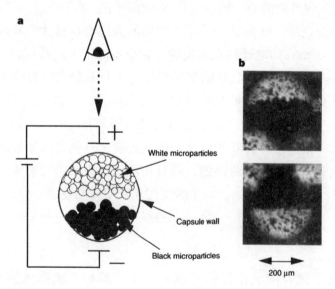

Fig. 1. Electrophoretic microcapsule. **a**, Schematic illustration of microencapsulated electrophoretic image display (white and black microparticles system). The top transparent electrode becomes positively charged, resulting in negatively-charged white microparticles migrating towards it. Oppositely charged black microparticles move towards the bottom electrode. **b**, Photomicrograph of an individual microcapsule addressed with a positive and negative field.

和纸上墨迹相近。相比较而言散射型液晶的典型折射率差值则小于 $\Delta n = 0.25$（参考文献 9）。

尽管具有这些有利的属性，现阶段的微粒显示器仍然存在许多缺点。其中包括，在双色拧转球系统中，由于在接近直角时偶极力下降，使其难以完全拧转（从而达到最大反差），此外还有制造上的困难。至于电泳系统，缺点包括胶体不稳定性[10]引起的寿命缩短。总之，至今为止，各种类型的双稳态显示媒体都不可能通过如印刷这样的简单过程来生产。（尽管用喷墨法在掺杂的电致发光聚合物薄膜上制备有机发光结构最近已有进展[11]，但这类发光非双稳态显示器并没有达到"电子纸张"的关键指标，即零能耗时信息的持久显示。）

为了实现可印刷的双稳态显示系统，我们合成了一种电泳墨水，通过将电泳分散的液滴微胶囊化，微胶囊的直径范围为 $30 \sim 300\ \mu m$。图 1a 显示了不同颜色的带电微粒微胶囊化系统的操作示意图，在该系统中一种或另一种粒子可以通过外加电场移向观察者。图 1b 显示了用正负电场编址的单一微胶囊的横截面显微照片。

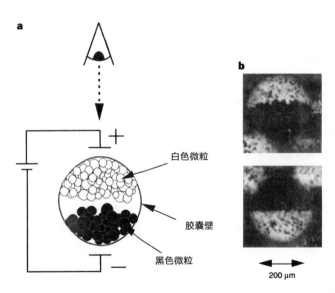

白色微粒

胶囊壁

黑色微粒

200 μm

图 1. 电泳微胶囊。**a**，电泳成像显示器微胶囊化的示意图（黑白微粒系统）。上方的透明电极带正电，引起带负电的白色微粒向它迁移。带相反电荷的黑色微粒向下方电极迁移。**b**，用正负电场编址单个微胶囊的显微照片。

The microencapsulated electrophoretic system was synthesized using the following process. The internal phase of the microcapsules was composed of a mixed dispersion of black and white microparticles in a dielectric fluid. To obtain white microparticles, a suspension of rutile titanium dioxide (specific gravity = 4.2) in molten low-molecular-weight polyethylene was atomized. A similar process was used to prepare black microparticles with an inorganic black pigment. The resulting particles were sieved to obtain a dry powder with an average diameter of 5 μm. Alternatively, smaller monodisperse particles, with a diameter less than 1 μm, were prepared chemically. The polyethylene serves to reduce the specific gravity of the particles (typically to ~1.5) and present a modified surface chemistry for charging purposes. The particles in suspension acquire a surface charge due to the electrical double layer[12]. The black and white particles have different zeta potential (and hence different mobility) due to the electroconductivity of the black pigment[13]. These microparticles are then dispersed in a mixture of tetrachloroethylene (specific gravity, 1.6) and an aliphatic hydrocarbon (specific gravity, 0.8) which is specific-gravity-matched to the manufactured particles. In the case where the particles are designed with charges of opposite sign, they are prevented from coagulation by providing a physical polymeric adsorbed layer on the surface of each particle which provides strong inter-particle repulsive forces[10]. Alternatively, a single particle system (white microparticles) dispersed in a dyed (Oil Blue N) dielectric fluid was prepared.

This suspension, which in typical electrophoretic displays would be interposed between two glass electrodes, was then emulsified into an aqueous phase and microencapsulated by means of an *in situ* polycondensation of urea and formaldehyde. This process produces discrete mechanically strong and optically clear microcapsules. The microcapsules are optionally filtered to obtain a desired size range, and are subsequently washed and dried.

Microcapsules (white particle in dye) were prepared and dispersed in a carrier (ultraviolet-curable urethane) and subsequently coated onto a transparent conductive film (indium tin oxide (ITO) on polyester). Rear electrodes printed from a silver-doped polymeric ink were then applied to the display layer. Figure 2 shows a series of microphotographs at different magnifications in which the letter "k" has been electronically addressed in the electronic ink. The sample shown was sieved to have a mean capsule size of 40 ± 10 μm yielding a capsule resolution of ~600 dots per inch. By going to a system using structured top electrodes (or address lines as opposed to a continuous electrode), we were able to fractionally address single microcapsules, yielding an addressable resolution of ~1,200 dots per inch.

微胶囊电泳系统经由以下工序合成。微胶囊的分散相由黑白微粒在介电流体的分散系混合而成。白色微粒是通过金红石型二氧化钛(比重 = 4.2)在熔化的低分子量聚乙烯的悬浮液中雾化而成。黑色微粒可用无机黑色颜料经过类似的过程制备。所得粒子经筛分得到平均直径为 5 µm 的干粉。此外直径小于 1 µm 的较小单分散粒子可用化学方法制备。聚乙烯用来降低粒子的比重(通常降至 1.5 左右)和提供经化学修饰的表面使之能够带电。悬浮液中粒子表面电荷来自双电层作用[12]。黑色颜料的导电性使黑白粒子具有不同的界面电位(进而有不同的迁移率)[13]。这些微粒被分散于四氯乙烯(比重 1.6)和一种脂肪烃(比重 0.8)的混合物中,混合物与所制备粒子的比重相匹配。在粒子按设计分别带有异性电荷的情况下,在粒子表面上通过物理吸附覆盖上一层高聚物,使得粒子间存在强互斥力[10]以避免发生聚沉。此外,还制备了分散在染色(油蓝 N)介电流体中的单粒子系统(白色微粒)。

该悬浮液置于典型电泳显示器的两块玻璃电极之间,在水相中乳化并通过尿素和甲醛的原位缩聚反应实现微胶囊化。该方法可制得高强度、光学性能清晰、分立的微胶囊。筛选后可获得所需尺寸范围的微胶囊,随之进行清洗和干燥。

在载体(紫外光固化的聚氨酯)中制备和分散微胶囊(染料中的白色粒子)后,随即覆以透明导电薄膜(载有氧化铟锡(ITO)的聚酯)。然后,用于显示层的背电极由银掺杂的聚合物墨水印制而成。图 2 显示了不同放大率下的一系列显微照片,为字母"k"在电子墨水中的电子编址。所示样品曾经过筛选,胶囊平均尺寸为 40±10 µm,得到的胶囊分辨率为每英寸约 600 点。通过运用结构化的面电极(或者编址线与连续电极方向相反)系统,我们可对部分单一微胶囊编址,得到的编址分辨率为每英寸约 1,200 点。

Fig. 2. Electrophoretic ink. Photomicrographs of 200-μm-thick film of electronic ink ("white particles in dye" type) with a capsule diameter of 40 ± 10 μm (top view). The electronically addressed letter "k" is white, other areas are blue.

In addition to making a flexible and printable system, microencapsulating the electrophoretic dispersion has solved a longstanding problem with electrophoretic displays, namely limited lifetime due to particle clustering, agglomeration and lateral migration[10]. This is because the dispersion is physically contained in discrete compartments, and cannot move or agglomerate on a scale larger than the capsule size. In ink samples that we have prepared to date, $> 10^7$ switching cycles have been observed with no degradation in performance. We have measured contrast ratios of 7:1 with 35% reflectance and a near-lambertian viewing characteristic. Using the same measurement system, we measured newspaper at 5:1 contrast and 55% reflectance. We have observed image storage times of several months, after which the material may be addressed readily.

In Fig. 3a we show maximum minus minimum reflected optical power (non-normalized contrast) versus drive frequency for several different applied fields; open symbols indicate data taken from low to high frequency, filled symbols show data from high to low frequency. Figure 3b shows plots of contrast versus applied field for different frequencies. The data in Fig. 3 were obtained for a 200-μm film of electronic ink (of the type "white particles in dye"), with capsules of diameter 40 ± 10 μm on ITO polyester with a silver-ink rear electrode.

图 2. 电泳墨水。胶囊直径为 40±10 μm，厚度为 200 μm 的电子墨水（"染料中白色粒子"型）薄膜显微
照片（俯视图）。电子编址的字母"k"为白色，其他区域为蓝色。

除了制作灵活的可印刷系统之外，微胶囊电泳分散法解决了电泳显示器的一个
存在已久的问题，即由于粒子簇集，聚沉，以及横向迁移带来的寿命受限问题[10]。
因为分散被物理地限制在分立的空间中，移动和聚集都不可能超出胶囊的尺寸。目
前为止，我们制备的墨水样品，经 10^7 次以上的切换循环，没有观测到性能有所下
降。测量得到的对比度为 7:1，反射率为 35%，并且观测到近朗伯显示特性。运用
相同的测量系统，我们测得报纸的对比度为 5:1，反射率为 55%。而且经几个月的
存放后，该材料仍然可以很容易地编址。

图 3a 中展示了反射光强的最大和最小值之差（对比度未归一化）与不同外加场
下驱动频率的关系；空心图形代表频率由低至高时采集的数据，实心图形代表频率
由高至低时所采集的数据。图 3b 为不同频率下对比度与外加电场的关系。图 3 中的
数据源于 200 μm 的电子墨水（"染料中白色粒子"型）薄膜，在具有银墨水背电极的
ITO 聚酯膜上的胶囊直径为 40±10 μm。

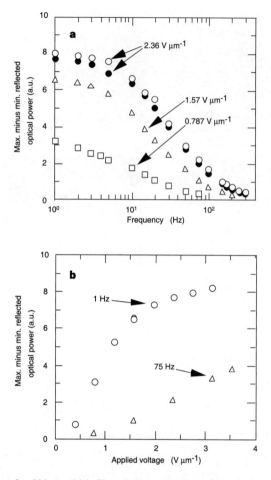

Fig. 3. Properties of a 200-μm-thick film of electronic ink ("white particles in dye" type) with capsule diameter of 40 ± 10 μm. **a**, Maximum minus minimum reflected optical power (contrast) versus drive frequency for several different applied fields (open symbols, data taken from low to high frequency; filled symbols, from high to low frequency). **b**, Plot of contrast versus applied field for different frequencies.

We may calculate the particle mobility, zeta potential and charge per particle as follows. The low-frequency asymptote in Fig. 3a indicates that at sufficiently slow driving frequency there is time for the internal particles to make a full traverse of the capsule, after which there is no further contribution to the contrast. Thus the cusp of the sigmoid in Fig. 3a may be taken to indicate the frequency (5 Hz) at which the particle just traverses a round trip (2×40 μm = 80 μm) in the microcapsule yielding a velocity of $v = 400$ μm s^{-1}. The Reynolds number is given by $\mathrm{Re} = \rho v r \eta^{-1}$ where ρ is the internal fluid density, r is the particle radius, η is the internal fluid viscosity and v is the particle velocity. For the present system we have $\rho = 1.6 \times 10^{-15}$ kg μm^{-3}, $v = 400$ μm s^{-1}, $r = 0.5$ μm, and $\eta = 0.8 \times 10^{-9}$ kg μm^{-1} s^{-1}; and $\mathrm{Re} = 0.0004 \ll 1$ and the flow is laminar[14]. The transient time to establish the laminar flow may be estimated from the Navier–Stokes equation to be $\tau_{ss} = (1/9)r^2 \rho \eta^{-1} = 55$ ns. Thus for time-scales of interest, the particle mobility is given as $\mu = v/E = \epsilon \zeta / 6\pi\eta = q/12\pi r \eta$ where ϵ is the dielectric constant of the internal fluid, ζ is the zeta potential of the particles and q is the charge per particle. Thus for our system we have

286

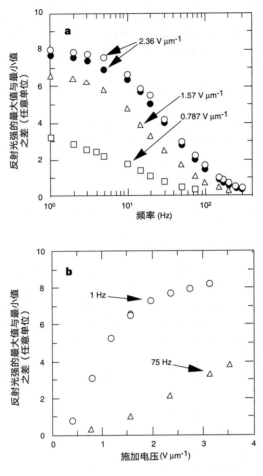

图 3. 胶囊直径为 40 ± 10 μm，厚度为 200 μm 的电子墨水（"染料中白色粒子"型）薄膜的属性。**a**，反射光强的最大值与最小值之差（对比度）与不同外场下驱动频率的关系（空心图形，频率由低至高采集的数据；实心图形，频率由高至低所采集的数据）。**b**，不同频率下对比度与外场的关系。

　　我们可以通过如下方法计算粒子的迁移率、界面电位和单位粒子的荷电量。图 3a 中低频渐近线意味着当驱动频率足够低时，内部粒子有时间完全横穿胶囊，之后对对比度没有进一步的贡献。因此图 3a 中 S 形的突起点可以视为粒子刚好在胶囊中以 5 Hz 的频率往返一周（2×40 μm $= 80$ μm）时的速率，为 $v = 400$ μm·s⁻¹。雷诺常数由 $\mathrm{Re} = \rho v r \eta^{-1}$ 给出，其中 ρ 为内部流体密度，r 为粒子半径，η 为内部流体黏度，v 为粒子速度。对于现在的系统，有 $\rho = 1.6 \times 10^{-15}$ kg·μm⁻³，$v = 400$ μm·s⁻¹，$r = 0.5$ μm，$\eta = 0.8 \times 10^{-9}$ kg·μm⁻¹·s⁻¹，$\mathrm{Re} = 0.0004 \ll 1$，并且流动为层流[14]。确定层流的瞬态时间可以从纳维–斯托克斯方程中估算出来，为 $\tau_{ss} = (1/9) r^2 \rho \eta^{-1} = 55$ ns。因此在我们所感兴趣的时间尺度之内，粒子迁移率可以表示为 $\mu = v/E = \epsilon \zeta / 6\pi\eta = q/12\pi r\eta$，其中 ϵ 为内部流体的介电常数，ζ 为粒子的界面电位，q 为单位粒子的电量。对于我们的系统，$\mu = 169$ μm²·V⁻¹·s⁻¹，$\epsilon = 2.5 \times 8.85 \times 10^{-6}$ kg·μm·s⁻²·V⁻²，$\zeta = 120$ mV，

$\mu = 169\ \mu\text{m}^2\,\text{V}^{-1}\,\text{s}^{-1}$, $\epsilon = 2.5 \times 8.85 \times 10^{-6}\ \text{kg}\ \mu\text{m}\ \text{s}^{-2}\,\text{V}^{-2}$, $\zeta = 120\ \text{mV}$, and $q = 2.6 \times 10^{-18}\ \text{C} = 16e^{-}$. The very small charge per particle indicates the mechanism for "field-off" bistability. Particle/capsule-wall and particle/particle binding dominate the particle/particle repulsion coming from the very small charge per particle. Other mechanisms for "field-off" bistability include charge redistribution and minimization.

Figure 3a indicates the compromise between switching speed and contrast; Fig. 3b shows the compromise between applied field and contrast. In both sets of curves, the high-contrast asymptote indicates the regime in which the particles have made full traverse of the capsule, and further time or applied field has no further consequence. Two-particle (black/white particles) systems, as opposed to "single particle in dye" systems, relax these trade-offs due to the absence of exponential light absorption in the dye case due to dye interstitially present between white particles.

In order to efficiently address a large number of pixels, matrix addressing is required. Matrix addressing in turn requires that either the contrast material (passive matrix) or an underlying electronic layer (active matrix) possess a threshold in order to prevent crosstalk between address lines. Typically, the display material may support passive matrix addressing for small numbers of pixels[15]. Larger numbers of pixels require the nonlinearity of an electronic element as implemented in active matrix addressing[16]. Active matrix structures, typically formed from polysilicon on glass, are expensive and would obviate many of the advantages of a flexible low-cost paper-like display. Our group has recently demonstrated an all-printed metal–insulator–metal (MIM) diode structure[17] capable of forming the active matrix for electrophoretic ink, which should enable us to drive high-pixel-count sheets of electrophoretic ink while maintaining the ease of fabrication and low-cost structure provided by the ink material itself. Work by other groups may also be of use in this endeavour[18]. We believe that such systems will open up a new field of research in printable microscopic electro-mechanical systems. Coupled with recent advances in printable logic, such technology offers the prospect of fundamentally changing the nature of printing, from printing form to printing function.

(**394**, 253-255; 1998)

Barrett Comiskey, J. D. Albert, Hidekazu Yoshizawa & Joseph Jacobson
Massachusetts Institute of Technology, The Media Laboratory, 20 Ames Street, Cambridge, Massachusetts 02139-4307, USA

Received 8 December 1997; accepted 18 May 1998.

References:

1. Pankove, J. I. *Color Reflection Type Display Panel* (Tech. Note No. 535, RCA Lab., Princeton, NJ, 1962).

2. Sheridon, N. K. & Berkovitz, M. A. The gyricon–a twisting ball display. *Proc. Soc. Information Display* **18(3,4)**, 289-293 (1977).

3. Sheridon, N. K. *et al.* in *Proc. Int. Display Research Conf.* (ed. Jay Morreale) L82–L85 (Toronto, 1997).

4. Ota, I., Honishi, J. & Yoshiyama, M. Electrophoretic image display panel. *Proc. IEEE* **61**, 832-836 (1973).

5. Dalisa, A. L. Electrophoretic display technology. *IEEE Trans. Electron Devices* **ED-24(7)**, 827-834 (1977).

6. Castellano, J. A. & Harrison, K. J. in *The Physics and Chemistry of Liquid Crystal Devices* (ed. Sprokel, G. J.) 263-288 (Plenum, New York, 1980).

$q = 2.6 \times 10^{-18}$ C $= 16e^-$。单位粒子带电量极低表示遵守"去场"双稳态机制。粒子/胶囊壁和粒子/粒子间的结合对来自带电量很小的微粒间的粒子/粒子互斥起主导作用。"去场"双稳态的其他机制包括电荷再分布以及最小化。

图 3a 指示出切换速度与对比度之间的制约关系；图 3b 显示了外场与对比度之间的制约关系。在两组曲线中，高对比度渐近线暗示着粒子完全穿越胶囊的机制，随后时间和外场都不再起作用。和"染料中的单粒子系统"不同的是，在双粒子（黑/白粒子）系统中，由于白粒子处于染料之中，而染料对光不发生指数型吸收，因而削弱了这种制约。

大量像素的有效编址要应用矩阵编址。矩阵编址又依次要求对比材料（被动矩阵）或者基本电层（主动矩阵）具有一个阈值以防止编址线之间的串扰。通常，显示材料可以支持少量像素的被动矩阵编址[15]。大量像素要求电子元件在实现主动矩阵编址时具有非线性[16]。主动矩阵结构，通常是由在玻璃上的多晶硅形成，昂贵而且不具备低成本的柔韧性类纸张显示器的许多优点。我们的团队近期提出了一种可印刷的金属-绝缘体-金属（MIM）的二极管结构[17]，可以组成电泳墨水的主动矩阵，应当有可能使驱动电泳墨水的高像素数印刷板成为现实，因为使用了这种墨水材料，保持了制造的简单易行和结构的低成本。其他团队的工作也可能应用到这项研究之中[18]。我们相信这种系统会开创可印刷微观机电系统的一个新研究领域。结合现阶段其他可印刷技术的思路，这种技术预期将从根本上改变印刷的特性，从印刷形式到印刷功能。

（崔宁 翻译；宋心琦 审稿）

7. Fitzhenry-Ritz, B. Optical properties of electrophoretic image displays. *IEEE Trans. Electron Devices* **ED-28**(6), 726-735 (1981).

8. Comiskey, B., Albert, J. D. & Jacobson, J. Electrophoretic ink: a printable display material. In *Digest of Tech. Papers* 75-76 (Soc. Information Display, Boston, 1997).

9. Okada, M., Hatano, T. & Hashimoto, K. Reflective multicolor display using cholesteric liquid crystals. In *Digest of Tech. Papers* 1019-1026 (Soc. Information Display, Boston, 1997).

10. Murau, P. & Singer, B. The understanding and elimination of some suspension instabilities in an electrophoretic display. *J. Appl. Phys.* **49**, 4820-4829 (1978).

11. Hebner, T. R., Wu, C. C., Marcy, D., Lu, M. H. & Sturm, J. C. Ink-jet printing of doped polymers for organic light emitting devices. *Appl. Phys. Lett.* **72**, 519-521 (1998).

12. Fowkes, F. M., Jinnai, H., Mostafa, M. A., Anderson, F. W. & Moore, R. J. in *Colloids and Surfaces in Reprographic Technology* (eds Hair, M. & Croucher, M. D.) (Am. Chem. Soc., Washington DC, 1982).

13. Claus, C. J. & Mayer, E. F. in *Xerography and Related Processes* (eds Dessauer, J. H. & Clark, H. E.) 341-373 (Focal, New York, 1965).

14. Landau, L. D. & Lifshitz, E. M. *Fluid Mechanics* (Pergamon, New York, 1959).

15. Ota, I., Sato, T., Tanka, S., Yamagami, T. & Takeda, H. Electrophoretic display devices. In *Laser 75* 145-148 (Optoelectronics Conf. Proc., Munich, 1975).

16. Shiffman, R. R. & Parker, R. H. An electrophoretic image display with internal NMOS address logic and display drivers. *Proc. Soc. Information Display* **25.2**, 105-115 (1984).

17. Park, J. & Jacobson, J. in *Proc. Materials Research Soc.* B8.2 (Mater. Res. Soc., Warrendale, PA, 1998).

18. Service, R. F. Patterning electronics on the cheap. *Science* **278**, 383 (1997).

Correspondence and requests for materials should be addressed to J.J. (e-mail: jacobson@media.mit.edu).

Asynchrony of Antarctic and Greenland Climate Change during the Last Glacial Period

T. Blunier *et al.*

Editor's Note

Clues to the mechanisms of natural climate shifts can be gleaned from looking at whether or not they happen synchronously across the globe. This is particularly pertinent for the rapid climate transitions seen in ice-core records of temperature during the last ice age. Here Thomas Blunier of the University of Bern and his coworkers compare these fast climate fluctuations between Greenland in the Northern Hemisphere and Antarctica in the Southern Hemisphere. They find that the Antarctic events happen about 1,000–2,500 years before those in Greenland. The researchers suspect that ocean circulation links the hemispheric climates: changes that originate in the south are transferred to the north by concomitant changes to circulation in the North Atlantic.

A central issue in climate dynamics is to understand how the Northern and Southern hemispheres are coupled during climate events. The strongest of the fast temperature changes observed in Greenland (so-called Dansgaard–Oeschger events) during the last glaciation have an analogue in the temperature record from Antarctica. A comparison of the global atmospheric concentration of methane as recorded in ice cores from Antarctica and Greenland permits a determination of the phase relationship (in leads or lags) of these temperature variations. Greenland warming events around 36 and 45 kyr before present lag their Antarctic counterpart by more than 1 kyr. On average, Antarctic climate change leads that of Greenland by 1–2.5 kyr over the period 47–23 kyr before present.

GREENLAND ice-core isotopic records[1,2] have revealed 24 mild periods (interstadials) of 1–3 kyr duration during the last glaciation[1] known as Dansgaard–Oeschger (D–O) events[3], where temperature increased by up to 15 °C compared with full glacial values[4,5]. The temperature over Greenland is essentially controlled by heat released from the surface of the North Atlantic Ocean[6]. This heat is brought to the North Atlantic by the thermohaline circulation which is driven by the North Atlantic Deep Water (NADW) formation[7]. NADW strongly influences the deep circulation on a global scale. Therefore climate change related to the North Atlantic thermohaline circulation is not restricted to the North Atlantic region but has global implications[8]. Data correlating with D–O events have been found, for examples in the southern Atlantic[9], the Pacific Ocean[10] and probably also the Indian Ocean[11]. On the continents, related events were found in North America where D–O events are manifested as wetter climate periods[12,13]. Also, the hydrological cycle

292

末次冰期南极和格陵兰气候变化的非同步性

布鲁尼尔等

编者按

有关自然气候变化机制的线索可以通过考察其是否在全球同步发生来收集。这尤其适合用来研究末次冰期期间冰芯温度记录所展现的气候快速变化问题。本文中，来自伯尔尼大学的托马斯·布鲁尼尔及其合作者比较了北半球格陵兰和南半球南极二者间的这些快速气候波动。他们发现南极的事件比格陵兰的对应事件早发生约1,000～2,500 年。研究者们猜测海洋环流联动了两个半球的气候：起源于南半球的变化，通过伴随的北大西洋环流的变化，传递至北半球。

气候动力学的核心问题是弄清在气候事件发生过程中南北半球之间如何耦合。在格陵兰观测到的末次冰期内最强的快速温度变化（所谓丹斯果–厄施格尔事件）和南极温度记录有一定的相似性。对比南极和格陵兰冰芯中记录的全球大气甲烷浓度，可以确定这些温度变化之间的位相关系（超前还是滞后）。距今 3.6 万年和 4.5 万年的格陵兰增暖比南极相应的温度变化滞后 1,000 年以上。总的说来，在距今 4.7 万年前到 2.3 万年前期间南极气候变化比格陵兰的超前 1,000～2,500 年。

格陵兰冰芯的同位素记录[1,2]揭示在末次冰期内[1]有 24 个时间长度约 1,000 年到 3,000 年、称之为丹斯果–厄施格尔事件（简称 D-O 事件）[3]的温暖时段（间冰阶），当时其温度与整个冰期的温度相比增高达 15℃[4,5]。格陵兰的温度基本上受控于北大西洋洋面所释放的热量[6]，这些热量是由北大西洋深层水（NADW）的生成所驱动的温盐环流带到北大西洋洋面的[7]。NADW 强烈地影响着全球范围海洋的深层环流，因此，和北大西洋温盐环流有关的气候变化并不局限于北大西洋地区，而是有着全球性的影响[8]。例如在南大西洋[9]、太平洋[10]，可能还包括印度洋[11]，都已发现与 D-O 事件有关的记录。在大陆上，有关的事件在北美也有发现，该地 D-O 事件表现为更湿润的气候特征[12,13]。热带水循环的变化似乎也和 D-O 事件同步出

in the tropics appears to have changed in parallel with D–O events. This is indicated by changes in atmospheric methane concentration, the main source of which is wetlands situated in the tropics during the last glaciation[14] .

Temperature variations in Greenland inferred from the $\delta^{18}O$ record are characterized by fast warmings and slow coolings during the last glaciation. Antarctic temperature variations (from $\delta^{18}O$ and δD records) show a different pattern with fewer, less pronounced and rather gradual warming and cooling events over the same period[15,16] .

One of the great challenges of climate research is therefore to find a mechanism that is able to reconcile the very different dynamical behaviour of high-latitude Northern and Southern hemispheres and their phase relationship during rapid climate change. A crucial test for our understanding of climate change is the relative phase of events in the two hemispheres. Using the global isotopic signal of atmospheric oxygen, Bender et al.[17] hypothesized that interstadial events start in the north and spread to the south provided that they last long enough (~ 2 kyr), postulating that northern temperature variations lead those in Antarctica. On the other hand, they did not exclude the possibility that corresponding events are out of phase, as suggested by Stocker et al.[18] on the basis of a simplified ocean–atmosphere model which shows that heat is drawn from the Antarctic region when NADW switches on at the onset of a D–O event. This leads to a cooling of surface air temperatures in the south and a corresponding warming in the north. Charles et al.[9] used stable-isotope ratios of benthic and planktonic foraminifera in a single core from the Southern Ocean to argue that the Northern Hemisphere climate fluctuations lagged those of the Southern Hemisphere by ~ 1.5 kyr. They conclude that a direct link of high-latitude climate is not promoted by deep-ocean variability.

Conclusive information regarding the phase lag of climate events can come only from high-resolution palaeoarchives which have either absolute or synchronized timescales with uncertainties smaller than 500 yr (for example, to distinguish the Younger Dryas from the Antarctic Cold Reversal[19], hereafter YD and ACR). This prerequisite is at present satisfied only in polar ice cores. Here we use fast variations in methane during the last glaciation for a synchronization that permits a better estimate of the relative timing of events in the two hemispheres.

The CH_4 records from two Antarctic ice cores (Byrd station 80° S, 120° W; Vostok 78.47° S, 106.80° E) and one Greenland core (GRIP ice core, Summit 72.58° N, 37.64° W) have been used to establish coherent timescales for the Antarctic cores with respect to the GRIP timescale over the period from 50 kyr BP to the Holocene epoch. The timescale for the Greenland reference core (GRIP) has been obtained by stratigraphic layer counting[20] down to 14,450 yr BP with an uncertainty of ± 70 yr at 11,550 yr BP (termination of the YD). For older ages, the chronology is based on an ice-flow model[21]. This timescale is not considered to be absolute; for instance, the model-based dating differs from the stratigraphic layer counting in the deeper part of the core by several thousand years. However, our conclusions do not depend on the particular timescale, as explained below.

现。这种末次冰期时热带水循环的改变，以源于热带湿地的大气甲烷浓度的变化为标志[14]。

从 $\delta^{18}O$ 记录中推断出格陵兰在末次冰期中的温度变化特征表现为快速增暖和缓慢变冷。南极在同一时期的温度变化（从 $\delta^{18}O$ 和 δD 记录推断）则显示出不同的模态：即较少的、不那么显著且较为渐进的增暖和变冷事件[15,16]。

因此，气候研究的最大挑战之一就是找出一种能够调和高纬度南北半球气候快速变化之间显著不同的动力行为以及位相关系的机制。对两半球发生的事件的相对位相的研究是认识气候变化的重要的检验方法之一。本德等[17]分析了全球大气中的氧同位素信息，提出猜想认为如果间冰阶事件持续时间足够长（约 2,000 年），那么这些事件开始于北半球，然后向南半球扩展，即假设北半球温度变化导致了南极温度的变化。另外，这些实验也不排除两半球对应事件异相的可能性，如同施托克尔等[18]从一个简化的海洋–大气模式中展示的那样：在 D–O 事件伊始，NADW 开启，热量从南极地区流出。这就使南半球地表气温降低，北半球温度对应升高。查尔斯等[9]应用南大洋一个岩芯中底栖和浮游有孔虫的稳定同位素比率资料，提出北半球的气候波动比南半球滞后 1,500 年左右。他们的结论是，两半球高纬度气候的直接联系并非由深海的变化促成。

关于气候事件之间位相滞后的关键信息只能来自高分辨率的古资料，它需要绝对的或同步化的时标，其不确定性要小于 500 年（以便区分如南极转冷期[19]（ACR）和新仙女木期（YD））。现在只有极地冰芯可以满足这一先决条件。我们应用了末次冰期内的甲烷快速变化以保证同步化的时标，这样就可以对两半球气候事件之间的相对时间有更好的估计。

应用两个南极冰芯（80°S，120°W 的伯德站冰芯和 78.47°S，106.80°E 的东方站冰芯）和一个格陵兰冰芯（72.58°N，37.64°W 的萨米特站的 GRIP 冰芯（GRIP 即格陵兰冰芯计划））的 CH_4 记录，建立距今 5 万年前到全新世时期南极冰芯相对于 GRIP 冰芯的同步时标。应用地层计数法[20]已经将格陵兰参考冰芯（GRIP 冰芯）的时标向下标定至距今 14,450 年，在距今 11,550 年（YD 结束时）存在 ±70 年的不确定性。对于更早的时期，其年代由冰流模型确定[21]。这样确定的时标并不是绝对的，例如，基于模型的定年和用地层计数法得到的冰芯深部年龄有数千年的差异。无论如何，如下所述，我们的结论并不依赖于求得的个别时标。

We are able to show that long-lasting Greenland warming events around 36 and 45 kyr BP (D–O events 8 and 12) lag their Antarctic counterpart by 2–3 kyr (comparing the starting points of corresponding warmings). On average, Antarctic climate change leads that of Greenland by 1–2.5 kyr over the period 47–23 kyr BP. This finding contradicts the hypothesis that Antarctic warmings are responses to events in the Northern Hemisphere[17]. The observed time lag also calls into question a coupling between northern and southern polar regions via the atmosphere, and favours a connection via the ocean. Indeed, climate models[18,22-24] suggest that heat is extracted from the Southern Hemisphere when NADW formation switches on. This interhemispheric coupling is clearly identified in interstadial events 8 and 12, as well as during the termination of the last glaciations.

Synchronization of Ice-core Records

Because the porous firn layer on the surface of the ice sheet exchanges air with the overlying atmosphere, air, which gets enclosed in bubbles at 50–100 m below the surface (present-day conditions), has a mean age which is younger than the surrounding ice[5]. Consequently, a timescale for the ice and a timescale for the air need to be known, both of which depend on local climate characteristics. For GRIP, the difference between the age of the ice and the mean age of the gas at close-off depth (Δage) has been calculated using a dynamic model[5] for firn densification and diffusional mixing of the air in the firn, including the heat transport in the firn and temperature dependence of the close-off density. For Summit, estimates of the main parameters—accumulation rate and temperature[4,25]—allow the calculation of Δage with an uncertainty of only ± 100 yr between 10 and 20 kyr BP, increasing to ± 300 yr back to 50 kyr BP (ref. 5). Temperature was deduced from the $\delta^{18}O$ profile as suggested from borehole temperature measurements for the glacial–interglacial transition[4], corresponding roughly to a linear dependence with a slope of 0.33‰ per °C. This relation is in contrast to today's relation[26] with a slope of 0.67‰ per °C. However, there is good evidence that the $\delta^{18}O$–temperature relation was close to the one derived from the borehole temperature profile throughout the past 40 kyr (ref. 5).

The model calculations are confirmed by an independent measurement of Δage on the GISP2 core, drilled 30 km to the west of the GRIP drill site. A Δage of 809 ± 20 yr was determined, on the basis of thermal fractionation of nitrogen at the termination of YD[27]. This is consistent with Δage of 775 ± 100 yr for the YD calculated with the same model used here[5].

Two steps are necessary to synchronize ice cores via a gas record: (1) correlation of the CH_4 records, and (2) calculation of the age of the ice by determining Δage. The initial GRIP CH_4 profile[14,28] was refined between 9 and 55 kyr BP by an additional 180 samples measured both in the Bern and Grenoble laboratories, resulting in a mean sampling resolution of about 150 yr in the period of interest. A number of new samples were measured on the Antarctic cores from Byrd and Vostok station; 179 and 56, respectively. These measurements cover the time period 9–47 kyr BP on Byrd and 30–43 kyr BP on Vostok,

我们能够说明的是，发生在大约距今 3.6 万年前和 4.5 万年前（D–O 事件 8 和 12）的持续较长的格陵兰增暖事件比南极的相应事件滞后 2,000～3,000 年（由对比相应的增暖事件的开始时间而得）。平均说来，在距今 4.7 万到 2.3 万年前的南极气候变化比格陵兰超前 1,000～2,500 年。这一发现反驳了南极增暖是对北半球对应事件的响应这一假设 [17]。所观测到的时间滞后使人们对南北半球极区间的耦合是通过大气来实现的观点产生怀疑，转而支持通过海洋产生关联的观点。实际上气候模式 [18,22-24] 显示，当 NADW 的形成开启，它将从南半球获得热量。这种半球之间的耦合关系可以从间冰阶事件 8 和 12 以及末次冰期的结束期中被清晰地辨认出来。

冰芯记录的同时化

由于冰盖表面覆盖的透气粒雪层和其上的大气间有空气交换，包裹在冰盖表面（指现代冰盖表面）以下 50～100 m 处气泡中的空气的平均年代比周围冰的年代要晚 [5]。因此需要知道冰的时标和空气的时标，它们二者都取决于当地的气候特点。对于 GRIP 冰芯，冰的年代和包裹在深处气泡中的气体的平均年代的差（Δage）已经可以用一个关于雪粒致密化和雪粒中空气扩散混合的动力学模式 [5] 计算出来，该模式考虑了雪粒中的热量输送与封闭密度对温度的关系。对于萨米特站，通过主要参数（累积率和温度 [4,25]）的估计可以算出 Δage，其不确定性在距今 1 万到 2 万年间仅为 ±100 年，到距今 5 万年时不确定性增加到 ±300 年（见参考文献 5）。就像对冰期–间冰期转换期间钻孔温度的测量 [4] 那样，温度可由 $\delta^{18}O$ 廓线推导出来，二者粗略符合斜率为 0.33‰/℃ 的线性关系。这一关系和现今斜率为 0.67‰/℃ 的关系 [26] 大不相同。不过，有充分的证据表明 $\delta^{18}O$–温度关系和从过去 4 万年钻孔温度廓线得来的关系是相近的（见参考文献 5）。

模式的计算结果已被 GRIP 冰芯钻孔点西面 30 km 处的 GISP2 冰芯（格陵兰冰盖计划 2）独立测量的 Δage 值证实。根据 YD 结束时氮的热力学分级 [27]，Δage 值为 809±20 年。这和应用同一模式对 YD 时期计算得到的 775±100 年的 Δage 值是一致的 [5]。

应用气体记录使冰芯资料同时化需要两个步骤：（1）CH_4 记录的相关性计算。（2）通过求 Δage 计算冰的年代。结合伯尔尼和格勒诺布尔的两个实验室测量的 180 个样本，初始的 GRIP 冰芯 CH_4 廓线 [14,28] 在距今 9,000 年到 5.5 万年段的精度得以提高，这使得其平均采样分辨率在所关心的时段可达到 150 年左右。在南极伯德站和东方站冰芯测得了大量新的样本，两站样本数分别为 179 和 56。这些观测值涵盖伯德站距今 9,000 年到 4.7 万年时期和东方站距今 3 万到 4.3 万年的时期，其平均采

with a mean sampling resolution of 200 and 250 yr, respectively. For the synchronization of the CH_4 records, a Monte Carlo method was used, searching for a maximal correlation between the records[5]. The fast CH_4 variations allow us to fit the records to ± 200 yr. The time period 25–17 kyr BP was excluded from the synchronization because only low CH_4 variations appear during that period, making the correlation uncertain.

The uncertainty in Δage for the Antarctic sites is dominated by the uncertainty in temperature and accumulation. Estimates of the glacial–interglacial temperature difference at Vostok and Byrd station range from 7 to 15 °C (refs 29, 30). This range is covered in our Δage calculations for Byrd station (see Methods). Δage for the Byrd core is one order of magnitude smaller than for the Vostok core, and comparable to the GRIP Δage. This makes the synchronization of isotopes more convincing for Byrd/GRIP than for Vostok/GRIP (see below and Methods).

To validate the resulting timescales, the distinct double peak of the cosmogenic radioisotope [10]Be around 40 kyr BP can be used. The GRIP [10]Be record[31] is plotted together with that of Byrd[32] on the new synchronized timescale in Fig. 1. Unfortunately, the resolution of the latter is much lower, which does not allow a firm validation. Nevertheless, the two timescales—resulting from assuming 7 °C or 15 °C glacial–interglacial temperature change—are both consistent with the [10]Be signal. As they represent extreme positions relative to the GRIP record around the peak position, we are confident that the true Byrd timescale (relative to GRIP) lies within our estimates. The relative uncertainty at the location of the [10]Be peak given by the two temperature models is about 300 yr.

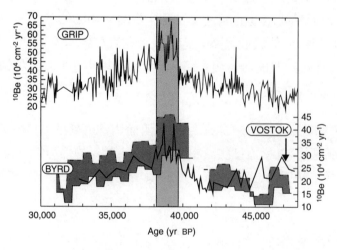

Fig. 1. [10]Be fluxes from Byrd[32], Vostok[33] and GRIP[31] ice cores. Values for Byrd were reduced by 15%, accounting for a difference in standards used for the measurements on the Byrd and the GRIP/Vostok cores (J. Beer, personal communication). Dating of the GRIP core is after ref. 21. For Byrd, two dating scenarios were used: lower temperature and accumulation (solid line); higher temperature and accumulation (dashed line; see Methods for details). Vostok was dated as described in Methods. The hatched area indicates the [10]Be peak area deduced from the GRIP core.

样分辨率分别为 200 年和 250 年。为了使 CH_4 记录同时化，应用蒙特卡罗方法寻求记录之间的最大相关性[5]。CH_4 的快速变化使得我们拟合记录的精度达到 ±200 年。距今 2.5 万到 1.7 万年段在同时化时被剔除，因为这一时期 CH_4 的变化较慢，使相关性变得不确定。

南极测站 Δage 值的不确定性来源于温度及累积率的不确定性。东方站和伯德站的冰期–间冰期温度差异范围为 7 ~ 15℃（参考文献 29 和 30），这一差异范围在我们对伯德站 Δage 计算的范围之内（详见方法部分）。伯德站冰芯的 Δage 值比东方站冰芯的值小一个量级，而与 GRIP 冰芯的 Δage 值接近。这就使得对于同位素记录的同时化，伯德站/GRIP 冰芯的对比相比东方站/GRIP 冰芯的对比更具说服力（详见下文和方法部分）。

可以利用宇宙线产生的放射性同位素 ^{10}Be 在距今 4 万年左右的显著双峰现象来确认所求出的时间标尺的有效性，在图 1 中 GRIP 冰芯的 ^{10}Be 的记录[31] 和伯德站冰芯的记录[32] 一起画在新的同步化时标上。遗憾的是，后者的分辨率很低，因而可用程度不高。不过，从假定的 7℃ 或 15℃ 的冰期–间冰期温度变化得到的两个时标都与 ^{10}Be 的信号一致。由于它们代表了相对于 GRIP 记录在峰值附近的极端位置，我们可以认为真实的伯德站冰芯时标（相对于 GRIP 冰芯时标）处于我们估计的范围内。两个温度模式给出的 ^{10}Be 峰值位置的相对不确定性大约为 300 年。

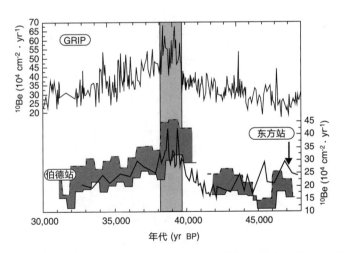

图 1. 伯德站[32]、东方站[33] 和 GRIP[31] 冰芯的 ^{10}Be 通量。伯德站的值减少了 15%，说明伯德站和 GRIP/东方站观测标准有差别（比尔，个人交流）。GRIP 冰芯的年代根据参考文献 21 确定。伯德站的定年运用了两种情景：较低的温度及累积率（实线）；较高的温度及累积率（虚线，详见方法部分）。东方站的定年在方法部分进行了描述。阴影区表示由 GRIP 冰芯推断的 ^{10}Be 的峰值区。

Main Features of Antarctic Climate

Vostok station has a low accumulation rate and low temperature; the present values are 1.7–2.6 cm of ice per year and −55°C, respectively[16]. This results in a large Δage, and makes Δage extremely sensitive to uncertainties in these parameters. Therefore, the Vostok CH_4 record alone is not well suited to correlate the Northern and the Southern hemispheres on a millennial scale. However, in combination with the detailed ^{10}Be record[31,33] a reasonable dating relative to Greenland can be achieved (for details, see Methods). The use of Vostok, in view of the relation between Greenland and Antarctic records, is to indicate which features seen in the Byrd core are representative for the climate in Antarctica (we limit our discussion to these features). They include the deglaciation and the ACR, a cold event occurring during the Bølling/Allerød[19,34,35], and warm events centring at 35.5 and 44.5 kyr BP (denoted as A1 and A2 in Fig 2).

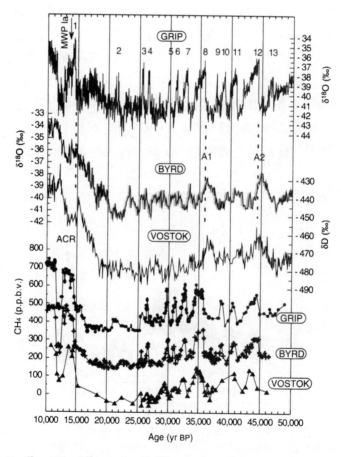

Fig. 2. GRIP[1], Byrd[15] and Vostok[16] isotopic and CH_4 records on the common timescale (GRIP timescale in years before 1989). For the data from Byrd, a solid line shows the lower-temperature scenario, and the shaded area indicates the range of Δage for the two temperature scenarios discussed in Methods. The CH_4 scale at lower left corresponds to the GRIP values; Byrd and Vostok values were lowered by 200 and 400 p.p.b.v., respectively, for better visibility. The numbers on top of the GRIP isotopic record

南极气候的主要特征

东方站冰的累积率低，温度也低：当前冰的累积率为每年 1.7～2.6 厘米，温度为 −55℃ [16]。这导致一个大的 Δage 值，且使 Δage 对这些参数的不确定性极端敏感。因此，该站的 CH₄ 记录不适合单独表示南北半球千年尺度的气候特征。但是，如果再联合详细的 ¹⁰Be 记录 [31,33] 就能得到相对于格陵兰冰芯记录的合理的年代值（详见方法部分）。鉴于格陵兰和南极记录的关系，应用东方站的资料可以告诉我们从伯德站冰芯中所见的哪些特征对于南极气候是有代表性的（此处讨论仅限于这些特征）。这些气候事件包括冰消期和 ACR，后者是发生在波令/阿勒罗德期间 [19,34,35] 的一个冷事件，前者是中心位于距今 3.55 万和 4.45 万年的暖事件（图 2 中用 A1 和 A2 表示）。

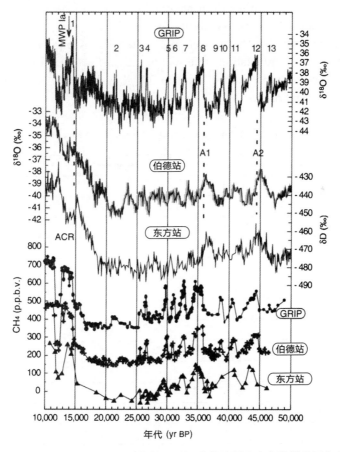

图 2. 同一时标下（1989 年以前的 GRIP 的时标）GRIP[1]、伯德站 [15] 和东方站 [16] 的同位素和 CH₄ 记录。对于伯德站的资料，实线表示低温的情景，阴影区域表示方法部分中讨论的两种情景的 Δage 值的范围。左下方的 CH₄ 标尺对应 GRIP 的值；伯德站和东方站的值分别在图示中降低 200 ppbv 和 400 ppbv，以使两站的结果看得更清楚。GRIP 的同位素记录顶部的数字表示 D–O 事件的位置；ACR 是茹泽尔等 [34]

indicate the location of Dansgaard–Oeschger events; ACR is the location of the Antarctic Cold Reversal as described by Jouzel *et al.*[34]; Antarctic warmings are indicated by A1 and A2; vertical dashed lines show the location of Greenland warmings 1, 8 and 12 in the Antarctic cores. The Byrd $\delta^{18}O$ variation between 25 and 17 kyr BP shows most likely local climate characteristics because: (1) it is not seen in the Vostok or in the Dome C record[51], and (2) it is not compatible with the climate mechanism seen during events A1, A2 and also the ACR present in all Antarctic records. The arrow marked MWP 1a indicates the position of the first meltwater pulse[45].

Phase Lag

In Fig. 2 the isotopic records from Byrd ($\delta^{18}O$), Vostok (δD) and GRIP ($\delta^{18}O$), which is representative of Greenland[2,21] and clearly registers Northern Hemispheric climate change[36], are plotted on the common timescale (Table 1) based on GRIP. Our synchronization clearly demonstrates that the Antarctic warmings (A1 and A2) are out of phase with the corresponding D–O events (8 and 12). Antarctic temperature variations A1 and A2 are characterized by a gradual rather than abrupt increase and decrease; D–O events by a fast increase and a slow decrease. While the Antarctic temperature is already increasing, Greenland is still cooling. The Antarctic temperature rise is interrupted once Greenland temperature jumps to an interstadial state within only a few decades[2,21]. Subsequently, the temperature decreases in both hemispheres to full glacial level, where Antarctica reaches this level earlier than Greenland.

Table 1. Corresponding depths for the three ice cores

Age of the ice (kyr BP)	Depth GRIP (m)	Depth Byrd*†(m)	Depth Vostok†(m)
12	1638.0	1074.4	261.1
13	1674.0	1124.2	278.6
14	1722.6	1161.4	293.1
15	1770.0	1198.6	307.6
16	1802.2	1235.8	322.8
17	1834.3	1269.5	336.4
18	1867.6	1294.4	348.9
19	1894.5	1319.4	360.1
20	1916.5	1344.3	371.0
21	1937.7	1369.3	381.9
22	1958.8	1394.2	392.9
23	1975.6	1419.1	404.3
24	1993.1	1444.1	415.9
25	2011.1	1469.0	427.3
26	2031.6	1493.9	438.9
27	2051.6	1518.7	451.3
28	2065.1	1539.9	464.2

描述的南极转冷事件；南极增温用 A1 和 A2 表示；垂直虚线表示南极冰芯中对应于格陵兰增温事件 1、8 和 12 的位置。距今 2.5 万到 1.7 万年之间伯德站 δ¹⁸O 的变化很可能表示局地气候特征，这是因为：(1)这些变化并未在东方站或冰穹 C 记录 [51] 中见到；(2) 它与事件 A1 和 A2 中所见的气候机制不兼容，且所有南极记录中都有 ACR 出现。MWP 1a 所示的箭头表示第一次融水脉冲事件 [45] 的位置。

相 位 滞 后

图 2 中给出的来自伯德站(δ¹⁸O)、东方站(δD)和 GRIP(δ¹⁸O)三个点的同位素记录共同标注在以 GRIP 冰芯为基础的时标上(参见表 1)。GRIP 代表格陵兰 [2,21]，清晰地记录了北半球的气候变化 [36]。我们的同时化结果清晰地表明，南极增暖(A1 和 A2)和对应的 D–O 事件(8 和 12)异相。南极的温度变化事件 A1 和 A2 以渐进变化为特征，而不是突然的增温或降温，而 D–O 事件的特点则是快速增温和缓慢降温。当南极开始增温时，格陵兰仍在降温。一旦格陵兰温度在短短几十年内突然转到间冰阶时 [2,21]，南极的温度上升就被打断了。之后，两半球的温度同时降低，进入完全的冰期状态，不过南极达到这一状态要比格陵兰早一些。

表 1. 三个冰芯的相应深度

冰的年代(距今千年)	GRIP 深度(米)	伯德站深度 *†(米)	东方站深度 †(米)
12	1638.0	1074.4	261.1
13	1674.0	1124.2	278.6
14	1722.6	1161.4	293.1
15	1770.0	1198.6	307.6
16	1802.2	1235.8	322.8
17	1834.3	1269.5	336.4
18	1867.6	1294.4	348.9
19	1894.5	1319.4	360.1
20	1916.5	1344.3	371.0
21	1937.7	1369.3	381.9
22	1958.8	1394.2	392.9
23	1975.6	1419.1	404.3
24	1993.1	1444.1	415.9
25	2011.1	1469.0	427.3
26	2031.6	1493.9	438.9
27	2051.6	1518.7	451.3
28	2065.1	1539.9	464.2

Continued

Age of the ice (kyr BP)	Depth GRIP (m)	Depth Byrd*†(m)	Depth Vostok†(m)
29	2079.9	1561.1	477.6
30	2097.5	1582.3	492.0
31	2112.6	1603.5	505.6
32	2129.5	1624.7	516.4
33	2149.9	1645.8	528.5
34	2162.9	1667.0	541.5
35	2182.7	1688.2	553.8
36	2203.6	1709.4	566.4
37	2214.7	1730.6	581.1
38	2227.5	1751.7	593.5
39	2242.8	1772.9	604.2
40	2255.1	1786.5	614.2
41	2272.0	1798.8	623.6
42	2282.6	1811.2	633.4
43	2297.6	1823.5	642.7
44	2315.9	1840.7	653.2
45	2330.2	1858.3	665.8
46	2340.2	1875.8	680.0
47	2353.2	1893.4	694.2

* The indicated depth corresponds to the higher Byrd temperature scenario (see Methods).

† We note that in the new Byrd, as well as in the new Vostok, age–depth relations, rather unrealistic changes in the annual layer thickness appear. They result because the synchronization is based on the GRIP timescale which is, although well dated, not an absolute timescale. Therefore care must be taken when using the Byrd and Vostok timescales with respect to annual layer or accumulation changes. Our conclusions do not depend on an absolute timescale, and are unaffected by such problems (see Methods).

We point out that our synchronization excludes the possibility that Antarctic and Greenland events are in phase. To bring the isotopic records in phase, the Byrd Δage would have to be reduced by more than 1,000 yr. As Δage is only about 500 yr during A1 and A2 (see Fig. 3) this would result in a negative Δage, which is impossible. Further, the relative position of A1 is also confirmed by the [10]Be peak recorded in the ice of the three investigated cores.

Antarctic isotopic events other than A1 and A2 can hardly be allocated individually with D–O events. However, one can calculate the mean time shift required to obtain the best correlation between Greenland and Antarctic isotopic records in the period 45–23 kyr BP. Best correlations are obtained when the Antarctic signal is shifted towards younger ages by 1–2.5 kyr. This, and the timing of the events (A1 and A2), confirms the chronology obtained by deep sea cores[9] and contradicts the hypothesis that long interstadials begin in the Northern Hemisphere and spread to Antarctica, creating a lag of Antarctic climate with respect to Greenland (at least for the past 50 kyr).

冰的年代（距今千年）	GRIP 深度（米）	伯德站深度 *†（米）	东方站深度 †（米）
29	2079.9	1561.1	477.6
30	2097.5	1582.3	492.0
31	2112.6	1603.5	505.6
32	2129.5	1624.7	516.4
33	2149.9	1645.8	528.5
34	2162.9	1667.0	541.5
35	2182.7	1688.2	553.8
36	2203.6	1709.4	566.4
37	2214.7	1730.6	581.1
38	2227.5	1751.7	593.5
39	2242.8	1772.9	604.2
40	2255.1	1786.5	614.2
41	2272.0	1798.8	623.6
42	2282.6	1811.2	633.4
43	2297.6	1823.5	642.7
44	2315.9	1840.7	653.2
45	2330.2	1858.3	665.8
46	2340.2	1875.8	680.0
47	2353.2	1893.4	694.2

* 所示深度对应于较高的伯德站冰芯温度情景（参见方法部分）。

† 我们注意到，在新的伯德站冰芯和东方站冰芯的年代–深度关系中，年层厚度出现了相当不切实际的变化。这是由于同时化是基于 GRIP 冰芯的时标，而这一时标虽然经过了很好的定年，但仍不是绝对的时标。因此，当使用与年层和冰累积量变化有关的伯德站和东方站时标时需要多加小心。我们的结论并不依赖于绝对的时标，也不受这一问题的影响（参见方法部分）。

需要指出，我们的同时化处理排除了南极和格陵兰发生的事件同相的可能性。因为为了使同位素记录同相，伯德站冰芯的 Δage 必须减少 1,000 年以上。而 Δage 在 A1 和 A2 期间大约只有 500 年（参见图 3），这将导致负的 Δage 值，这是不可能的。而且，A1 的相对位置也是由研究所用的三个冰芯中 [10]Be 的峰值所确认的。

南极的同位素事件除 A1 和 A2 以外，很难单独由 D–O 事件确定。然而我们可以通过计算所需的平均时间偏移得到距今 4.5 万到 2.3 万年期间格陵兰和南极同位素记录的最佳相关关系。当南极的信号向现今移动 1,000～2,500 年时，可以得到最佳相关关系。这与事件 A1 和 A2 的定年结果一起证实了由深海岩芯得到的年表[9]，并且反驳了长间冰阶始于北半球并向南半球扩张，导致南半球的气候变化（至少在过去 5 万年期间）滞后于格陵兰这一假说。

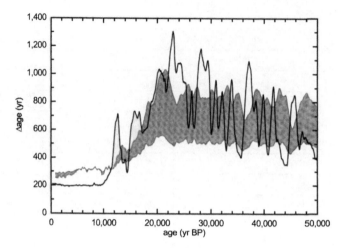

Fig. 3. Plot of Δage versus ice age. The solid line shows data from GRIP[5], the shaded area gives the range of Δage for the Byrd core covered by the temperature scenarios discussed in Methods, calculated with the same model as the GRIP Δage.

Implication for North–South Connection

A connection between the hemispheres can take place over the ocean or over the atmosphere. Charles *et al.*[9] suggested that both high northern and southern temperature could be driven over the atmosphere or the ocean by tropical temperature. Whereas coupling between Antarctic and tropical temperatures would be immediate, northern temperature would lag a tropical temperature change due to the thermal influence of the ice sheets or by the ice sheets' influence on salt balance of the North Atlantic surface layer. An immediate response of the Southern Hemisphere to tropical temperature would tend to synchronize the CH_4 (mainly related to tropical moisture changes) and the Antarctic isotope signal, whereas we observe a synchronism of CH_4 and Greenland $\delta^{18}O$ (refs 5, 14). This makes a coupling via atmosphere improbable, and points instead to a dominant role of the ocean controlling the past climate of both polar regions.

It is supposed that changes in the Atlantic thermohaline circulation, the direct cause of temperature changes in central Greenland[6], are closely linked to ice-rafting events and associated melt water which occur simultaneously in the North Atlantic region[37]. Melting reduces the deep-water formation and initiates cooling in the North Atlantic region. The cooling then slows the melting and permits re-establishment of the NADW formation, rapidly bringing heat to the North Atlantic[6]. The temperature increase then leads again to a freshwater input into the North Atlantic and to gradual shut down of NADW formation[38]. Model simulations demonstrate the NADW's sensitivity to freshwater input in the North Atlantic[24,39], and they also indicate that the resumption of the "conveyor belt" circulation after a shut-down is a rapid process[39,40].

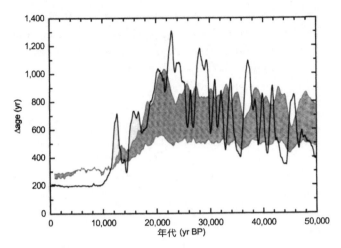

图 3. Δage 和冰年代的关系图。实线表示 GRIP[5] 的数据，阴影区表示方法部分中讨论的伯德站冰芯的温度情景的 Δage 范围，其计算模式和 GRIP 冰芯的 Δage 计算模式相同。

南北半球联系的启示

两半球间的联系可以通过海洋，也可以通过大气发生。查尔斯等[9] 认为，南北半球高纬度地区的温度都可以由热带温度通过大气或海洋驱动。然而南极和热带温度之间的耦合是即时的，由于冰盖的热力影响或冰盖对北大西洋表层盐平衡的影响，北半球的温度变化滞后于热带温度的变化。南半球对热带温度的即时响应致使 CH_4（主要和热带的湿度变化有关）和南极同位素信号的变化同步，而我们观测到了 CH_4 和格陵兰 $\delta^{18}O$ 变化之间的同步性（参考文献 5 和 14）。这就使得通过大气的耦合不大可能发生，取而代之的是海洋对过去两极气候的控制起主导作用。

可以推测，作为格陵兰中部气温变化直接原因的大西洋温盐环流变化[6] 与同时发生在北大西洋地区的冰筏事件和融水事件[37] 紧密相连。冰的融化减少了深层水的形成，引起北大西洋地区降温。降温过程的发生减慢了冰的融化，使 NADW 的形成重新建立，并迅速将热量带至北大西洋[6]。而温度的上升又再次导致淡水进入北大西洋，使 NADW 的形成逐渐关闭[38]。模式模拟指出了 NADW 对北大西洋地区淡水输入的敏感性[24,39]，同时也指出，"输送带"环流关闭后的重启是一个快速过程[39,40]。

D–O events and corresponding ice raftings occur more frequently than Antarctic warmings. Therefore, ice-rafting events can only partly, possibly through an increase in sea level, be initiated by Antarctic warmings. For the D–O events that have no counterpart in Antarctica, the trigger mechanism for ice-rafting events thus originates probably from an internal process of one of the ice sheets ablating into the North Atlantic. The Laurentide ice sheet can be excluded as it has a time constant of about 7 kyr (refs 41, 42), longer than the required 2–3 kyr between ice-rafting events[37]. Bond and Lotti[37] proposed instead ice sheets in Greenland, the Barents Sea or Scandinavia as candidates for such a trigger mechanism.

Central to the dynamics that we reconstruct on the basis of the ice cores is the role of the thermohaline circulation in the North Atlantic. Activation of the thermohaline circulation in the North Atlantic tends to cool the Southern Hemisphere[22,43]. This is exactly what happens during D–O events 12 and 8 (corresponding to A1 and A2), but also during D–O event 1, which is followed by the ACR[19]. Further, the YD/ACR out-of-phase relationship between Northern and Southern hemispheres is consistent with model simulations from ref. 18, recently reproduced with a three-dimensional coupled atmosphere–ocean general circulation model[24].

We may speculate that Antarctica is asynchronously coupled to the Northern Hemisphere during the whole glacial period investigated. Indeed, some of the smaller D–O events (for example, events 7 or 11) might have a concomitant cooling in Antarctica. The smaller amplitude of these coolings compared to the main events A1 and A2 might point to a positive correlation between the duration of a D–O event and the amplitude of the successive Antarctic cooling. However, it is also possible that the small Antarctic events may merely be noise in the isotopic records. In that case, the heat would only be drawn from Antarctica when Antarctica is in a warm phase. This leads to the speculation that under full glacial conditions Greenland and Antarctic temperature, and hence also thermohaline circulation and Antarctic circumpolar sea surface temperature[9], are only weakly coupled. A weaker coupling of the hemispheres is achieved when the Atlantic thermohaline circulation exchanges less water with the Southern Ocean, and so has a reduced influence on the heat balance of the southern latitudes. The strength of this circulation also determines its stability to freshwater flux perturbations: a weaker thermohaline circulation is more likely to shut down for a given perturbation[44]. If a weak Atlantic thermohaline circulation switches off, then the north will still experience strong cooling, but with a reduced influence on southern climate. We hypothesize that this is the case during the short D–O events which have no corresponding signal in Antarctic records. D–O events 12 and 8 and the Bølling/Allerød/YD sequence, on the other hand, would be examples of strong coupling between the hemispheres.

Warming in Antarctica tends to reduce sea surface density and hence the density of newly formed deep water. This favours NADW formation[43], and hence brings the ocean circulation closer to its strong mode with a tighter coupling of the circumpolar sea surface temperature to the Atlantic thermohaline circulation. A switch-on of the thermohaline circulation after ice-rafting events leads to the observed cooling in Antarctica, which results

D–O 事件和相应的冰筏事件相比南极增温发生得更为频繁。因此，冰筏事件可能只是部分由北大西洋增温引起的海平面上升导致。在南极，没有类似 D–O 事件的过程，因而触发冰筏事件的机制可能源于一块冰盖融化到北大西洋中的内部过程。可以排除劳伦太德冰盖，因为它的冰筏事件间隔是一个大约 7,000 年的固定时间（参考文献 41 和 42），这一数字比冰筏事件中所需要的 2,000～3,000 年要长 [37]。邦德和洛蒂 [37] 提出格陵兰、巴伦支海或者斯堪的纳维亚的冰盖可以作为这样一种触发机制。

根据冰芯建立的动力学的核心问题是北大西洋温盐环流的作用。北大西洋温盐环流的激活导致南半球降温 [22,43]。这确实在 D–O 事件 12 和 8(相当于 A1 和 A2) 期间发生过，但也同样在 ACR 之前的 D–O 事件 1 期间发生过 [19]。另外，南北半球 YD/ACR 的异相关系和参考文献 18 的模式模拟结果一致，该模拟结果最近由一个三维大气–海洋耦合环流模式重现 [24]。

我们推测，在我们所研究的整个冰期，南极和北半球的耦合是非同步的。某些较小的 D–O 事件 (如事件 7 或 11) 确实可以在南极出现伴随的降温，这些降温的幅度与主要事件 A1 和 A2 相比要小，这可能指示 D–O 事件的持续时间和随之而来的南极降温幅度之间的正相关。但是，小的南极事件也可能仅仅是同位素记录中的噪音。在这种情况下，当南极处于暖位相时，热量仅来源于南极地区。由此可以推测，在全冰期尺度下，格陵兰和南极的温度以及温盐环流和南极绕极海面温度 [9] 都仅有较弱的耦合。当大西洋的温盐环流和南大洋的海水交换较少并因此对南纬区域热平衡影响减小时，半球之间的耦合就更弱。这一环流的强度也决定了在淡水输入造成扰动的情况下其表现出的稳定性：弱的温盐环流更容易由于扰动的影响而消失 [44]。如果弱的大西洋温盐环流消失了，北半球将仍处于强烈的降温状态，但南半球气候受到的影响会减小。我们猜想，那些在南极记录中缺乏对应信号的短 D–O 事件就属于这种情况。另一方面，D–O 事件 12 和 8 及波令/阿勒罗德/YD 序列则是两半球间强耦合的例子。

南极升温会导致海水密度的减小，并因此使得新形成的深层水的密度减小。这有利于 NADW 的形成 [43]，从而使海洋环流更接近其强模态，在此模态下，绕极海面温度和大西洋温盐环流之间的耦合更紧密。冰筏事件之后温盐环流的开启导致观测到的南极降温，结果再次导致数千年范围内两半球间更弱的耦合。这也和事件

again in a weaker coupling of the hemispheres for several thousand years. This is also consistent with an increased heat transport to high northern latitudes during A1 and A2 leading to the observed prolonged D–O events 8 and 12.

During the deglaciation, the climate system shows very similar behaviour. During the first part of the deglaciation the Antarctic warming is not interrupted as no ice-rafting events in the North Atlantic region are taking place: perhaps the lowered sea level makes it impossible for an ice sheet to initiate significant surging of other North Atlantic ice sheets. Then a sudden temperature increase in Greenland (14.5 kyr BP) initiates the same pattern as in the glaciation, namely a cooling in Antarctica (ACR). The first meltwater pulse (MWP 1A)[45] reduces NADW, at first gradually and then completely, allowing further warming in Antarctica. This makes the deglaciation a complicated process involving Milankovitch forcing, internal ice-sheet processes, and ocean circulation changes.

Further data from Antarctica which are synchronized with the existing records, together with model simulations, will help us disentangle the complicated pattern of climate change between Northern and Southern hemispheres for the entire glacial period. More, and better, data should be available from current European, Japanese and American ice-core drilling programmes.

Methods

To calculate Δage, the accumulation rate and temperature need to be known. The temperature at the time of precipitation can be determined from the $\delta^{18}O$ signal recorded in the ice. Assuming that today's spatial dependence between $\delta^{18}O$ and temperature was also valid in the past, the glacial–interglacial temperature increase in Antarctica is $\sim$7–10 °C (refs 16, 29). On the other hand, recent borehole temperature measurements at Vostok suggest[30] a temperature change of up to 15 °C.

The amount of snow falling over the year depends on the mean temperature of the inversion layer. This temperature (T_I) can be calculated for the surface temperature (T_G) according to[46]:

$$T_I = 0.67 \ T_G - 1.2\,°C \tag{1}$$

where T_I and T_G are in °C. Assuming a linear relation between accumulation rate and the derivative of the water vapour partial pressure (P) with respect to temperature (T) (ref. 16), the accumulation rate which has led to a layer now found at depth z is given by:

$$a(z) = a_0 \left(\left.\frac{\partial P}{\partial T}\right|_{T_0} \right)^{-1} \left.\frac{\partial P}{\partial T}\right|_{T_I(z)} \tag{2}$$

where T_0 and $T_I(z)$ are the inversion layer temperature for today's accumulation rate a_0 and for the past accumulation rate $a(z)$, respectively.

A1、A2 期间向北半球高纬度热量输送的增加是一致的，热量输送的增加导致了观测到的延长的 D–O 事件 8 和 12。

在冰消期，气候系统表现出非常相似的行为。冰川消退的开始阶段，南极的升温并未中断，因为在北大西洋没有冰筏事件发生：或许最低的海平面使得一个北大西洋冰盖无法令其他北大西洋冰盖发生明显涌动。此后格陵兰的突然增温（距今 1.45 万年）启动了与冰期相同的模态（称为南极转冷事件，即 ACR）。第一次融水脉冲事件（MWP 1A）[45] 减弱了 NADW，一开始逐渐地、之后彻底地引起南极的进一步升温。这就使冰川消退成为涉及米兰科维奇驱动、冰盖内部过程和海洋环流变化的复杂过程。

和已有记录进行过同时化处理的更多资料和模式模拟结果将帮助我们弄清整个冰期南北半球之间气候变化的复杂过程。从现在欧洲、日本和美洲的冰芯钻探计划可以获得更多更好的资料。

方　法

为了计算 Δage 需要知道冰的累积率和温度。降水时的温度可以从冰的 $\delta^{18}O$ 信号求得。假设现今的 $\delta^{18}O$ 和温度的空间关系在过去也是成立的，则南极冰期和间冰期之间的温度增加大约为 $7 \sim 10$℃（参考文献 16 和 29）。另外，由最近东方站的钻孔温度测量得知[30]，温度变化可以高达 15℃。

年降雪量取决于逆温层的平均温度。这一温度（T_I）可由地表温度（T_G）按下式计算[46]：

$$T_I = 0.67\, T_G - 1.2\text{℃} \tag{1}$$

式中 T_I 和 T_G 的单位用℃表示。假定冰的累积率和水汽分压（P）对温度（T）的导数是线性关系（参考文献 16），则深度 z 的累积率用下式表示：

$$a(z) = a_0 \left(\left.\frac{\partial P}{\partial T}\right|_{T_0} \right)^{-1} \left.\frac{\partial P}{\partial T}\right|_{T_{I(z)}} \tag{2}$$

其中 T_0 和 $T_I(z)$ 分别是现今累积率 a_0 情况下和过去累积率 $a(z)$ 情况下逆温层的温度。

Byrd. As the accumulation is calculated from the temperature and present accumulation rate, its uncertainty is essentially linked to the uncertainty in both parameters. For Byrd we have calculated a new age of the ice for two temperature scenarios. For the warmer scenario, the $\delta^{18}O$–temperature relation from ref. 29 was used (equation (3)); for the colder scenario[30] we use the linear relation given in equation (4):

$$T = (1.01\,°C\ per\ ‰)\,\delta^{18}O + 6.57\,°C \tag{3}$$

$$T = (2.07\,°C\ per\ ‰)\,\delta^{18}O + 42.7\,°C \tag{4}$$

From the two temperature scenarios, accumulation was calculated according to equations (1) and (2) with $a_0 = 11.2$ cm per yr ice equivalent[47]. Thereafter, Δage was calculated using the dynamical model (Fig. 3) leading, in combination with the CH_4 synchronization, to a timescale for the warmer- and the colder-temperature scenario.

The accumulation rate is calculated from these temperatures and therefore also covers a wide range of possible accumulations (mean accumulation rates for the glacial are 7.9 and 5.5 cm yr^{-1} for the higher- and the lower-temperature scenario, respectively). Alternatively, the accumulation rate can be calculated from the annual layer thickness, if the ice flow pattern is known. The flow pattern in the Byrd station area is very complicated (due to bedrock topography and melting at the bottom), which results in uncertainties in the calculated accumulation rate[15,48]. However, our accumulation range over the glacial is consistent with measured annual layer thickness[47], assuming a linear thinning of annual layers with depth deduced from the Holocene period.

Vostok. As Vostok is a low accumulation site, the ice from the time period considered here is found in the upper 740 m of the ice sheet. There, ice flow is well known and was described by a glaciological model[49]. Thus, it should in principle be possible to get coherency between the timescales for the ice and the occluded gas, accumulation and temperature. In a first approach the ^{10}Be peak around 40 kyr was synchronized between GRIP[31] and Vostok[33]. For this purpose, accumulation rates were calculated from equations (1) and (2) from the isotopic temperature[16]. Δage was obtained using this accumulation and the isotopic temperature where the present-day accumulation rate a_0 was set to 2.1 cm yr^{-1}, within the uncertainty for today's accumulation rate[50]. To obtain a timescale from the accumulation rate, annual layer thickness was calculated using the thinning function from the glaciological model[49]. With this approach the main features in the isotopic record, as well as the Vostok CH_4 signal, appear shifted relative to the Byrd record. Reducing the Vostok temperature (suggested by Salamatin *et al.*[30]) and therefore also accumulation rate would make the match even worse. In order to get the Vostok CH_4 record in agreement with its GRIP and Byrd counterpart, Vostok and/or the GRIP/Byrd timescales must be squeezed and stretched.

In a first case, we consider Vostok as reference timescale and adapt the GRIP/Byrd timescales. Consequently accumulation has to be changed in both cores and with the same amplitude and sign; therefore this does not change the relative phasing of GRIP and Byrd. This scenario brings event A2 in to agreement between Byrd and Vostok, but it creates fluctuations in the annual layer thickness of GRIP, which are not expected from the exponential relation between accumulation and $\delta^{18}O$ (ref. 25).

312

伯德站 由于过去冰的累积率是用温度和现在的累积率计算的，因此其不确定性本质上和这两个参数的不确定性有关。对伯德站，我们用两个温度情景计算了新的冰的年代值。对较暖的情景，$\delta^{18}O$ 和温度的关系取自参考文献 29(即 (3) 式)，对较冷的情景 [30]，应用 (4) 式给出线性关系：

$$T = (1.01\,℃/‰)\,\delta^{18}O + 6.57\,℃ \tag{3}$$

$$T = (2.07\,℃/‰)\,\delta^{18}O + 42.7\,℃ \tag{4}$$

应用这两个温度情景按式 (1) 和 (2) 取 $a_0 = 11.2$ 厘米/年(冰当量) [47] 计算了冰的累积率。然后使用经同时化处理的 CH_4 资料应用动力学模式在冷暖两个温度情景下计算了 Δage(见图 3)。

从上述温度情景计算的冰的累积率覆盖了一个较宽的可能累积率范围(冰期平均累积率在高温情景和低温情景下分别为 7.9 厘米/年和 5.5 厘米/年)。另外，如果知道冰流型，也可以从冰的年层厚度计算冰的累积率。然而，由于基岩地形和底部融化，伯德站区域的冰流型非常复杂，这就导致所计算的累积率有相当的不确定性 [15,48]。不过，如果假定在全新世时期冰的年层厚度随深度线性减少的话，我们计算的整个冰期累积率的范围和冰的年层厚度的测量结果是一致的 [47]。

东方站 由于东方站是一个低累积率站点，这里所讨论的时期内的冰位于冰盖上部 740 米范围。这里冰流的情况已为人们所熟知，并已被冰川学模型所描述 [49]。因此，原则上我们有可能得到冰的时标与封闭在其中的气体、冰的累积率和温度的一致性关系。初步结果说明，^{10}Be 在 4 万年附近的峰值同步出现在 GRIP 冰芯 [31] 和东方站冰芯 [33]。为此，从式 (1) 和 (2) 应用同位素温度 [16] 计算出冰的累积率。应用这一累积率和同位素温度，求得了 Δage，此处现今累积率 a_0 设为 2.1 厘米/年，该值在现今累积率的不确定性范围内 [50]。为了从累积率获得时标，应用从冰川学模型 [49] 得到的减薄函数计算了冰的年层厚度。应用这一方法，同位素记录和东方站 CH_4 信号的主要特征表现出相对于伯德站记录的偏移。降低东方站温度(像萨拉马京等人 [30] 建议的那样)和累积率后将会使吻合程度变得更差。为了使东方站的 CH_4 记录与 GRIP 冰芯、伯德站冰芯吻合，东方站冰芯和(或) GRIP 冰芯/伯德站冰芯的时标必须缩短或拉长。

在第一种情况下，我们把东方站冰芯当作参考时标，调整 GRIP 冰芯/伯德站冰芯的时标。因此这两个冰芯的冰累积率必须做出正负相同、幅度一致的改变，这样就不会改变 GRIP 冰芯和伯德站冰芯的相对位相。这种情景使得事件 A2 的时间在伯德站和东方站保持一致，但是它会产生 GRIP 冰芯中年层厚度的波动，而这与冰累积率和 $\delta^{18}O$ 之间的指数关系不甚符合(参考文献 25)。

In a second case, we consider GRIP as the reference timescale. The Vostok gas timescale is squeezed and stretched in order to fit the GRIP CH_4 record. The Vostok ice timescale is treated in the same way which is changing the accumulation rate. Thus Δage is re-calculated with the accumulation rates consistent with the modified timescale (using the thinning function from the glaciological model[49]). The resulting change in the gas chronology was found to be negligible; this means that the Vostok CH_4 record is still in agreement with the GRIP CH_4 chronology. This scenario also brings event A2 into agreement between Byrd and Vostok, but now creates unexpected fluctuations in the annual layer thickness of Vostok. $\delta^{18}O$, ^{10}Be and CH_4 data are presented using this chronology in Figs 1 and 2.

In both cases, maintaining consistency between ^{10}Be peaks and CH_4 records, we obtain good correlation between the isotope records of Byrd and Vostok, and are therefore confident that the characteristic features reflect identical climate events. We are left with the unsolved problem of annual layer fluctuation either in the Vostok or in the GRIP ice core. However, the synchronization of the Byrd and GRIP core is robust to this problem.

Three cores from Vostok are involved in this study. The isotopic profile was obtained on the 3G core[16] while ^{10}Be was measured on the 4G core[33]. The occurrence of a dust horizon in both cores at around 550 m depth (J. R. Petit, personal communication) allows an exact synchronization of the cores (4G depth increased by 3.41 m). The CH_4 profile was measured on the 5G core, which is of better quality than the 3G and 4G cores. The dust horizon at 550 m depth was not found in the 5G core. We thus performed additional CH_4 measurements on the 3G core over the CH_4 increase associated with D–O event 5, allowing us to estimate the depth offset between 3G and 5G (5G depth decreased by 3 m).

(**394**, 739-743; 1998)

T. Blunier[*], J. Chappellaz[†], J. Schwander[*], A. Dällenbach[*], B. Stauffer[*], T. F. Stocker[*], D. Raynaud[†], J. Jouzel[‡†], H. B. Clausen[§], C. U. Hammer[§] & S. J. Johnsen[§‖]

[*] Climate and Environmental Physics, Physics Institute, University of Bern, Sidlerstrasse 5, CH-3012 Bern, Switzerland

[†] CNRS Laboratoire de Glaciologie et Géophysique de l'Environnement (LGGE), BP 96, 38402 St Martin d'Hères Cedex, Grenoble, France

[‡] Laboratoire des Sciences du Climat et de l'Environnement, UMR CEA-CNRS 1572, CEA Saclay, Orme des Merisiers, 91191 Gif sur Yvette, France

[§] Department of Geophysics (NBIfAPG), University of Copenhagen, Juliane Maries Vej 30, 2100 Copenhagen O, Denmark

[‖] Science Institute, Department of Geophysics, University of Iceland, Dunhaga 3, Is-107 Reykjavik, Iceland

Received 1 September 1997; accepted 13 February 1998.

References:

1. Dansgaard, W. *et al.* Evidence for general instability of past climate from a 250-kyr ice-core record. *Nature* **364,** 218-220 (1993).

2. Grootes, P. M., Stuiver, M., White, J. W C., Johnsen, S. & Jouzel, J. Comparison of oxygen isotope records from the GISP2 and GRIP Greenland ice cores. *Nature* **366,** 552-554 (1993).

3. Oeschger, H *et al.* in *Climate Processes and Climate Sensitivity* (eds Hansen, J. E. & Takahashi, T.) 299-306 (Vol. 29, Geophys. Monogr. Ser., Am. Geophys. Union, Washington DC, 1984).

4. Johnsen, S., Dahl-Jensen, D., Dansgaard, W. & Gundestrup, N. Greenland palaeotemperatures derived from GRIP bore hole temperature and ice core isotope profiles. *Tellus B* **47,** 624-629 (1995).

5. Schwander, J. *et al.* Age scale of the air in the summit ice: Implication for glacial-interglacial temperature change. *J. Geophys. Res.* **102,** 19483-19494 (1997).

314

在第二种情况下，我们把 GRIP 冰芯当作参考时标。将东方站的气体时标缩短或拉长，以适应 GRIP 冰芯的 CH_4 记录。东方站冰的时标也用同一种方法处理，即改变冰的累积率。此时，使用与修正过的时标相符的冰累积率重新计算 Δage(应用冰川学模型 [49] 得到的减薄函数)。我们发现这样做引起的气体年代值的改变是可以忽略的；这就是说，东方站的 CH_4 记录仍然符合 GRIP 冰芯的 CH_4 记录的年代。这种情景也使东方站和伯德站的 A2 事件的年代保持一致，不过它导致了东方站年层厚度的意外波动。我们在图 1、图 2 中给出的 $\delta^{18}O$、^{10}Be 和 CH_4 资料便是用的这种年代计算方法。

在上述两种情况下，我们保持 ^{10}Be 峰值和 CH_4 记录之间的一致性，从而得到了伯德站和东方站同位素记录较好的相关性，也因此可以确信独有的特征反映了相同的气候事件。尚未解决的问题是，要么东方站冰芯，要么 GRIP 冰芯，会出现年层厚度波动的问题。无论如何，伯德站冰芯和 GRIP 冰芯的同时化对于这个问题而言是稳健的。

本研究用了东方站的三个冰芯。同位素廓线来自 3G(第 3 代)冰芯 [16]，而 ^{10}Be 则由 4G 冰芯测得 [33]。在两个冰芯约 550 米深处(珀蒂，个人交流)出现的尘埃层顶提供了冰芯之间的精确同时化(4G 冰芯的厚度增加了 3.41 米)。CH_4 的廓线测量自 5G 冰芯，它比 3G 冰芯和 4G 冰芯的质量要好。但在 550 米深度发现的尘埃面并未出现在 5G 冰芯中。因此我们对 3G 冰芯中与 D–O 事件 5 有关的 CH_4 增长进行了补充测量，使我们可以估计 3G 和 5G 冰芯之间的深度偏差(5G 冰芯的深度减少 3 米)。

(周家斌 翻译；李崇银 审稿)

6. Bond, G. *et al.* Correlations between climate records from North Atlantic sediments and Greenland ice. *Nature* **365,** 143-147 (1993).

7. Roemmich, D. Estimation of meridional heat flux in the North Atlantic by inverse methods. *J. Phys. Oceanogr.* **10,** 1972-1983 (1981).

8. Broecker, W. S. & Denton, G. H. The role of ocean-atmosphere reorganizations in glacial cycles. *Geochim. Cosmochim. Acta* **53,** 2465-2501 (1989).

9. Charles, C. D., Lynch-Stieglitz, J., Ninnemann, U. S. & Fairbanks, R. G. Climate connections between the hemisphere revealed by deep sea sediment core/ice core correlations. *Earth Planet. Sci. Lett.* **142,** 19-27 (1996).

10. Behl, R. J. & Kennett, J. P. Brief interstadial events in the Santa Barbara basin, NE Pacific, during the past 60 kyr. *Nature* **379,** 243-379 (1996).

11. Bard, E., Rostek, F. & Sonzogni, C. Interhemispheric synchrony of the last deglaciation inferred from alkenone palaeothermometry. *Nature* **385,** 707-710 (1997).

12. Grimm, E. C., Jacobson, G. L. Jr, Watts, W. A., Hansen, B. C. S. & Maasch, K. A. A 50,000-year record of climate oscillations from Florida and its temporal correlation with the Heinrich events. *Science* **261,** 198-200 (1993).

13. Benson, L. V. *et al.* Climatic and hydrologic oscillations in the Owens Lake basin and adjacent Sierra Nevada, California. *Science* **274,** 746-749 (1996).

14. Chappellaz, J. *et al.* Synchronous changes in atmospheric CH_4 and Greenland climate between 40 and 8 kyr BP. *Nature* **366,** 443-445 (1993).

15. Johnsen, S. J., Dansgaard, W., Clausen, H. B. & Langway, C. C. Oxygen isotope profiles through the Antarctic and Greenland ice sheets. *Nature* **235,** 429-434 (1972).

16. Jouzel, J. *et al.* Vostok ice core: A continuous isotope temperature record over the last climatic cycle (160,000 years). *Nature* **329,** 403-408 (1987).

17. Bender, M. *et al.* Climate correlations between Greenland and Antarctica during the past 100,000 years. *Nature* **372,** 663-666 (1994).

18. Stocker, T. F., Wright, D. G. & Mysak, L. A. A zonally averaged, coupled ocean-atmosphere model for paleoclimate studies. *J. Clim.* **5,** 773-797 (1992) .

19. Blunier, T. *et al.* Timing of the Antarctic Cold Reversal and the atmospheric CO_2 increase with respect to the Younger Dryas event. *Geophys. Res. Lett.* **24,** 2683-2686 (1997).

20. Hammer, C. U. *et al. Report on the Stratigraphic Dating of the GRIP Ice Core* (Spec. Rep. of the Geophysical Dept, Niels Bohr Institute for Astronomy, Physics and Geophysics, Univ Copenhagen, in the press).

21. Johnsen, S. J. *et al.* Irregular glacial interstadials recorded in a new Greenland ice core. *Nature* **359,** 311-313 (1992).

22. Crowley, T. J. North Atlantic deep water cools the Southern Hemisphere. *Paleoceanography* **7,** 489-497 (1992) .

23. Stocker, T. F. & Wright, D. G. Rapid changes in ocean circulation and atmospheric radiocarbon. *Paleoceanography* **11,** 773-796 (1996).

24. Schiller, A., Mikolajewicz, U. & Voss, R. The stability of the thermohaline circulation in a coupled ocean-atmosphere general circulation model. *Clim. Dyn.* **13,** 325-348 (1997).

25. Dahl-Jensen, D., Johnsen, S. J., Hammer, C. U., Clausen, H. B. & Jouzel, J. in *Ice in the Climate System* (ed. Peltier, W. R.) 517-532 (Springer, Berlin, (1993).

26. Johnsen, S. J., Dansgaard, W. & White, J. W. C. The origin of Arctic precipitation under present and glacial conditions. *Tellus B* **41,** 452-468 (1989).

27. Severinghaus, J. P., Sowers, T., Brook, E. J., Alley, R. B. & Bender, M. L. Timing of abrupt climate change at the end of the Younger Dryas interval from thermally fractionated gases in polar ice. *Nature* **391,** 141-146 (1998).

28. Blunier, T., Chappellaz, J., Schwander, J., Stauffer, B. & Raynaud, D. Variations in atmospheric methane concentration during the Holocene epoch. *Nature* **374,** 46-49 (1995).

29. Robin, G. de Q. in *The Climatic Record in Polar Ice Sheets* (ed. Robin, G. de Q.) 180-184 (Cambridge Univ. Press, London, 1983).

30. Salamatin, A. N. *et al.* Ice core age dating and paleothermometer calibration based on isotope and temperature profiles from deep boreholes at Vostok Station (East Antarctica). *J. Geophys. Res.* **103,** 8963-8978 (1998).

31. Yiou, F. *et al.* Beryllium 10 in the Greenland Ice Core Project ice core at Summit, Greenland. *J. Geophys. Res.* **102,** 26783-26794 (1997).

32. Beer, J. *et al. The Last Deglaciation: Absolute and Radiocarbon Chronologies* (eds Bard, E. & Broecker, W. S.) 141-153 (NATO ASI Ser. I 2, Springer, Berlin, 1992) .

33. Raisbeck, G. M. *et al.* in *The Last Deglaciation: Absolute and Radiocarbon Chronologies* (eds Bard, E. & Broecker, W. S.) 127-140 (NATO ASI Ser. I 2, Springer, Berlin, 1992).

34. Jouzel, J. *et al.* The two-step shape and timing of the last deglaciation in Antarctica. *Clim. Dyn.* **11,** 151-161 (1995).

35. Sowers, T. & Bender, M. Climate records covering the last deglaciation. *Science* **269,** 210-214 (1995).

36. Siegenthaler, U., Eicher, U., Oeschger, H. & Dansgaard, W. Lake sediments as continental $\delta^{18}O$ records from the transition of glacial-interglacial. *Ann. Glaciol.* **5,** 149-152 (1984).

37. Bond, G. C. & Lotti, R. Iceberg discharges into the North Atlantic on millennial time scales during the last deglaciation. *Science* **267,** 1005-1010 (1995).

38. Stocker, T. F. & Wright, D. G. The effect of a succession of ocean ventilation changes on radiocarbon. *Radiocarbon* **40,** 359-366 (1998).

39. Stocker, T. F. & Wright, D. G. Rapid transitions of the ocean's deep circulation induced by changes in surface water fluxes. *Nature* **351,** 729-732 (1991).

40. Wright, D. G. & Stocker, T. F. in *Ice in the Climate System* (ed Peltier, W. R.) 395-416 (NATO ASI Ser .I 12, Springer, Berlin, 1993).

41. MacAyeal, D. R. A low-order model of the Heinrich event cycle. *Paleoceanography* **8,** 767-773 (1993).

42. MacAyeal, D. R. Binge/purge oscillations of the Laurentide ice sheet as a cause of the North Atlantic's Heinrich events. *Paleoceanography* **8,** 775-784 (1993).

43. Stocker, T F., Wright, D. G. & Broecker, W. S. The influence of high-latitude surface forcing on the global thermohaline circulation. *Paleoceanography* **7,** 529-541 (1992).

44. Tziperman, E. Inherently unstable climate behaviour due to weak thermohaline ocean circulation. *Nature* **386,** 592-595 (1997).

45. Bard, E. *et al.* Deglacial sea-level record from Tahiti corals and the timing of global meltwater discharge. *Nature* **382,** 241-244 (1996).

46. Jouzel, J. & Merlivat, L. Deuterium and oxygen 18 in precipitation: modeling of the isotopic effects during snow formation. *J. Geophys. Res.* **89,** 11749-11757 (1984).

47. Hammer, C. U., Clausen, H. B. & Langway, C. C. Jr. Electrical conductivity method (ECM) stratigraphic dating of the Byrd Station ice core, Antarctica. *Ann. Glaciol.* **20,** 115-120 (1994).

48. Whillans, I. M. Ice flow along the Byrd station strain network, Antarctica. *J. Glaciol.* **24,** 15-28 (1979).

49. Ritz, C. *Un Modèle Thermo-Mécanique d'évolution Pour le Bassin Glaciaire Antarctique Vostok-glacier Byrd : Sensiblité aux Valeurs des Paramètres Mal Connus.* Thesis, Univ. J. Fourier (1992).

50. Jouzel, J. *et al.* Extending the Vostok ice-core record of palaeoclimate to the penultimate glacial period. *Nature* **364,** 407-412 (1993).

51. Lorius, C., Merlivat, L., Jouzel, J. & Pourchet, M. A 30,000-yr isotope climatic record from Antarctic ice. *Nature* **280,** 644-648 (1979).

Acknowledgements. This work, in the frame of the Greenland Ice Core Project (GRIP), was supported by the University of Bern, the Swiss National Science Foundation, the Federal Department of Energy (BFE), the Schwerpunktprogramm Umwelt (SPPU) of the Swiss National Science Foundation, the EC program "Environment and Climate 1994-1998", the Fondation de France and the Programm National de Dynamique du Climat of CNRS. We thank F. Finet for the CH_4 measurements on Vostok and part of GRIP, C. Rado and J. R. Petit for ice sampling at Vostok station, C. C. Langway for providing us with additional Byrd samples and F. Yiou, G. Raisbeck and J. Beer for the ^{10}Be data.

Correspondence and requests for materials should be addressed to T.B. (e-mail: blunier@climate.unibe.ch).

Energy Implications of Future Stabilization of Atmospheric CO$_2$ Content

M. I. Hoffert *et al.*

Editor's Note

Mitigating climate change caused by human-induced emissions of greenhouse gases will, if it is not to impede economic growth, require extensive recourse to "clean" energy-generating technologies that reduce or eliminate these emissions. Here Martin Hoffert and his colleagues provide a sobering assessment of the scale of that challenge. They estimate that even relatively modest goals for stabilizing carbon dioxide emissions will demand much more emissions-free energy production than is anticipated for the year 2050 without any governmental intervention. This implies that "massive" investment in the research that could enable such a change is urgently needed: the researchers suggest that this be attributed the kind of importance associated with the Manhattan Project and the Apollo space programme.

The United Nations Framework Convention on Climate Change[1] calls for "stabilization of greenhouse-gas concentrations in the atmosphere at a level that would prevent dangerous anthropogenic interference with the climate system ...". A standard baseline scenario[2,3] that assumes no policy intervention to limit greenhouse-gas emissions has 10 TW (10×10^{12} watts) of carbon-emission-free power being produced by the year 2050, equivalent to the power provided by all today's energy sources combined. Here we employ a carbon-cycle/energy model to estimate the carbon-emission-free power needed for various atmospheric CO$_2$ stabilization scenarios. We find that CO$_2$ stabilization with continued economic growth will require innovative, cost-effective and carbon-emission-free technologies that can provide additional tens of terawatts of primary power in the coming decades, and certainly by the middle of the twenty-first century, even with sustained improvement in the economic productivity of primary energy. At progressively lower atmospheric CO$_2$-stabilization targets in the 750–350 p.p.m.v. range, implementing stabilization will become even more challenging because of the increasing demand for carbon-emission-free power. The magnitude of the implied infrastructure transition suggests the need for massive investments in innovative energy research.

INTERGOVERNMENTAL Panel on Climate Change (IPCC) working groups developed six scenarios for greenhouse-gas emissions based on socioeconomic projections[2,3]. Scenario IS92a incorporated widely accepted projections and the then-current consensus on population, economic development and energy technology to generate projections of greenhouse-gas emissions for the twenty-first century assuming "no new climate change policies". This baseline IS92a scenario has been dubbed "business as usual".

318

未来使大气中 CO_2 含量稳定的能源要求

霍费尔特等

编者按

要在不阻碍经济增长的情况下缓解人为温室气体排放所造成的气候变化，就需要使用大量的"清洁"能源生成技术来减少或消除这些温室气体的排放。本文中，马丁·霍费尔特和他的同事们对这一挑战的规模进行了不容乐观的评估。他们估计，如果没有政府的干预，即使设置相对比较适度的稳定二氧化碳排放量的目标，也需要大量的、高于 2050 年预期目标的无碳排放能源的生产。这意味着迫切需要进行"大规模"的研究投入以实现这一改变：研究人员认为，这一研究的重要性堪比曼哈顿计划和阿波罗太空计划。

联合国气候变化框架公约 [1] 呼吁"把大气中温室气体浓度稳定在一定水平上，以防止人类活动对气候系统产生危险的干扰……"。在假定没有政策措施来限制温室气体排放的标准基线排放情景下 [2,3]，到 2050 年可生产的无碳排放的能量将达10 TW(10×10^{12} 瓦特)，这相当于当前所有能源所提供能量的总和。在本文中，我们采用碳循环/能量模型来估计保持各种大气 CO_2 含量稳定值情景下所需要的无碳排放的能源。研究发现，在接下来的几十年，也就是到二十一世纪中叶，随着经济的增长，即使一次能源的经济生产力持续增长，要使 CO_2 含量稳定，也必须寻找创新型、有成本效益的无碳排放技术来满足额外的几十太瓦的一次能源需求。在 750 ~ 350 ppmv 的范围内，随着 CO_2 稳定含量目标值的逐步降低，对无碳排放能量的需求将不断增长，这使得碳稳定化的实施变得更加严峻。其中隐含的基础设施的转型规模也表明，需要加大对创新能源的研究投入。

联合国政府间气候变化专门委员会(IPCC)工作组根据社会经济预估，提出了六种温室气体排放情景 [2,3]。情景 IS92a 综合各种广受认可的预估方案以及当时的人口、经济发展状况和能源技术水平，最终形成了假定"没有关于气候变化的新政策出台"情况下二十一世纪的温室气体排放预估。该排放基线 IS92a 情景被称为"照常排放"。

In general, the rate at which carbon is emitted (as CO_2) by energy production is given by the Kaya identity[4-6];

$$\dot{M}_c \equiv \mathcal{N}(GDP/\mathcal{N})(\dot{E}/GDP)(C/E) \qquad (1)$$

expressing emissions as the product of population ($\mathcal{N}$), per capita gross domestic product ($GDP/\mathcal{N}$), primary energy intensity ($\dot{E}/GDP$) and carbon intensity (C/E). Here we express the rate of primary energy consumption from all fuel sources (the "burn rate") in watts (W) and the gross domestic product in (1990 US) \$ yr^{-1} so their ratio, the energy intensity, has units of W yr \$$^{-1}$. Carbon intensity, the weighted average of the carbon-to-energy emission factors of all energy sources, has the units kg C W^{-1} yr^{-1}. For example, from equation (1) fossil-fuel CO_2 emissions in 1990 were $\dot{M}_c = 5.3 \times 10^9$ persons $\times$ 4,100 \$ per person $yr^{-1} \times$ 0.49 W yr \$$^{-1} \times$ 0.56 kg C W^{-1} $yr^{-1} \approx 6.0$ Gt C yr^{-1} (1 Gt C = 10^{12} kg C).

To illustrate the relative contributions of the factors in equation (1) historically and under IS92a, we evaluated each of them globally over the 210-year period from 1890 to 2100 (Fig. 1) from historical data[7-9] before 1990, and from documents defining IPCC scenarios[2,10-12] after 1990. Although some of this information is implicit in IPCC documents, it has not been previously presented in this way.

Although illustrating the "big picture", aggregating Kaya decomposition terms globally can mask regional developmental differences[13]. Data for individual nations indicate that $\dot{E}/GDP$ typically increases during economic development, reaching a maximum as infrastructure investments peak, and declining only after some lag as economic productivity rises and the economy shifts structurally to less-energy-intensive activities (for example, services)[14]. Apparently declining, energy intensities of India and China today are still 2 to 5 times the global mean[8]. To focus on energy supply issues we provisionally accept the IS92a projections of 1% yr^{-1} improvement in global $\dot{E}/GDP$, recognizing that achieving this will depend crucially on the technology and structural changes adopted by developing nations[15].

Another opportunity for emission reductions is continuation of the "decarbonization" of the past 100 years reflected in decreasing carbon intensity of the global energy mix[8]. Under IS92a, the global mean C/E continues to decrease monotonically over the next century (Fig. 1d). Indeed, the evolving global energy mix based on assumed declining costs of nuclear and carbon-free energy relative to fossil fuels has global C/E dropping to that of natural gas by 2030; and it declines even more thereafter. Such rapid decarbonization is possible only by the massive introduction of carbon-free power, $\sim$10 TW by 2050. Even with this much carbon-free power and sustained 1% yr^{-1} improvements in energy intensity, the net effect of the factors in equation (1) more than doubles 1990 CO_2 emissions by 2050.

通常，能源生产排放碳（以 CO_2 的形式）的速率可根据茅阳一（Kaya）碳排放公式 [4-6] 得出：

$$\dot{M}_c \equiv N(GDP/N)(\dot{E}/GDP)(C/E) \tag{1}$$

以人口总数（N）、人均国内生产总值（GDP/N）、一次能源消耗强度（$\dot{E}/GDP$）和碳排放强度（C/E）的乘积来表示排放量。本文中我们用瓦特（W）来表示所有燃料能源的一次能源消耗速率（"燃速"），用（1990 年美元）\$ · yr^{-1} 作为国内生产总值的单位，那么两者的比值（即能源消耗强度）的单位就为 W · yr · $\$^{-1}$。碳排放强度是所有能源的由碳到能量的转化排放系数的加权平均值，单位为 kg C · W^{-1} · yr^{-1}。例如，根据公式（1），1990 年化石燃料的 CO_2 排放系数为：$\dot{M}_c$ = 5.3×10^9 人 × 每人 4,100 \$ · yr^{-1} × 0.49 W · yr · $\$^{-1}$ × 0.56 kg C · W^{-1} · yr^{-1} ≈ 6.0 Gt C · yr^{-1}（1 Gt C = 10^{12} kg C）。

为了说明公式（1）中各因子在历史上以及 IS92a 情景中的相对贡献，我们研究了 1890 年至 2100 年这 210 年内每个因子在全球范围内的变化（图 1），其中 1990 年以前的数据是根据历史资料 [7-9] 得到的，而 1990 年后的数据则根据定义 IPCC 情景的文献 [2,10-12] 得来。尽管 IPCC 文献中隐含了上述有关信息，但本文首次清晰地展示了这些信息。

茅阳一碳排放公式虽然有利于说明整体状况，但对该公式的各项进行全球累加却有可能掩盖区域发展的不同 [13]。个别国家的数据显示，经济发展过程中 $\dot{E}/GDP$ 将不断增大，在基础建设投资达峰值时 $\dot{E}/GDP$ 达到最大值，只有在经济生产力提高且经济结构向低能耗经济活动（如服务业）转型一段时间之后，$\dot{E}/GDP$ 才开始降低 [14]。虽然能耗强度呈明显下降趋势，但如今印度和中国的能耗强度仍为世界平均值的 2~5 倍 [8]。为突出能源供给问题，我们暂时采用 IS92a 预估值，即全球 $\dot{E}/GDP$ 每年改善 1%。必须考虑到，要实现这一目标，发展中国家要采取的技术和结构的转型至关重要 [15]。

另一个减少排放的机会是过去 100 年"脱碳"的继续，这体现在全球能源结构中碳排放强度的减少 [8]。根据 IS92a 情景，下个世纪全球平均 C/E 值将持续递减（图 1d）。实际上，全球能源结构的优化是基于核能与无碳能源成本相对于化石能源成本的降低这一假设而实现的，这将使 2030 年全球 C/E 值降低到天然气的碳排放强度，并且之后还会继续降低。只有大量引入无碳能源（到 2050 年约达 10 TW）才能使含碳能源的使用快速减少。即使有这样多的无碳能源，并且能耗强度每年稳定改善 1%，公式（1）中各因子的净作用仍将使 CO_2 排放量在 2050 年时达到 1990 年的两倍以上。

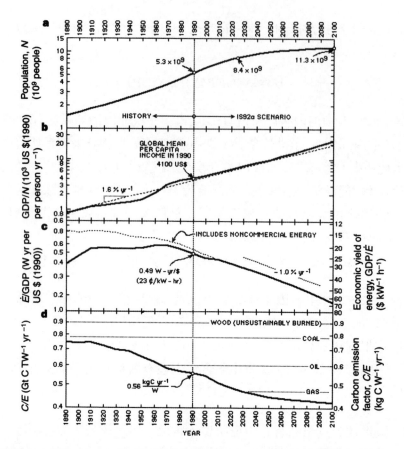

Fig. 1. Evolution of factors governing the rate of global fossil-fuel carbon emissions in the Kaya identity: $\dot{M}_c \equiv N(\text{GDP}/N)(\dot{E}/\text{GDP})(C/E)$. Historical curves (1890–1990) are from archival data[7-9]; future projections (1990–2100) are computed for the IPCC IS92a scenario[2,10-12]. GDP is inflation-corrected to 1990 US dollars. **a**, Global population; **b**, Per capita GDP; **c**, primary energy intensity ($\dot{E}/\text{GDP}$: left hand scale) and economic productivity of energy (GDP/$\dot{E}$: right hand scale); **d**, carbon intensity of the energy mix; the horizontal lines are emission individual carbonaceous fuels. Global population, N, shown after 1990 is the UN mid-range projection made at the time the IS92 scenarios were developed. It reaches 11.3 billion by 2100. Per capita global mean GDP continues its monotonic rise at ~1.6% yr^{-1} over the entire twenty-first century. The primary energy intensity incorporates both engineering energy efficiency and structural changes in the economy governing the material content of economic growth.

Figure 2 shows carbon emissions (Fig. 2a), primary power levels (Fig. 2b) and carbon-free primary power (Fig. 2c) required over the twenty-first century for IS92a and CO$_2$ stabilization scenarios. Even the optimistic decline of the last two factors in the Kaya identity cannot prevent emissions from increasing from 6 Gt C yr^{-1} in 1990 to ~20 Gt C yr^{-1} by 2100 under "business as usual" (Fig. 2a). Also shown as differently shaded zones are the relative contributions of natural gas, oil and coal to emissions. A feature of IS92a worth noting is that the share of carbon-intensive coal, relative to less-carbon-intensive natural gas and oil, rises after 2025, but the carbon intensity (C/E) of the fuel mix declines overall, a feature possible only by the massive introduction of carbon-free energy sources.

322

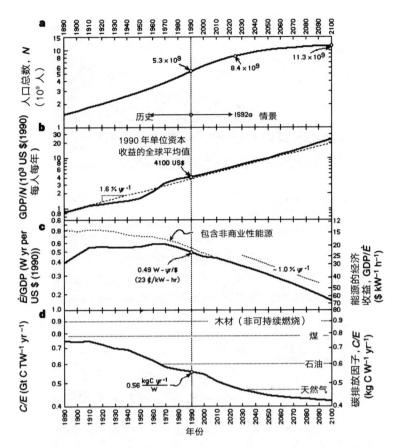

图 1. 茅阳—碳排放公式 $\dot{M}_c \equiv N(GDP/N)(\dot{E}/GDP)(C/E)$ 中控制全球化石燃料 CO_2 排放速率的各因子的变化情况。历史曲线（1890～1990）是根据档案资料[7-9]得到的，未来规划（1990～2100）则是根据 IPCC 的 IS92a 情景[2,10-12]计算而来的。GDP 被修正至 1990 年美元的通胀水平。**a**，全球人口总数；**b**，人均 GDP；**c**，一次能源消耗强度（$\dot{E}/GDP$：左侧刻度）与能源的经济生产力（$GDP/\dot{E}$：右侧刻度）；**d**，各类能源的碳排放强度，水平线为各种含碳燃料的排放因子。1990 年后的全球人口总数 N 是设计 IS92 情景时，联合国制定的人口预估的中值。到 2100 年该值将达 113 亿。整个二十一世纪，人均全球平均 GDP 以每年 1.6% 左右的速度持续递增。一次能源消耗强度综合考虑了控制经济增长物质组成的工程能源效率和经济结构的变化。

图 2 为 IS92a 和 CO_2 稳定情景下，二十一世纪要求的碳排放量（图 2a）、一次能源水平（图 2b）和无碳一次能源（图 2c）情况。倘若"照常排放"，即使最乐观地估计茅阳—放碳排放公式中的后两个因子的下降速度，也无法阻止排放量由 1990 年的 6 Gt C · yr^{-1} 升至 2100 年的约 20 Gt C · yr^{-1}（图 2a）。图中还用不同的颜色带表示出了天然气、石油和煤对排放量的贡献。IS92a 中一个值得注意的特征就是，在 2025 年后，相比于含碳量较低的天然气和石油，含碳量较高的煤所占比重有所上升，而燃料结构的碳排放强度（C/E）总体上是下降的。这一特征只有在大量引入无碳能源的情况下才可能出现。

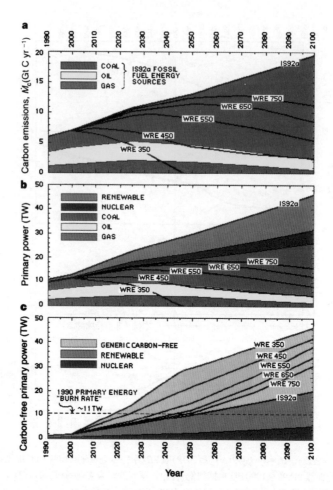

Fig. 2. Fossil-fuel carbon emissions and primary power in the twenty-first century for IPCC IS92a and WRE stabilization scenarios. **a**, Carbon emissions; **b**, primary power and **c**, carbon-free primary power. Coloured areas are gas, oil, coal, nuclear and renewable components of IS92a from the energy economics model of Pepper et al.[12]. Carbon emissions for WRE scenarios[16] are outputs of our inverse carbon-cycle model[17]. Fossil-fuel power for WRE scenarios is based on IS92a burn rates of gas, oil and coal employed sequentially in descending priority, and limited by total carbon emission caps. Carbon-free primary power is total primary power less fossil-fuel carbon power.

The curves in Fig. 2a are allowable emission levels over time which ultimate stabilize atmospheric CO$_2$ at 750, 650, 550, 450 and 350 p.p.m.v. computed for the Wigley, Richels and Edmonds stabilization paths[16,17]. These delayed stabilization paths, which follow IS92a emissions early on, with large reductions later (hereafter, WRE 350, 450, ..., 750 scenarios), could buy time to attain CO$_2$ stabilization goals. This is possible because a given atmospheric CO$_2$ concentration goal depends roughly on cumulative carbon emissions and can be approached by an infinite number of paths, some of which constrain emissions early, some later. However, Wigley, Richels and Edmonds did not consider whether their paths were a realistic transition from the present fossil-fuel system. They emphasized that their "results should not be interpreted as a 'do nothing' or 'wait and see' policy", calling for prompt and sustained commitment to research, development and

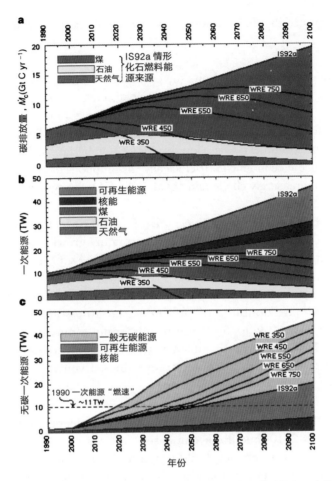

图 2. 在 IPCC 的 IS92a 和 WRE 稳定情景下，二十一世纪化石燃料的碳排放量与一次能源。**a**，碳排放量；**b**，一次能源；**c**，无碳一次能源。彩色部分为根据佩珀等 [12] 的能源经济模型得出的 IS92a 情景下天然气、石油、煤、核能及其他可再生能源等组分。WRE 情景的碳排放量 [16] 是根据我们的大气二氧化碳反演模式 [17] 得到的结果。WRE 情景的化石燃料能源是根据 IS92a 中天然气、石油和煤的燃速按照优先顺序得出的，并受总排放量限制。无碳一次能源是用一次能源总量减去化石燃料量得到的。

    图 2a 中的曲线是根据威格利、里歇尔斯和埃德蒙兹的稳定路径 [16,17] 计算出的（当最终大气 CO₂ 浓度分别稳定在 750 ppmv、650 ppmv、550 ppmv、450 ppmv 和 350 ppmv 的水平上时）允许排放水平随时间的变化情况。上述延迟稳定路径是：早期依照 IS92a 的排放路径，后期大幅地降低排放（自那以后分别为 WRE 350，450，…，750 情景），来为达到 CO₂ 的稳定目标争取时间。这是可以实现的，因为一个给定的大气 CO₂ 浓度目标能否实现大致取决于累计碳排放量，可以通过部分在早期、部分在晚期进行限制排放的诸多方案来达到该目标。然而，威格利、里歇尔斯和埃德蒙兹并未考虑到，在当前的化石燃料体系下其路径能否实现。他们强调，"这一结果不应该被解释为'无所作为'或者'等等看'的策略"，而是号召人们立刻开始并不断致

demonstration to ensure that low-carbon and carbon-free energy alternatives are available when needed[16].

The black topmost curve in Fig. 2b shows the evolution of total primary power required to meet the economic goals of IS92a, with gas, oil, coal, nuclear and renewable components shown as coloured areas. Also shown are allowable primary power levels from fossil-fuel sources computed for WRE stabilization scenarios. The difference between the IS92a total primary power, $\dot{E}$, and fossil-fuel power allowable for CO_2 stabilization, $\dot{E}_f$, must be provided by carbon-free sources if the economic and "efficiency" assumptions of IPCC "business as usual" are maintained; an increasingly challenging goal as the CO_2 concentration target is lowered.

Figure 2c shows carbon-free power, $\dot{E}_{cf} = \dot{E} - \dot{E}_f$, required for IS92a and for CO_2 stabilization via WRE 350–750 paths in a world economy growing as IS92a and with the same improvement rate in $\dot{E}/GDP$. In that case stabilizing CO_2 concentrations at 1990 levels according to WRE 350 will require ~10 TW of carbon-free primary power by 2018, equal to the total 1990 primary power of the world economy. WRE 550, which leads to an approximate doubling of pre-industrial atmospheric CO_2, requires this much carbon-free power by 2035. These results imply a massive transition from the present fossil-fuel-dominated energy infrastructure to carbon-free sources in the coming decades to stabilize CO_2 at reasonable target values. Even IS92a requires ~10 TW carbon-free power by 2050.

Studies by US DOE laboratories in support of the Kyoto negotiations to reduce carbon emissions in the 1998–2010 time frame emphasize demand side reductions from improved energy efficiency in motor vehicles, buildings and electrical appliances[18]. Our analysis indicates that beyond 2010 new carbon-free primary power technologies will increasingly be needed even with significant improvements in the ability to convert primary power into GDP.

Figure 3 shows the trade-off between increases in carbon-free power and energy efficiency improvements, expressed as the annual percentage improvement in $\dot{E}/GDP$ required to achieve a $2 \times$ pre-industrial CO_2 goal by the WRE 550 path. Sustained improvement of energy intensity of order $\pm 1.0\%$ yr^{-1} relative to the base case (1% yr^{-1}) creates increasingly large differences in carbon-free power required as the twenty-first century progresses (Fig. 3). For a 2% yr^{-1} compounded growth, the carbon-free power required remains modest even by the year 2100. But 2% yr^{-1} may be almost impossible to sustain over the next century, as that would imply growing the world population and economy significantly at nearly constant primary power. If there is no improvement in energy intensity, some 40 carbon-free terawatts could be needed by 2050.

力于研究、发展和论证，以保证在需要时有低碳和无碳能源替代品可用[16]。

图 2b 最上面的黑色曲线为达到 IS92a 的经济目标所需要的一次能源需求总量，其中天然气、石油、煤、核能以及其他可再生能源成分以彩色区域标注。图中还给出了根据 WRE 稳定情景计算出的、允许使用的来自于化石燃料能源的一次能源水平。倘若考虑到经济和"效率"而保留 IPCC"照常排放"的假设，IS92a 中一次能源总量 $\dot{E}$ 和 CO_2 稳定目标下允许的化石燃料能源 $\dot{E}_f$ 之差，必须由无碳能源来提供。随着 CO_2 目标浓度的降低，这是一个越来越严峻的挑战。

图 2c 所示为 IS92a 情景下和满足 WRE350～750 路径下实现 CO_2 稳定所需要的无碳能源 $\dot{E}_{cf} = \dot{E} - \dot{E}_f$，这些是建立在全球经济按照 IS92a 情景中的速率增长且 $\dot{E}$/GDP 有相同的改善效率的假设前提下的。那样的话，根据 WRE350，要想将 CO_2 浓度稳定在 1990 年的水平上，到 2018 年无碳能源就必须达到 10 TW 左右，这相当于 1990 年世界经济发展消耗的一次能源的总量。WRE550 的目标是使大气 CO_2 含量稳定在工业化以前水平的两倍左右，这要求到 2035 年时无碳能源达到 10 TW。这也意味着，未来几十年，为使 CO_2 含量稳定在合理的目标值，必然要从现今的以化石燃料为主的基本能源结构向无碳能源的大规模转变。即便是 IS92a 情景也要求，到 2050 年无碳能源需达到 10 TW 左右。

美国能源部的国家实验室为支持京都协议，对 1998～2010 年间如何减少碳排放量做了研究，研究强调可以通过提高机动车辆、建筑以及电器的能效来减少（对于能源的）需求[18]。分析结果表明，在 2010 年以后，即使一次能源转化为 GDP 的能力得到明显提高，对无碳一次能源新技术的需求仍将日益迫切。

图 3 给出了在增加无碳能源与提高能效之间的权衡，图中在 WRE550 路径将 CO_2 含量稳定在工业化以前水平的两倍的目标下，以每年 $\dot{E}$/GDP 所需要改善的百分点这一情况来表示。到二十一世纪，随着时间的推移，即使能耗强度相对于基线（1% yr^{-1}）的改善均为 ±1.0% yr^{-1} 左右，每年对无碳能源需求的差别也是逐渐增大的（图 3）。而当复合增长率为 2% yr^{-1} 时，即使到 2100 年，对无碳能源的需求将相对较少。但到下个世纪要保持 2% yr^{-1} 几乎是不可能实现的。因为，那将意味着，在一次能源近乎不变的情况下，世界人口与经济均明显增长。倘若不改善能耗强度，到 2050 年，将需要约 40 TW 的无碳能源。

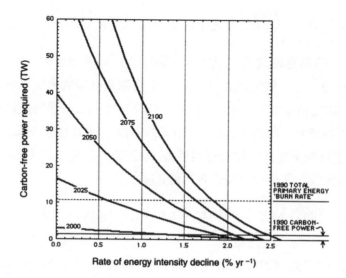

Fig. 3. Twenty-first century trade-offs, between carbon-free power required and "energy efficiency", to stabilize at twice the pre-industrial CO_2 concentration. Carbon-free primary power is plotted versus the annual rate of decline in energy intensity. The latter is defined as $-[d(\dot{E}/GDP)/dt](\dot{E}/GDP)^{-1}100\%$. All cases assume WRE 550 CO_2 stabilization paths and GDP growth of 2.9% yr^{-1} to 2025, 2.3% yr^{-1} thereafter. The rate of energy intensity decline in IS92a and in our CO_2 stabilization base cases is ~1.0% yr^{-1}.

The authors of the IPCC central "business as usual" scenario (IS92a) believed that an exponential decline in $\dot{E}$/GDP of 1% yr^{-1} would be sustainable over the next century employing only those emission-control policies internationally agreed to at the 1992 Rio climate treaty negotiations[1,2]. But either deploying 10–30 TW carbon-free power or improving $\dot{E}$/GDP by 2% yr^{-1} will be very difficult. The need to push hard on both improving energy supply and reducing demand is demonstrated by our analysis, as well as the fact that there are real trade-offs—more of one can significantly diminish the need for the other.

Stabilizing atmospheric CO_2 at twice pre-industrial levels while meeting the economic assumptions of "business as usual" implies a massive transition to carbon-free power, particular in developing nations. There are no energy systems technologically ready at present to produce the required amounts of carbon-free power[6]. Some suggest the answer may be integrated energy systems based on fossil fuels[19] in which carbon dioxide[20] or solid carbon[21] is sequestered in reservoirs isolated from the atmosphere, or "geoengineering" compensatory climate changes[22-24]. Despite potentially serious environmental and cost problems, these approaches would allow fossil fuel, increasingly coal, to continue its historic rise as the primary energy source of the next century.

It is time now to look hard at the engineering feasibility of transformative technologies that can change the way primary energy itself is produced. It is within the range of climate change and impact projections that stabilization of atmospheric CO_2 at some level below the IS92a baseline is necessary to mitigate large adverse effects on global economies and ecosystems[3]. In that case, a massive infusion of new energy-producing technologies at the

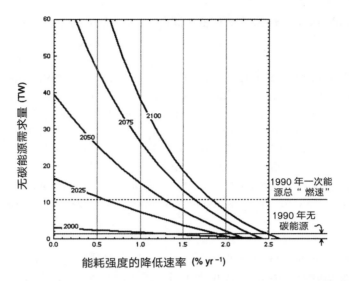

图 3. 21 世纪，为使 CO_2 含量稳定在工业化以前水平的两倍，需在无碳能源与"能效"间做出权衡。图中纵坐标为无碳一次能源需求量，横坐标为能耗强度年降低速率。后者用 $-[d(\dot{E}/GDP)/dt](\dot{E}/GDP)^{-1}100\%$ 表示。所有情形均以 WRE550 的 CO_2 稳定路径为准，且到 2025 年 GDP 的年增长率为 2.9%，2025 年以后为 2.3%。而在 IS92a 以及我们的 CO_2 稳定基线中，能耗强度的降低速率约为 1% yr^{-1}。

制定 IPCC 最重要的"照常排放"情景（IS92a）的学者认为，下个世纪，只要各国的控制排放政策符合 1992 年的里约气候条约[1,2]的规定，使 $\dot{E}$/GDP 每年指数递减 1% 是可行的。但无论是要开发 10～30 TW 的无碳能源还是使 $\dot{E}$/GDP 每年改善 2% 都不容易。通过分析我们已经证明，需要在大力推动能源供给提高的同时减少对能源的需求，而且，实际中确实存在着这样的权衡关系——某一项的增多将大大降低对另一项的需求。

要使 CO_2 含量稳定在工业化以前水平的两倍，同时还要达到经济正常发展的目标，就意味着要大量改用无碳能源，尤其是在发展中国家。目前还没有任何技术成熟的能源系统可以生产出那么多的无碳能源[6]。有人提议，可通过整合目前以化石燃料为基础的能源体系[19]，使 CO_2[20] 或固态碳[21] 保存在某个地方，与大气隔离开来；或实施"地球工程"来补偿气候的变化[22-24]。倘若不考虑一系列潜在的严峻的环境和成本问题，利用这些方法，下个世纪化石燃料（尤其是煤）将延续其在一次能源中占主要地位的历史。

现在是时候重视改造技术的工程可行性问题了，倘若改造技术实现，就可以改变一次能源本身的生产方式。在预估的气候变化与影响范围内，有必要将大气 CO_2 的稳定目标值定得比 IS92a 基线低些，以减轻给全球经济和生态系统带来的大量不利影响[3]。那样的话，就需要引入大量的新能源生产技术以防止"人类活动对气候系

329

required scale could be needed to prevent "dangerous anthropogenic interference with the climate system" (ref. 1).

Analyses of carbon emissions targets thus far have quite reasonably emphasized market economics theory. Some suggest that market forces alone, perhaps supplemented by carbon taxes, are sufficient to stimulate adequate levels of innovation in emission-reducing technologies. However, market inefficiencies may preclude timely development of such technologies at the required scale[25]. There are renewable[26,27], fission[28] and fusion[29] concepts incorporating innovative technological ideas at early research and development stages that could, in principle, provide the needed carbon-free power. But without policy incentives to overcome socioeconomic inertia these could take more than 50 years to penetrate to their market potential[30].

These results underscore the pitfalls of "wait-and-see". This past century, accelerated technology development from wartime and postwar research produced commercial aviation, radar, computer chips, lasers and the Internet, among other things. Researching, developing and commercializing carbon-free primary power technologies capable of 10–30 TW by the mid-twenty-first century could require efforts, perhaps international, pursued with the urgency of the Manhattan Project or the Apollo space programme. The roles of governments and market entrepreneurs in the eventual deployment of such technologies need to be considered more comprehensively than we can do here. But the potentially adverse effect of humanity on the Earth's climate could well stimulate new industries in the twenty-first century, as did the Second World War and the "cold war" in this century.

(**395**, 881-884; 1998)

Martin I. Hoffert[*], **Ken Caldeira**[†], **Atul K. Jain**[‡], **Erik F. Haites**[§], **L. D. Danny Harvey**[‖], **Seth D. Potter**[*¶], **Michael E. Schlesinger**[‡], **Stephen H. Schneider**[#], **Robert G. Watts**[*], **Tom M. L. Wigley**[**] & **Donald J. Wuebbles**[‡]

[*] Department of Physics, New York University, 4 Washington Place, New York, New York 10003-6621, USA
[†] Lawrence Livermore National Laboratory, Livermore, California 94550, USA
[‡] Department of Atmospheric Sciences, University of Illinois, Urbana, Illinois 61801, USA
[§] Margaree Consultants, Toronto, M5H 2X6, Canada
[‖] Department of Geography, University of Toronto, Toronto, M5S 3G3, Canada
[#] Department of Biological Sciences, Stanford University, Stanford, California 94305, USA
[*] Department of Mechanical Engineering, Tulane University, New Orleans, Louisiana 70118, USA
[**] National Center for Atmospheric Research, Boulder, Colorado 80307, USA
[¶] Present address: Boeing, Saal Beach, California 90740-7644, USA.

Received 10 August; accepted 8 October 1998.

References:

1. *United Nations Framework Convention on Climate Change (Text)* (UNEP/WMO, Climate Change Secretariat, Geneva, 1992).

2. Leggett, J. *et al.* in *Climatic Change 1992: the Supplementary Report to the IPCC Scientific Assessment* (eds Houghton, J. T. *et al.*) 69-95 (Cambridge Univ. Press, 1992).

3. Houghton, J. T. *et al.* (eds) *Climate Change 1995: the Science of Climate Change* (Cambridge Univ. Press, 1996).

4. Kaya, Y. Impact of carbon dioxide emission control on GNP growth: Interpretation of proposed scenarios (IPCC Response Strategies Working Group Memorandum, 1989).

统产生危险的干扰"(参考文献 1)。

基于迄今对碳排放目标的分析，我们完全有理由强调市场经济理论的作用。有人认为，仅凭市场力量本身，或许再辅以一定的排碳税，就足以促进减排技术的创新达到所需水平。然而，市场效率的低下可能会妨碍减排技术及时发展到所需的规模 [25]。可再生的 [26,27]、核裂变 [28] 及核聚变 [29] 概念结合创新技术观念正处于早期研发阶段，理论上讲，它们应该可以满足无碳能源的需求。但是，如果没有政策的激励以克服社会经济发展的惯性，要看到其市场潜能，还将需要 50 年以上的时间 [30]。

上述结果再次强调了"等等看"策略的隐患。在过去的 100 年中，战时与战后的研究加速了技术的发展，诞生了商用飞机、雷达、计算机芯片、激光以及互联网等等科技产物。到二十一世纪中叶，要使无碳能源技术的研究、发展以及商业化达到 $10 \sim 30$ TW 的水平，还需要国际社会的共同努力，能源的研究发展如同曼哈顿计划或阿波罗太空计划的实施一样迫切。相比于我们文中所讨论的情况，应该更全面地考虑政府和市场中企业家在减排技术的最终部署中的作用。正如本世纪第二次世界大战和"冷战"一样，人类对地球气候的潜在危害将有效地促进二十一世纪新工业的诞生。

（齐红艳 翻译；周天军 审稿）

5. Nakićenović, N., Victor, D., Grübler, A. & Schrattenholtzer, L. Long term strategies for mitigating global warming: Introduction. *Energy* **18,** 403-409 (1993).

6. Hoffert, M. I. & Potter, S. D. in *Engineering Response to Global Climate Change* (ed. Watts, R. G.) 205-260 (Lewis, Boca Raton, FL, 1997).

7. de Vries, B. & van den Wijngaart, R. *The Targets/IMage Energy (TIME) 1.0 Model* (GLOBO Rep. Ser. No. 16, National Institute of Public Health and the Environment (RIVM), Bilthoven, Netherlands, 1995).

8. Nakićenović, N. Freeing energy from carbon. *Daedalus* **125**(3), 95-112 (1996).

9. Boden, T. A. *et al.* (eds) *Trends 93: A Compendium of Data on Global Change* 501-584 (ORNL/CDIA- 65, Carbon Dioxide Information Analysis Center, Oak Ridge National Lab., Oak Ridge, TN, 1994).

10. Alcamo, J. *et al.* in *Climate Change 1994* (eds Houghton, J. T. *et al.*) 247-304 (Cambridge Univ. Press, Cambridge, UK, 1995).

11. Alcamo, J. (ed.) *Image 2.0: Integrated Modeling of Global Climate Change* (Kluwer, Netherlands, 1994).

12. Pepper, W. J. *et al. Emission Scenarios for the IPCC—An Update* (US Environmental Protection Agency, Washington DC, 1992).

13. Yang, C. & Schneider, S. H. Global carbon dioxide scenarios: sensitivity to social and technological factors in three regions. *Mitigat. Adapt. Strat. for Global Change* **2,** 373-404 (1998).

14. Reddy, A. K. N. & Goldemberg, J. in *Energy for Planet Earth* (ed. Piel, J.) 59-71 (Freeman, New York, 1991).

15. Goldemberg, J. Energy needs in developing countries and sustainability. *Science* **269,** 1058-1059 (1995).

16. Wigley, T. M. L., Richels, R. & Edmonds, J. A. Economic and environmental choices in the stabilization of atmospheric CO$_2$ concentration. *Nature* **379,** 240-243 (1996).

17. Jain, A. K., Kheshgi, H. S., Hoffert, M. I. & Wuebbles, D. J. Distribution of radiocarbon as a test of global carbon cycle models. *Glob. Biogeochem. Cycles* **9,** 153-166 (1995).

18. Romm, J., Levine, M., Brown, M. & Peterson, E. A road map for U.S. carbon reductions. *Science* **279,** 669-670 (1998).

19. National Academy of Sciences *Policy Implications of Greenhouse Warming* 340-375 (National Academy Press, Washington DC, 1992).

20. Parson, E. A. & Keith, D. W. Fossil fuels without CO$_2$ emissions: Progress, prospects and policy implications. *Science* (in the press).

21. Steinberg, M. & Dong, Y. in *Global Climate Change: Science, Policy and Mitigation Strategies* (eds Mathai, C. V. & Stensland, C.) 858-873 (Air & Waste Management Assoc., Pittsburgh, PA, 1994).

22. Marland, G. (ed.) Special section on geoengineering. *Clim. Change* **33**(3), 275-336 (1996).

23. Flannery, B. P. *et al.* in *Engineering Response to Global Climate Change* (ed. Watts, R. G.) 379-427 (Lewis, Boca Raton, FL, 1997).

24. Fogg, M. J. *Terraforming: Engineering Planetary Environments* (SAE International, Warrendale, PA, 1995).

25. Schneider, S. H. & Goulder, L. H. Achieving low-cost emission targets. *Nature* **389,** 13-14 (1997).

26. Johansson, T. M. *et al.* (eds) *Renewable Energy* (Island, Washington DC, 1993).

27. Glaser, P. E. *et al.* (eds) *Solar Power Satellites: A Space Energy System for Earth* (Wiley-Praxis, Chichester, 1998).

28. Weinberg, A. M. in *Technologies For a Greenhouse-Constrained Society* (eds Kuliasha, M. A. *et al.*) 227-237 (Lewis, Boca Raton, FL, 1992).

29. Fowler, T. K. *The Fusion Quest* (Johns Hopkins, Baltimore, 1997).

30. Grübler, A. in *Technological Trajectories and the Human Environment* (eds Ausubel, J. H. & Langford, H. D.) 14-32 (National Academy Press, Washington DC, 1997).

Acknowledgements. We thank DOE, NASA and NSF for partial support of this work. We also thank the Aspen Global Change Institute for discussions during the 1998 summer workshop "Innovative Energy Systems and CO$_2$ Stabilization".

Correspondence and requests for materials should be addressed to M.I.H. (e-mail: marty.haffert@nyu. edu).

Jefferson Fathered Slave's Last Child

E. A. Foster *et al.*

Editor's Note

This paper reports one of the more dramatic examples of the ability, developed over the previous decade or so, to make genetic comparisons between individuals by looking at specific "markers" in their genomes. For many years there had been suggestions that Thomas Jefferson, one of the Founding Fathers of the United States of America and its third president, had fathered one or more children by one of his slaves, named Sally Hemings. The paper reports a genetic analysis of the male-line descendants of Hemings' youngest and eldest sons (Thomas Woodson and Eston Hemings Jefferson, each of whom will have received a Y chromosome in direct paternal succession from these two respective men), which is compared with that of male-line descendants of Jefferson's paternal uncle. The results support the view that Eston, but not Thomas Woodson, was fathered by Thomas Jefferson.

THERE is a long-standing historical controversy over the question of US President Thomas Jefferson's paternity of the children of Sally Hemings, one of his slaves[1-4]. To throw some scientific light on the dispute, we have compared Y-chromosomal DNA haplotypes from male-line descendants of Field Jefferson, a paternal uncle of Thomas Jefferson, with those of male-line descendants of Thomas Woodson, Sally Hemings' putative first son, and of Eston Hemings Jefferson, her last son. The molecular findings fail to support the belief that Thomas Jefferson was Thomas Woodson's father, but provide evidence that he was the biological father of Eston Hemings Jefferson.

In 1802, President Thomas Jefferson was accused of having fathered a child, Tom, by Sally Hemings[5]. Tom was said to have been born in 1790, soon after Jefferson and Sally Hemings returned from France where he had been minister. Present-day members of the African–American Woodson family believe that Thomas Jefferson was the father of Thomas Woodson, whose name comes from his later owner[6]. No known documents support this view.

Sally Hemings had at least four more children. Her last son, Eston (born in 1808), is said to have borne a striking resemblance to Thomas Jefferson, and entered white society in Madison, Wisconsin, as Eston Hemings Jefferson. Although Eston's descendants believe that Thomas Jefferson was Eston's father, most Jefferson scholars give more credence to the oral tradition of the descendants of Martha Jefferson Randolph, the president's daughter. They believe that Sally Hemings' later children, including Eston, were fathered by either

杰斐逊是其奴隶最小孩子的生父

福斯特等

编者按

这篇论文报道了一种非常引人注目的、在过去十年发展起来的科学手段，该手段通过在不同个体基因组中寻找特异性的"标记"来进行遗传比对。许多年以来，一直有人认为托马斯·杰斐逊——美国的开国之父之一以及第三任总统，与他的一个奴隶莎丽·海明斯，生下了一个或者多个孩子。这篇论文报道了海明斯最大的和最小的儿子（托马斯·伍德森和艾斯顿·海明斯·杰斐逊）的男性后裔（他们分别从这两个人直接遗传一条父系 Y 染色体）的遗传分析，并与托马斯·杰斐逊的伯父的男性后裔的基因进行比较。结果支持托马斯·杰斐逊是艾斯顿而非托马斯·伍德森的父亲这一观点。

关于美国总统托马斯·杰斐逊与其奴隶之一莎丽·海明斯的孩子之间的血缘关系，长期以来一直是历史上争论不休的一个问题 [1-4]。从科学的角度来看待和分析这个争论，我们比较了不同来源的 Y 染色体 DNA 的单体型。其中一些标本来自托马斯·杰斐逊的伯父菲尔德·杰斐逊的男性后裔，另一些标本来自托马斯·伍德森的男性后裔，他被认定为莎丽·海明斯的第一个儿子，以及来自她最小的儿子艾斯顿·海明斯·杰斐逊的男性后裔的样本。分子层面的比对结果无法支持我们相信托马斯·杰斐逊是托马斯·伍德森的生父，但是提供了证据表明他是艾斯顿·海明斯·杰斐逊生物学上的生父。

1802 年，总统托马斯·杰斐逊被指控与莎丽·海明斯生下了一个孩子，名叫汤姆 [5]。据说汤姆生于 1790 年，就在杰斐逊和莎丽·海明斯从法国返回不久，杰斐逊曾经在法国做过部长。现在的非裔美籍伍德森家族成员都相信托马斯·杰斐逊是托马斯·伍德森的生父，后者的名字来自于他之后的主人 [6]。但是，没有任何已知的文件支持这种观点。

莎丽·海明斯至少有四个孩子。她最小的孩子艾斯顿（生于 1808 年），据说与托马斯·杰斐逊惊人得相似，他以艾斯顿·海明斯·杰斐逊的身份进入威斯康星州麦迪逊城的白人社会中。尽管艾斯顿的男性后裔相信托马斯·杰斐逊是艾斯顿的生父，但是研究杰斐逊的绝大多数学者更愿意相信总统的女儿玛莎·杰斐逊·伦道夫的后裔的口口相传。他们认为莎丽·海明斯所生的孩子，包括艾斯顿在内，他们的生父

Samuel or Peter Carr, sons of Jefferson's sister, which would explain their resemblance to the president.

Because most of the Y chromosome is passed unchanged from father to son, apart from occasional mutations, DNA analysis of the Y chromosome can reveal whether or not individuals are likely to be male-line relatives. We therefore analysed DNA from the Y chromosomes of: five male-line descendants of two sons of the president's paternal uncle, Field Jefferson; five male-line descendants of two sons of Thomas Woodson; one male-line descendant of Eston Hemings Jefferson; and three male-line descendants of three sons of John Carr, grandfather of Samuel and Peter Carr (Fig. 1a). No Y-chromosome data were available from male-line descendants of President Thomas Jefferson because he had no surviving sons.

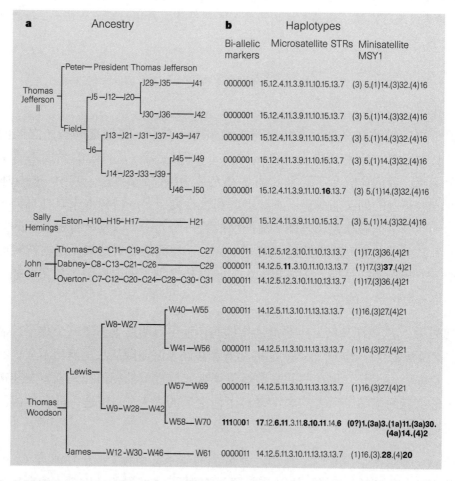

Fig. 1. Male-line ancestry and haplotypes of participants. **a,** Ancestry. Numbers correspond to reference numbers and names in more detailed genealogical charts for each family. **b,** Haplotypes. Entries in bold highlight deviations from the usual patterns for the group of descendants. **Bi-allelic markers**. Order of loci: YAP-SRYm8299-sY81-LLY22g-Tat-92R7-SRYm1532. 0, ancestral state; 1, derived state. **Microsatellite short tandem repeats (STRs)**. Order of loci: 19-388-389A-389B-389C-389D-390-391-

是塞缪尔或者彼得·卡尔——杰斐逊妹妹的儿子，这样也可以解释为什么他们的染色体会和杰斐逊总统之间有着相似之处。

绝大多数的 Y 染色体都会毫无改变地由生父传给儿子，除非偶然发生突变，因此，对 Y 染色体的 DNA 分析可以揭示不同个体之间是否具有某种亲缘关系。我们分析了来自不同人的 Y 染色体上的 DNA：总统的伯父菲尔德·杰斐逊的两个儿子的五个男性后裔；托马斯·伍德森的两个儿子的五个男性后裔；艾斯顿·海明斯·杰斐逊的男性后裔；约翰·卡尔的三个儿子的三个男性后裔，约翰·卡尔是塞缪尔和彼得·卡尔的祖父（见图 1a）。样本中没有来自于总统托马斯·杰斐逊的男性后裔的数据，因为他没有活下来的儿子。

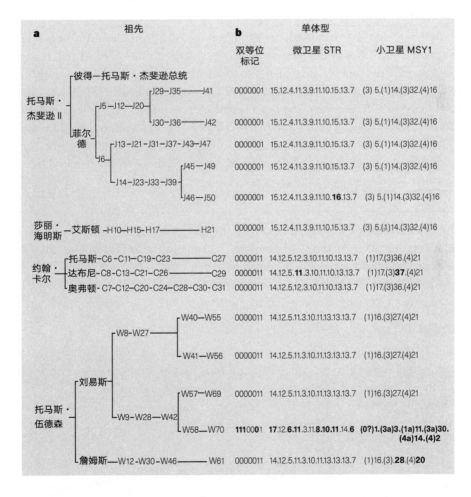

图 1. 男性族谱以及参与者的单体型。**a**，祖先。数字对应相应的人的编码，名字则是更为具体的每个家族的宗族谱图。**b**，单体型。粗体用来标注与整个家族正常遗传情况的差异。**双等位基因标记**。位点的顺序：YAP-SRYm8299-sY81-LLY22g-Tat-92R7-SRYm 1532。0，祖先状态；1，派生状态。**微卫星短串联重复序列（STR）**。位点顺序：19-388-389A-389B-389C-389D-390-391-392-393-dxys156y。在每一个位点

392-393-dxys156y. The number of repeats at each locus is shown. **Minisatellite MSY1.** Each number in brackets represents the sequence type of the repeat unit; the number after it is the number of units with this sequence type. For example, J41 has 5 units of sequence type 3, 14 units of sequence type 1, 32 units of sequence type 3, and 16 units of sequence type 4.

Seven bi-allelic markers (refs 7–12), eleven microsatellites (ref. 13) and the minisatellite MSY1 (ref. 14) were analysed (Fig. 1b). Four of the five descendants of Field Jefferson shared the same haplotype at all loci, and the fifth differed by only a single unit at one microsatellite locus, probably a mutation. This haplotype is rare in the population, where the average frequency of a microsatellite haplotype is about 1.5 per cent. Indeed, it has never been observed outside the Jefferson family, and it has not been found in 670 European men (more than 1,200 worldwide) typed with the microsatellites or 308 European men (690 worldwide) typed with MSY1.

Four of the five male-line descendants of Thomas Woodson shared a haplotype (with one MSY1 variant) that was not similar to the Y chromosome of Field Jefferson but was characteristic of Europeans. The fifth Woodson descendant had an entirely different haplotype, most often seen in sub-Saharan Africans, which indicates illegitimacy in the line after individual W42. In contrast, the descendant of Eston Hemings Jefferson did have the Field Jefferson haplotype. The haplotypes of two of the descendants of John Carr were identical; the third differed by one step at one microsatellite locus and by one step in the MSY1 code. The Carr haplotypes differed markedly from those of the descendants of Field Jefferson.

The simplest and most probable explanations for our molecular findings are that Thomas Jefferson, rather than one of the Carr brothers, was the father of Eston Hemings Jefferson, and that Thomas Woodson was not Thomas Jefferson's son. The frequency of the Jefferson haplotype is less than 0.1 per cent, a result that is at least 100 times more likely if the president was the father of Eston Hemings Jefferson than if someone unrelated was the father.

We cannot completely rule out other explanations of our findings based on illegitimacy in various lines of descent. For example, a male-line descendant of Field Jefferson could possibly have illegitimately fathered an ancestor of the presumed maleline descendant of Eston. But in the absence of historical evidence to support such possibilities, we consider them to be unlikely.

(**396**, 27-28; 1998)

上重复的次数显示在图上。**小卫星 MSY1**。每一个括号中的数字代表重复单元的序列类型；在它之后的数字代表该序列类型单元的数目。例如，J41 个体具有 5 个单元的序列类型 3，14 个单元的序列类型 1，32 个单元的序列类型 3，以及 16 个单元的序列类型 4。

我们对七个双等位基因标记（参考文献 7~12）、十一个微卫星遗传标记（参考文献 13）以及小卫星 MSY1（参考文献 14）进行了分析研究（图 1b）。来自菲尔德·杰斐逊家族五个男性后裔中的四个人在所有位点上都具有同样的单体型，第五个人只是在其中一个微卫星遗传位点上的一个单元上略有不同，这可能是一个突变。这个单体型在人群中出现的概率是很小的，每个微卫星单体型出现的平均概率大概是百分之一点五。事实上，这种单体型从来没有在杰斐逊家族以外被发现过，在微卫星标记的 670 个欧洲男性（在全世界范围内超过 1,200 个）或者 308 个欧洲男性（全世界范围内 690 个男性）中都未曾发现过 MSY1 位点标记。

来自托马斯·伍德森五个男性后裔中有四个具有相同的单体型（有一个 MSY1 位点突变），这个单体型与来自菲尔德·杰斐逊的 Y 染色体之间并不相似，但却是欧洲人的特征。来自伍德森男性后裔的第五个后裔具有完全不同的单体型，这种单体型在亚撒哈拉非裔中最常见，这说明个体 W42 之后子嗣中存在私生情况。与之相反，艾斯顿·海明斯·杰斐逊的后代确实具有与菲尔德·杰斐逊相同的单体型。来自约翰·卡尔的两个后代的单体型是一样的；第三个后代的差别其中之一位于微卫星位点上，另一个位于 MSY1 编码上。来自卡尔男性后裔的单体型与来自菲尔德·杰斐逊男性后裔的单体型有着显著的差别。

针对我们分子研究的发现，最简单和最有可能的解释是艾斯顿·海明斯·杰斐逊的生父是托马斯·杰斐逊，而不是卡尔兄弟中的某一个。另外，托马斯·伍德森不是托马斯·杰斐逊的儿子。杰斐逊单体型出现的频率低于百分之零点一，这个结果说明，总统作为艾斯顿·海明斯·杰斐逊生父的可能性是一个毫不相干的人作为他的生父的概率的 100 倍。

基于不同谱系的异常情况，我们无法完全排除针对我们发现的其他解释。例如，菲尔德·杰斐逊的一个男性后裔有可能非法生下了一个孩子，而他是我们假定的艾斯顿后裔中的祖先。但缺少历史证据来支持这样一种假设，我们认为这是不可能发生的。

（刘振明 翻译；胡松年 审稿）

Eugene A. Foster[*], **M. A. Jobling**[†], **P. G. Taylor**[†], **P. Donnelly**[‡], **P. de Knijff**[§], **Rene Mieremet**[§], **T. Zerjal**[¶], **C. Tyler-Smith**[¶]

[*] 6 Gildersleeve Wood, Charlottesville, Virginia 22903, USA e-mail: eafoster@aol.com

[†] Department of Genetics, University of Leicester, Adrian Building, University Road, Leicester LE1 7RH, UK

[‡] Department of Statistics, University of Oxford, South Parks Road, Oxford OX1 3TG, UK

[§] MGC Department of Human Genetics, Leiden University, PO Box 9503, 2300 RA Leiden, The Netherlands

[¶] Department of Biochemistry, University of Oxford, South Parks Road, Oxford OX1 3QU, UK

References:

1. Peterson, M. D. *The Jefferson Image in the American Mind* 181-187 (Oxford Univ. Press, New York, 1960).

2. Malone, D. *Jefferson the President: First Term, 1801–1805* Appendix II, 494-498 (Little Brown, Boston, MA, 1970).

3. Brodie, F. M. *Thomas Jefferson: An Intimate History* (Norton, New York, 1974).

4. Ellis, J. J. *American Sphinx: The Character of Thomas Jefferson* (Knopf, New York, 1997).

5. Callender, J. T. *Richmond Recorder* 1 September 1802. [Cited in: Gordon-Reed, A. *Thomas Jefferson and Sally Hemings: An American Controversy* (Univ. Press of Virginia, Charlottesville, 1997)].

6. Woodson, M. S. *The Woodson Source Book* 2nd edn (Washington, 1984).

7. Hammer, M. F. *Mol. Biol. Evol.* **11,** 749-761 (1994).

8. Whitfield, L. S., Sulston, J. E. & Goodfellow, P. N. *Nature* **378,** 379-380 (1995).

9. Seielstad, M. T. *et al. Hum. Mol. Genet.* **3,** 2159-2161 (1994).

10. Zerjal, T. *et al. Am. J. Hum. Genet.* **60,** 1174-1183 (1997).

11. Mathias, N., Bayes, M. & Tyler-Smith, C. *Hum. Mol. Genet.* **3,** 115-123 (1994).

12. Kwok, C. *et al. J. Med. Genet.* **33,** 465-468 (1996).

13. Kayser, M. *et al. Int. J. Legal Med.* **110,** 125-133 (1997).

14. Jobling, M. A., Bouzekri, N. & Taylor, P. G. *Hum. Mol. Genet.* **7,** 643-653 (1998).

Light Speed Reduction to 17 Metres per Second in an Ultracold Atomic Gas

L. V. Hau *et al.*

ditor's Note

Light moves more slowly in physical media than in empty space—but generally only by a little. Yet interference effects associated with quantum physics can in principle be used to slow the speed of light dramatically in a specially designed medium. Here, Lene Vestergaard Hau and colleagues demonstrate this effect in an ultracold gas of sodium atoms. They use a laser beam travelling at right angles to a test beam to prepare the atoms such that a fraction of them are in particular excited electronic states. Light in the test beam, being repeatedly absorbed and re-emitted by these atoms, travels at a speed of only 17 metres per second, some 20 million times slower than light in a vacuum.

Techniques that use quantum interference effects are being actively investigated to manipulate the optical properties of quantum systems[1]. One such example is electromagnetically induced transparency, a quantum effect that permits the propagation of light pulses through an otherwise opaque medium[2-5]. Here we report an experimental demonstration of electromagnetically induced transparency in an ultracold gas of sodium atoms, in which the optical pulses propagate at twenty million times slower than the speed of light in a vacuum. The gas is cooled to nanokelvin temperatures by laser and evaporative cooling[6-10]. The quantum interference controlling the optical properties of the medium is set up by a "coupling" laser beam propagating at a right angle to the pulsed "probe" beam. At nanokelvin temperatures, the variation of refractive index with probe frequency can be made very steep. In conjunction with the high atomic density, this results in the exceptionally low light speeds observed. By cooling the cloud below the transition temperature for Bose–Einstein condensation[11-13] (causing a macroscopic population of alkali atoms in the quantum ground state of the confining potential), we observe even lower pulse propagation velocities (17 m s^{-1}) owing to the increased atom density. We report an inferred nonlinear refractive index of 0.18 cm^2 W^{-1} and find that the system shows exceptionally large optical nonlinearities, which are of potential fundamental and technological interest for quantum optics.

T HE experiment is performed with a gas of sodium atoms cooled to nanokelvin temperatures. Our atom cooling set-up is described in some detail in ref. 14. Atoms emitted from a "candlestick" atomic beam source[15] are decelerated in a Zeeman slower and loaded into a magneto-optical trap. In a few seconds we collect a cloud of 10^{10} atoms

342

光速在超冷原子气中降低至 17 米每秒

豪等

编者按

光在物质媒介中的传播速度比在真空中的慢，但通常只慢一点点。然而，与量子物理相关的干涉效应原则上可以使光速在特别设计的介质中显著变慢。本文中，莱娜·韦斯特戈·豪和其同事们在超冷钠原子气体中展示了这种效应：使用与测试光束成直角传播的激光束来制备原子，使得其中的一部分处在特定的电子激发态。测试光束中的光，被这些原子反复吸收和重新发射，以每秒 17 米的速度传播，大约为真空中光速的 2,000 万分之一。

基于量子干涉效应的技术被积极研究用来操纵量子系统中的光学性质[1]。其中的一个例子就是电磁感应透明，一种允许光脉冲在其他状况下不透明的介质中传播的量子现象[2-5]。在此，我们报道了在超冷钠原子气中电磁感应透明的实验验证，光脉冲以在真空中光速的 2,000 万分之一的速度在该介质中传播。气体通过激光和蒸发制冷被降温至纳开（10^{-9} K）[6-10]。控制介质光学性质的量子干涉通过与脉冲"探测"光束成直角传播的"耦合"激光束来建立。在纳开温度下，折射率随探测频率可以发生非常急剧的变化。与高原子密度相结合，就会导致所观察到的极低光速。我们将原子云冷却至玻色-爱因斯坦凝聚临界温度[11-13]以下（引起处于囚禁势量子基态的碱金属原子的宏观布居），由于原子数密度增大，我们观测到更低的脉冲传播速度（$17 \text{ m} \cdot \text{s}^{-1}$）。经过测算，本文报道了该体系的非线性折射率系数为 $0.18 \text{ cm}^2 \cdot \text{W}^{-1}$，发现该系统显示出非常大的光学非线性，这对于量子光学的研究而言具有潜在的基础和技术吸引力。

本实验是在降温至纳开的钠原子中进行的。文献 14 中对我们的原子冷却装置有一些细节性的描述。从"烛台"原子束源[15]发出的原子在塞曼减速器中被减速，然后被加载到一个磁光阱中。若干秒后我们在 1 mK 温度，$6 \times 10^{11} \text{ cm}^{-3}$ 密度下采集到

at a temperature of 1 mK and a density of 6×10^{11} cm^{-3}. The atoms are then polarization gradient cooled for a few milliseconds to 50 μK and optically pumped into the $F = 1$ ground state with an equal population of the three magnetic sublevels. We then turn all laser beams off and confine the atoms magnetically in the "4 Dee" trap[14]. Only atoms in the $M_F = -1$ state, with magnetic dipole moments directed opposite to the magnetic field direction (picked as the quantization axis), are trapped in the asymmetric harmonic trapping potential. This magnetic filtering results in a sample of atoms that are all in a single atomic state (state $|1\rangle$ in Fig. 1b) which allows adiabatic optical preparation of the atoms, as described below, and minimal heating of the cloud.

Next we evaporatively cool the atoms for 38 s to the transition temperature for Bose–Einstein condensation, T_c. The magnetic fields are then adjusted to adiabatically soften the trap. The resulting trapping potential has a frequency of $f_z = 21$ Hz along the symmetry (z) axis of the 4 Dee trap, and transverse frequencies $f_x = f_y = 69$ Hz. The bias field, parallel to the z axis, is 11 G. When we cool well below T_c, we are left with 1–2 million atoms in the condensate. For these parameters the transition occurs at a temperature of $T_c = 435$ nK and a peak density in the cloud of 5×10^{12} cm^{-3} .

We now apply a linearly polarized laser beam, the coupling beam, tuned to the transition between the unpopulated hyperfine states $|2\rangle$ and $|3\rangle$ (Fig. 1b). This beam couples states $|2\rangle$ and $|3\rangle$ and creates a quantum interference for a weaker probe laser beam (left circularly polarized) which is tuned to the $|1\rangle \rightarrow |3\rangle$ transition. A stable eigenstate (the "dark state") of the atom in the presence of coupling and probe lasers is a coherent superposition of the two hyperfine ground states $|1\rangle$ and $|2\rangle$. The ratio of the probability amplitudes is such that the contributions to the atomic dipole moment induced by the two lasers exactly cancel. The quantum interference occurs in a narrow interval of probe frequencies, with a width determined by the coupling laser power.

一个由 10^{10} 个原子组成的原子云。这些原子随后被偏振梯度冷却若干毫秒至 50 μK，并被光泵抽运至 $F = 1$ 的基态，原子平均布居于三个磁子能级上。然后我们关闭所有的激光束，并用磁场将原子限制在"4 Dee"阱 [14] 中。只有处于 $M_F = -1$ 态的原子，其磁偶极矩与磁场方向（被选作量子化轴）相反，被捕获于非对称的谐振势阱中。这个磁过滤使所有的样品原子都处于同一原子态（图 1b 中的量子态 $|1\rangle$），这就使下文所述的原子绝热光学制备和对原子云最低限度的加热成为可能。

接下来我们将原子蒸发制冷 38 秒至玻色–爱因斯坦凝聚的临界转变温度 T_c。调整磁场，绝热地减小势阱的梯度。这样所获得的势阱沿 4 Dee 势阱对称轴（z）方向的频率 $f_z = 21$ Hz，横向频率 $f_x = f_y = 69$ Hz。偏置场平行于 z 轴，为 11 G。当温度完全低于 T_c，我们只剩下处于凝聚态的一百万到两百万个原子。在这些参数下，转变发生在 $T_c = 435$ nK，原子云峰值密度为 5×10^{12} cm^{-3}。

我们现在来加一线偏振激光束，即耦合光束，将之调谐至无粒子分布的超精细态 $|2\rangle$ 和 $|3\rangle$ 之间的跃迁（图 1b）。该光束将态 $|2\rangle$ 和 $|3\rangle$ 相耦合，从而对一更弱的调谐在态 $|1\rangle \rightarrow |3\rangle$ 跃迁的探测激光光束（左旋圆偏振）产生了量子干涉。在耦合激光和探测激光同时存在时，原子的稳定本征态（"暗"态）是两个超精细基态 $|1\rangle$ 和 $|2\rangle$ 的相干叠加。概率幅比值被设置为两激光对所激发的原子偶极矩的贡献恰好相抵消。量子干涉发生在探测频率的一狭窄的范围内，宽度由耦合激光功率决定。

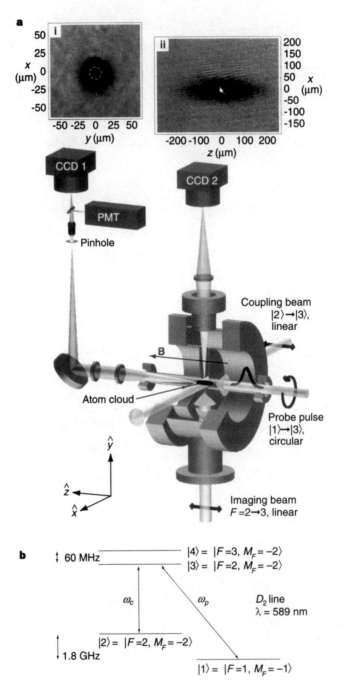

Fig. 1. Experimental set-up. A "coupling" laser beam propagates along the x axis with its linear polarization along the 11-G bias field in the z direction. The "probe" laser pulse propagates along the z axis and is left-circularly polarized. With a flipper mirror in front of the camera CCD 1, we direct this probe beam either to the camera or to the photomultiplier (PMT). For pulse delay measurements, we place a pinhole in an external image plane of the imaging optics and select a small area, 15 μm in diameter, of the probe beam centred on the atom clouds (as indicated by the dashed circle in inset (i)). The pulse delays are measured with the PMT. The imaging beam propagating along the y axis is used

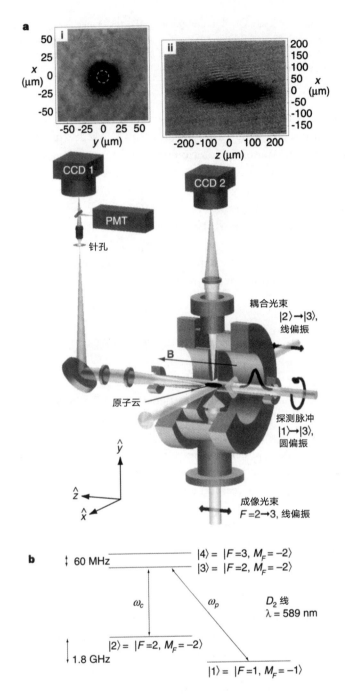

$|4\rangle = |F=3, M_F = -2\rangle$

$|3\rangle = |F=2, M_F = -2\rangle$

60 MHz

ω_c ω_p D_2 线
$\lambda = 589$ nm

$|2\rangle = |F=2, M_F = -2\rangle$

1.8 GHz

$|1\rangle = |F=1, M_F = -1\rangle$

图 1. 实验装置图。一"耦合"激光束沿 x 轴传播，其线偏振方向 z 轴上加有 11 G 的偏置场。"探测"激光脉冲沿 z 轴传播，为左旋圆偏振。利用摄像机 CCD1 前的转向镜，我们可以引导探测光速进入摄像机或者进入光电倍增管（PMT）。对于脉冲延迟的测量，我们在成像光路中一个外部的成像平面上放置一个针孔，并选择探测光束在原子云中心直径为 15 μm 的小区域（插图（i）中虚线圆所示）。脉冲延迟通过 PMT 测量。沿 y 轴传播的成像光束被用来使原子云在 CCD2 上成像，以获得原子云在沿脉冲传播方向

to image atom clouds onto camera CCD 2 to find the length of the clouds along the pulse propagation direction (z axis) for determination of light speeds. Inset (ii) shows atoms cooled to 450 nK which is 15 nK above T_c. (Note that this imaging beam is never applied at the same time as the probe pulse and coupling laser). The position of a cloud and its diameter in the two transverse directions, x and y, are found with CCD 1. Inset (i) shows an image of a condensate.

Figure 2a shows the calculated transmission of the probe beam as a function of its detuning from resonance for parameters which are typical of this work. In the absence of dephasing of the $|1\rangle \rightarrow |2\rangle$ transition, the quantum interference would be perfect, and at line centre, the transmission would be unity. Figure 2b shows the refractive index for the probe beam as a function of detuning. Due to the very small Doppler broadening of the $|1\rangle \rightarrow |2\rangle$ transition in our nanokelvin samples, application of very low coupling intensity leads to a transparency peak with a width much smaller than the natural line width of the $|1\rangle \rightarrow |3\rangle$ transition. Correspondingly, the dispersion curve is much steeper than can be obtained by any other technique, and this results in the unprecedented low group velocities reported here. The group velocity v_g for a propagating electromagnetic pulse is[16-19]:

$$v_g = \frac{c}{n(\omega_p) + \omega_p \dfrac{dn}{d\omega_p}} \approx \frac{\hbar c \epsilon_0}{2\omega_p} \frac{|\Omega_c|^2}{|\mu_{13}|^2 N} \tag{1}$$

Here $n(\omega_p)$ is the refractive index at probe frequency ω_p (rad s^{-1}), $|\Omega_c|^2$ is the square of the Rabi frequency for the coupling laser and varies linearly with intensity, μ_{13} is the electric dipole matrix element between states $|1\rangle$ and $|3\rangle$, N is the atomic density, and ϵ_0 is the permittivity of free space. At line centre, the refractive index is unity, and the second term in the denominator of equation (1) dominates the first. An important characteristic of the refractive index profile is that on resonance the dispersion of the group velocity is zero (see ref. 16), that is, $d^2 n/d\omega_p^2 = 0$, and to lowest order, the pulse maintains its shape as it propagates. The established quantum interference allows pulse transmission through our atom clouds which would otherwise have transmission coefficients of e^{-110} (below T_c), and creates a steep dispersive profile and very low group velocity for light pulses propagating through the clouds.

We note that the centres of the curves in Fig. 2 are shifted by 0.6 MHz from probe resonance. This is due to a coupling of state $|2\rangle$ to state $|4\rangle$ through the coupling laser field, which results in an a.c. Stark shift of level $|2\rangle$ and a corresponding line shift of the $2 \rightarrow 3$ transition. As the transparency peak and unity refractive index are obtained at two-photon resonance, this leads to a refractive index at the $1 \rightarrow 3$ resonance frequency which is different from unity. The difference is proportional to the a.c. Stark shift and hence to the coupling laser intensity, which is important for predicting the nonlinear refractive index as described below.

（z 轴）上的长度，并以此来确定光速。插图（ii）为降温至 450 nK 的原子，即在临界温度 T_c 以上 15 nK。（注意该成像束不能与探测光束或者耦合光束同时打开）。原子云的位置和在 x 和 y 两横向上的直径由 CCD1 测得。插图（i）为一凝聚态的成像图。

基于本文典型参数所确定的谐振频率，图 2a 为计算所得的探测光束透射率随其频率失谐的变化曲线。当 $|1\rangle \rightarrow |2\rangle$ 跃迁不存在失相时，量子干涉是完美的，在线中心，透射率应该是 1。图 2b 给出了探测光束的折射率与失谐的函数关系。由于我们的纳开温度样品中 $|1\rangle \rightarrow |2\rangle$ 跃迁时非常小的多普勒展宽，施加非常小的耦合强度都会导致一个宽度比 $|1\rangle \rightarrow |3\rangle$ 跃迁的自然线宽窄很多的透明度峰。相应地，色散曲线比通过其他任何技术所获得的都更陡，这也就导致了本文所述的前所未有的极低群速度。电磁脉冲传播的群速度 v_g 为 [16-19]：

$$v_g = \frac{c}{n(\omega_p) + \omega_p \dfrac{dn}{d\omega_p}} \approx \frac{\hbar c \epsilon_0}{2\omega_p} \frac{|\Omega_c|^2}{|\boldsymbol{\mu}_{13}|^2 N} \tag{1}$$

其中，$n(\omega_p)$ 为探测频率 $\omega_p (\mathrm{rad \cdot s^{-1}})$ 时的折射率，$|\Omega_c|^2$ 为耦合激光的拉比频率的平方，它随光强线性变化，$\boldsymbol{\mu}_{13}$ 为态 $|1\rangle$ 和 $|3\rangle$ 间的电偶极矩矩阵元，N 为原子数密度，ϵ_0 为自由空间的介电常数。在线中央，折射率为 1，等式（1）分母中的第二项的影响超过第一项，占主导地位。折射率图谱的一个重要特性是共振时群速度色散为 0（参见文献 16），即 $d^2 n/d\omega_p^2 = 0$，就最低阶来说，脉冲传播过程中保持波形不变。所建立的量子干涉允许脉冲传输通过在其他状态下透射系数仅为 e^{-110}（T_c 以下）的原子云，从而产生了极陡的色散图谱和极低的群速度。

我们注意到图 2 中的曲线中心与探测共振频率偏离了 0.6 MHz。这是由于耦合激光场使态 $|2\rangle$ 和 $|4\rangle$ 发生了耦合，从而能级 $|2\rangle$ 有交流斯塔克位移和与之相应的 $2 \rightarrow 3$ 跃迁的谱线位移。由于透明度峰值和等于 1 的折射率是在两光子共振时获得的，这就导致 $1 \rightarrow 3$ 跃迁共振频率下的折射率不为 1。二者之间的差别与交流斯塔克位移成比例，也因此与耦合激光强度成比例，这点对下文所述的预估非线性折射率至关重要。

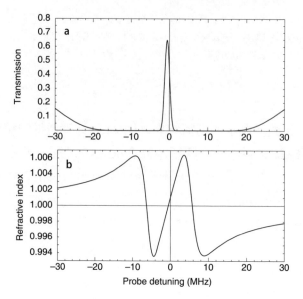

Fig. 2. Effect of probe detuning. **a**, Transmission profile. Calculated probe transmission as a function of detuning from the $|1\rangle \rightarrow |3\rangle$ resonance for an atom cloud cooled to 450 nK, with a peak density of 3.3 $\times 10^{12}$ cm^{-3} and a length of 229 μm (corresponding to the cloud in inset (ii) of Fig. 1a). The coupling laser is resonant with the $|2\rangle \rightarrow |3\rangle$ transition and has a power density of 52 mW cm^{-2}. **b**, Refractive index profile. The calculated refractive index is shown as a function of probe detuning for the same parameters as in **a**. The steepness of the slope at resonance is inversely proportional to the group velocity of transmitted light pulses and is controlled by the coupling laser intensity. Note that as a result of the a.c. Stark shift of the $|2\rangle \rightarrow |3\rangle$ transition, caused by a coupling of states $|2\rangle$ and $|4\rangle$ through the coupling laser field, the centre of the transmission and refractive index profiles is shifted by 0.6 MHz. The shift of the refractive index profile results in the nonlinear refractive index described in the text.

A diagram of the experiment is shown in Fig. 1a. The 2.5-mm-diameter coupling beam propagates along the x axis with its linear polarization parallel to the **B** field. The 0.5-mm-diameter, σ^- polarized probe beam propagates along the z axis. The size and position of the atom cloud in the transverse directions, x and y, are obtained by imaging the transmission profile of the probe beam after the cloud onto a charge-coupled-device (CCD) camera. An image of a condensate is shown as inset (i). A 55 mW cm^{-2} coupling laser beam was present during the 10-μs exposure of the atoms to a 5 mW cm^{-2} probe beam tuned close to resonance. The $f/7$ imaging optics are diffraction-limited to a resolution of 7 μm.

During the pulse delay experiments, a pinhole (placed in an external image plane of the lens system) is used to select only the part of the probe light that has passed through the central 15 μm of the atom cloud where the column density is the greatest. The outline of the pinhole is indicated with the dashed circle in inset (i).

Both coupling and probe beams are derived from the same dye laser. The frequency of the coupling beam is set by an acousto-optic modulator (AOM) to the $|2\rangle \rightarrow |3\rangle$ resonance. Here we take into account both Zeeman shifts and the a.c. Stark shift described above.

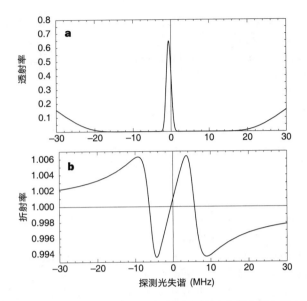

图 2. 探测光失谐的影响。**a**，透射率曲线。计算所得的探测透射率与激光频率相对原子云从 $|1\rangle \rightarrow |3\rangle$ 共振失谐的函数关系，该原子云被降温至 450 nK，峰值密度为 3.3×10^{12} cm^{-3}，长度为 229 μm（对应图 1a 中插图（ii）的原子云）。耦合激光与 $|2\rangle \rightarrow |3\rangle$ 的跃迁共振，功率密度为 52 mW·cm^{-2}。**b**，折射率曲线。计算所得的折射率与 **a** 中相同参数下探测光失谐的函数关系。达到共振时斜率的陡度与投射光脉冲的群速度成反比，并且由耦合激光强度所控制。要注意到，通过耦合激光场，态 $|2\rangle$ 和 $|4\rangle$ 产生耦合所造成的 $|2\rangle \rightarrow |3\rangle$ 跃迁的交流斯塔克位移的存在，导致透射率曲线的中心与折射率曲线的中心相差 0.6 MHz。折射率曲线的偏移引起了文中所述的非线性折射率现象。

实验图示如图 1a 所示。直径 2.5 mm 的耦合光束沿 x 轴传播，其线偏振方向平行于磁场 **B**。探测光束直径 0.5mm、σ$^-$ 偏振，沿 z 轴传播。原子云在横向 x，y 方向的尺寸和位置由探测光束通过原子云后成像在电荷耦合器件（CCD）摄像机上的透射率剖面获得。插图（i）为一凝聚态的图像。当原子在调谐至与共振频率接近的 5 mW·cm^{-2} 的探测光束中暴露 10 μs 的过程中，另一 55 mW·cm^{-2} 的耦合激光光束同时存在。f/7 成像光学受衍射限制分辨率为 7 μm。

在脉冲延迟的实验中，利用一针孔（置于透镜系统的外成像面）来只选择那些通过原子云中心 15 μm 的部分探测光束，这部分原子柱的数密度最大。针孔的轮廓在插图（i）中以虚线圆标记出。

耦合光束和探测光束都源于相同的染料激光。耦合激光器的频率由一声光调制器（AOM）设定为 $|2\rangle \rightarrow |3\rangle$ 的共振频率。这里我们将塞曼位移和上文所提到的交流斯塔克位移都考虑在内。

The corresponding probe resonance is found by measuring the transmission of the probe beam as a function of its frequency. We apply a fast frequency sweep, across 32 MHz in 50 μs, and determine resonance from the transmission peak. The sweep is controlled by a separate AOM. The frequency is then fixed at resonance, and the temporal shape of the probe pulse is generated by controlling the r.f. drive power to the AOM. The resulting pulse is approximately gaussian with a full-width at half-maximum of 2.5 μs. The peak power is 1 mW cm^{-2} corresponding to a Rabi frequency of $\Omega_p = 0.20\,A$, where the Einstein A coefficient is 6.3×10^7 rad s^{-1}. To avoid distortion of the pulse, it is made of sufficient duration that its Fourier components are contained within the transparency peak.

Probe pulses are launched along the z axis 4 μs after the coupling beam is turned on (the coupling field is left on for 100 μs). Due to the magnetic filtering discussed above, all atoms are initially in state $|1\rangle$ which is a dark state in the presence of the coupling laser only. When the pulse arrives, the atoms adiabatically evolve so that the probability amplitude of state $|2\rangle$ is equal to the ratio $\Omega_p/(\Omega_p^2 + \Omega_c^2)^{1/2}$, where Ω_p is the probe Rabi frequency. To establish the coherent superposition state, energy is transferred from the front of the probe pulse to the atoms and the coupling laser field. At the end of the pulse, the atoms adiabatically return to the original state $|1\rangle$ and the energy returns to the back of the probe pulse with no net energy and momentum transfer to the atomic cloud. Because the refractive index is unity, the electric field is unchanged as the probe pulse enters the medium. As the group velocity is decreased, the total energy density must increase so as to keep constant the power per area. This increase is represented by the energy stored in the atoms and the coupling laser field during pulse propagation through the cloud.

The pulses are recorded with a photomultiplier (3-ns response time) after they penetrate the atom clouds. The output from the photomultiplier is amplified by a 150-MHz-bandwidth amplifier and the waveforms are recorded on a digital scope. With a "flipper" mirror in front of the camera we control whether the probe beam is directed to the camera or to the photomultiplier.

The result of a pulse delay measurement is shown in Fig. 3. The front pulse is a reference pulse obtained with no atoms present. The pulse delayed by 7.05 μs was slowed down in an atom cloud with a length of 229 μm (see Fig. 1a, inset (ii)). The resulting light speed is 32.5 m s^{-1}. We used a coupling laser intensity of 12 mW cm^{-2} corresponding to a Rabi frequency of $\Omega_c = 0.56\,A$. The cloud was cooled to 450 nK (which is 15 nK above T_c), the peak density was 3.3×10^{12} cm^{-3}, and the total number of atoms was 3.8×10^6. From these numbers we calculate that the pulse transmission coefficient would be e^{-63} in the absence of the coupling laser. The probe pulse was indeed observed to be totally absorbed by the atoms when the coupling beam was left off. Inhomogeneous broadening due to spatially varying Zeeman shifts is negligible (~20 kHz) for the low temperatures and correspondingly small cloud sizes used here.

　　相应的探测共振频率由测量探测光束的透射率与频率的函数关系得出。我们运用一快速频率扫描，在 50 μs 内扫描 32 MHz，通过透射峰值来确定共振。扫描受另一 AOM 控制。然后将频率固定在共振频率，探测脉冲的时域波形就由控制 AOM 的无线电频率驱动来产生。所得脉冲是一个近高斯分布，半高宽为 2.5 μs。拉比频率 $\Omega_p = 0.20A$ 时的峰值功率为 1 mW·cm^{-2}。这里取爱因斯坦自激系数 A 为 6.3×10^7 rad·s^{-1}。为避免脉冲失真，我们给予其足够的持续时间来使其傅里叶分量包含在透明度峰之中。

　　在耦合光束打开 4 μs 之后探测脉冲沿 z 轴发射（耦合场保持打开 100 μs）。由于上文所讨论的磁过滤，所有的原子最初都处于 $|1\rangle$ 态，只有耦合激光存在时，这是一个暗态。当接受到光脉冲时，原子绝热演化，导致 $|2\rangle$ 态的概率幅与 $\Omega_p/(\Omega_p^2 + \Omega_c^2)^{1/2}$ 相等，其中 Ω_p 是探测光拉比频率。为了实现相干叠加态，能量从探测光脉冲的前沿传递到原子和耦合激光场。在探测光脉冲末端，原子绝热地回到原始态 $|1\rangle$，能量返回到探测脉冲后端，但没有净能量和动量转移到原子云。由于折射率等于 1，在探测脉冲进入介质时电场不变。随着群速度降低，总能量密度必须提高才能保证单位面积功率不变。这种提高就体现在脉冲在原子云中传播时，能量在原子以及耦合激光场的储存。

　　这些脉冲在穿过原子云后被一光电倍增管（响应时间为 3 ns）所记录。光电倍增管的输出信号被 150 MHz 带宽的放大器放大，波形显示在一数字示波器上。在摄像机前放置一"转向"镜，我们就可以控制探测光束是射入摄像机还是射入光电倍增管中。

　　探测光脉冲延迟测量的结果如图 3 所示。前面的脉冲是没有原子存在时的参考脉冲。脉冲穿过长度为 229 μm 的原子云时被减慢，延迟了 7.05 μs（见图 1a，插图 (ii)）。由此可得光速为 32.5 m·s^{-1}。实验中我们使用的耦合激光强度为 12 mW·cm^{-2}，对应拉比频率为 $\Omega_c = 0.56A$。原子云被冷却至 450 nK（临界温度 T_c 以上 15 nK）。峰值密度为 3.3×10^{12} cm^{-3}，原子总数为 3.8×10^6。根据这些数据我们可以计算得出在没有耦合激光时，脉冲透射系数为 e^{-63}。实验中确实观察到当耦合激光关闭时，探测脉冲完全被原子所吸收。本实验在低温下进行且原子云尺寸相对较小，由空间变化的塞曼位移引起的非均匀展宽（约 20 kHz）可以忽略。

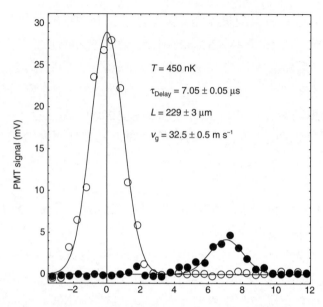

Fig. 3. Pulse delay measurement. The front pulse (open circles) is a reference pulse with no atoms in the system. The other pulse (filled circles) is delayed by 7.05 μs in a 229-μm-long atom cloud (see inset (ii) in Fig. 1a). The corresponding light speed is 32.5 m s⁻¹. The curves represent gaussian fits to the measured pulses.

The size of the atom cloud in the z direction is obtained with another CCD camera. For this purpose, we use a separate 1 mW cm^{-2} laser beam propagating along the vertical y axis and tuned 20 MHz below the $F = 2 \rightarrow 3$ transition. The atoms are pumped to the $F = 2$ ground state for 10 μs before the imaging which is performed with an exposure time of 10 μs. We image the transmission profile of the laser beam after the atom cloud with diffraction-limited $f/5$ optics. An example is shown in Fig. 1a, inset (ii), where the asymmetry of the trap is clear from the cloud's elliptical profile. We note that the imaging laser is never applied at the same time as the coupling laser and probe pulse, and for each recorded pulse or CCD picture a new cloud is loaded.

We measured a series of pulse delays and corresponding cloud sizes for atoms cooled to temperatures between 2.5 μK and 50 nK. From these pairs of numbers we obtain the corresponding propagation velocities (Fig. 4). The open circles are for a coupling power of 52 mW cm^{-2} ($\Omega_c = 1.2\ A$). The light speed is inversely proportional to the atom density (equation (1)) which increases with lower temperatures, with an additional density increase when a condensate is formed. The filled circles are for a coupling power of 12 mW cm^{-2}. The lower coupling power is seen to cause a decrease of group velocities in agreement with equation (1). We obtain a light speed of 17 m s^{-1} for pulse propagation in an atom cloud initially prepared as an almost pure Bose–Einstein condensate (condensate fraction is ⩾ 90%). Whether the cloud remains a condensate during and after pulse propagation is an issue that is beyond the scope of this Letter.

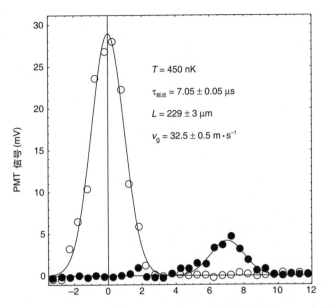

图 3. 脉冲延迟测量。前面的脉冲（空心圆）为系统中没有原子的参考脉冲。另一个脉冲（实心圆）在 229 μm 长的原子云中被延迟 7.05 μs（见图 1a 插图（ii））。相应的光速为 32.5 m·s⁻¹。曲线为测量脉冲的高斯拟合。

沿 z 方向的原子云尺寸通过另一 CCD 摄像机捕获。为了达到这个目的，我们运用了另一独立的 1 mW·cm⁻² 的激光束沿垂直的 y 轴传播，并调谐到比 $F = 2 \to 3$ 跃迁低 20 MHz。原子在 10 μs 的曝光成像之前，先被泵到 $F = 2$ 的基态 10 μs。我们使用衍射极限 $f/5$ 的光学器件在原子云之后对激光束的纵向透射率剖面进行成像。一个例子如图 1a 插图（ii）所示，其中由云的椭圆轮廓可清楚看见磁阱的不对称性。值得注意的是，成像激光不会与耦合激光及探测脉冲同时施加，并且对于每个记录的脉冲或 CCD 图片，都加载一新的原子云。

我们在原子冷却到温度 2.5 μK 和 50 μK 之间测量了一系列脉冲延迟和相应的云尺寸。从这些数据对中，我们获得了相应的传播速度（图 4）。空心圆对应耦合光强 52 mW·cm⁻²（$\Omega_c = 1.2A$）。光速与原子数密度成反比（等式（1）），而原子数密度随温度降低而增大，当凝聚态形成时原子数密度进一步增加。实心圆对应耦合光强 12 mW·cm⁻²。与等式（1）一致，可以看到耦合光强的降低引起群速度的减小。在最初制备的几乎纯净的玻色－爱因斯坦凝聚态（凝聚比率大于等于 90%）原子云中，我们获得了 17 m·s⁻¹ 的光脉冲传播速度。在脉冲传播过程中和传播后原子云是否保持凝聚不在本文讨论范围之内。

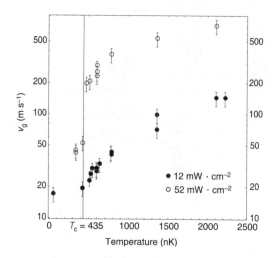

Fig. 4. Light speed versus atom cloud temperature. The speed decreases with temperature due to the atom density increase. The open circles are for a coupling power of 52 mW cm^{-2} and the filled circles are for a coupling power of 12 mW cm^{-2}. The temperature T_c marks the transition temperature for Bose–Einstein condensation. The decrease in group velocity below T_c is due to a density increase of the atom cloud when the condensate is formed. From imaging measurements we obtain a maximum atom density of 8×10^{13} cm^{-3} at a temperature of 200 nK. Here, the dense condensate component constitutes 60% of all atoms, and the total atom density is 16 times larger than the density of a non-condensed cloud at T_c. The light speed measurement at 50 nK is for a cloud with a condensate fraction $\geqslant 90\%$. The finite dephasing rate due to state $|4\rangle$ does not allow pulse penetration of the most dense clouds. This problem could be overcome by tuning the laser to the D_1 line as described in the text.

Transitions from state $|2\rangle$ to state $|4\rangle$, induced by the coupling laser (detuned by 60 MHz from this transition), result in a finite decay rate of the established coherence between states $|1\rangle$ and $|2\rangle$ and limit pulse transmission. The dephasing rate is proportional to the power density of the coupling laser and we expect, and find, that probe pulses have a peak transmission that is independent of coupling intensity and a velocity which reduces linearly with this intensity. The dephasing time is determined from the slope of a semi-log plot of transmission versus pulse delay[19]. At a coupling power of 12 mW cm^{-2}, we measured a dephasing time of 9 µs for atom clouds just above T_c.

Giant Kerr nonlinearities are of interest for areas of quantum optics such as optical squeezing, quantum nondemolition, and studies of nonlocality. It was recently proposed that they may be obtained using electromagnetically induced transparency[20]. Here we report the first (to our knowledge) measurement of such a nonlinearity. The refractive index for zero probe detuning is given by $n = 1 + (n_2 I_c)$ where I_c is the coupling laser intensity, and n_2 the cross phase nonlinear refractive index. As seen from Fig. 2b, the nonlinear term $(n_2 I_c)$ equals the product of the slope of the refractive index at probe resonance and the a.c. Stark shift of the $|2\rangle \rightarrow |3\rangle$ transition caused by the coupling laser. We can then express n_2 by the formula (see equation (1));

$$n_2 = \frac{\Delta\omega_S}{I_c} \frac{dn}{d\omega_p} \approx \frac{1}{2\pi} \frac{\Delta\omega_S}{I_c} \frac{\lambda}{v_g} \tag{2}$$

where $\Delta\omega_S$ is the a.c. Stark shift, proportional to I_c, and λ the wavelength of the probe

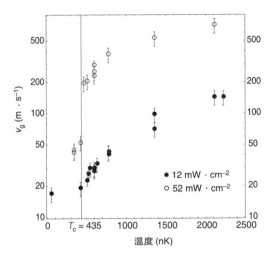

图 4. 光速与原子云温度的关系。由于原子数密度的增大，光速随温度的降低而降低。空心圆对应耦合光强 52 mW·cm⁻²，实心圆对应耦合光强 12 mW·cm⁻²。温度 T_c 标记出了玻色–爱因斯坦凝聚的临界转变温度。T_c 以下的群速度减小是由形成凝聚态时原子云密度增大而引起。通过成像测量，我们得到 200 nK 时最大的原子数密度为 8×10^{13} cm⁻³。密集凝聚态成分占所有原子的 60%，并且总原子数密度是 T_c 时非凝聚态云密度的 16 倍。50 nK 时的光速测量是在凝聚分数 $\geqslant 90\%$ 的云中进行的。由于态 $|4\rangle$ 造成的有限的退相率不允许脉冲穿过最密集的云。这个问题可以通过文中所述的调谐激光至 D_1 线的方法来解决。

由耦合激光激发的 $|2\rangle \rightarrow |4\rangle$ 态的跃迁（耦合激光与该跃迁之间存在 60 MHz 失谐），使得所建立的 $|1\rangle$ 和 $|2\rangle$ 态的相干具有一个有限衰变率，并限制了探测光脉冲的透射。退相率与耦合激光的功率密度成比例，我们预期并发现，探测脉冲具有一个透射率峰值，该值不受耦合强度以及随此强度线性减小的速度的影响。退相时间由透射率–脉冲延迟的半对数曲线的斜率决定[19]。在耦合光强为 12 mW·cm⁻² 时，对于温度刚刚超过 T_c 的原子云，我们测量到退相时间为 9 μs。

巨克尔非线性在量子光学领域具有很大的吸引力，例如光学压缩、量子非破坏性以及非定域性等方面的研究。近期的研究还指出，这些现象很可能通过电磁感应透明来获得[20]。本文报道了（据我们所知）首次对这样一种非线性的测量。零失谐探测光的折射率表达式为 $n = 1 + (n_2 I_c)$，其中 I_c 为耦合激光强度，n_2 为交叉相位非线性折射率。正如图 2b 所示，非线性项 $n_2 I_c$ 等于探测光在共振时的折射率曲线斜率与耦合激光激发的 $|2\rangle \rightarrow |3\rangle$ 跃迁的交流斯塔克位移之积。因此 n_2 可以用以下公式表式（见等式（1））：

$$n_2 = \frac{\Delta\omega_S}{I_c} \frac{dn}{d\omega_p} \approx \frac{1}{2\pi} \frac{\Delta\omega_S}{I_c} \frac{\lambda}{v_g} \tag{2}$$

其中 $\Delta\omega_S$ 为交流斯塔克位移，与 I_c 成比例，λ 为探测跃迁的光波长。当耦合激光强

transition. We measured an a.c. Stark shift of 1.3×10^6 rad s^{-1} for a coupling laser intensity of 40 mW cm^{-2}. For a measured group velocity of 17 m s^{-1} (Fig. 4), we obtain a nonlinear refractive index of 0.18 cm^2 W^{-1}. This nonlinear index is $\sim 10^6$ times greater than that measured in cold Cs atoms[21].

With a system that avoids the $|1\rangle - |2\rangle$ dephasing rate described above (which can be obtained by tuning to the D_1 line in sodium), the method used here could be developed to yield the collision-induced dephasing rate of the double condensate which is generated in the process of establishing electromagnetically induced transparency (see also refs 22, 23). In that case, the square of the probability amplitude for state $|3\rangle$ could be kept below 10^{-5} during pulse propagation, with no heating of the condensate as a result. With improved frequency stability of our set-up and lower coupling intensities, even lower light speeds would be possible, perhaps of the order of centimetres per second, comparable to the speed of sound in a Bose–Einstein condensate. Under these conditions we expect phonon excitation during light pulse propagation through the condensate. By deliberately tuning another laser beam to the $|2\rangle \rightarrow |4\rangle$ transition, it should be possible to demonstrate optical switching at the single photon level[24]. Finally, we note that during propagation of the atom clouds, light pulses are compressed in the z direction by a ratio of c/v_g. For our experimental parameters, that results in pulses with a spatial extent of only 43 μm.

(**397**, 594-598; 1999)

Lene Vestergaard Hau[*†], **S. E. Harris**[‡], **Zachary Dutton**[*†] & **Cyrus H. Behroozi**[*§]

[*] Rowland Institute for Science, 100 Edwin H. Land Boulevard, Cambridge, Massachusetts 02142, USA

[†] Department of Physics, [§] Division of Engineering and Applied Sciences, Harvard University, Cambridge, Massachusetts 02138, USA

[‡] Edward L. Ginzton Laboratory, Stanford University, Stanford, California 94305, USA

Received 3 November; accepted 21 December 1998.

References:

1. Knight, P. L., Stoicheff, B. & Walls, D. (eds) Highlights in quantum optics. *Phil. Trans. R. Soc. Lond. A* **355**, 2215-2416 (1997).

2. Harris, S. E. Electromagnetically induced transparency. *Phys. Today* **50(7)**, 36-42 (1997).

3. Scully, M. O. & Zubairy, M. S. *Quantum Optics* (Cambridge Univ. Press, 1997).

4. Arimondo, E. in *Progress in Optics* (ed. Wolf, E.) 257-354 (Elsevier Science, Amsterdam, 1996).

5. Bergmann, K., Theuer, H. & Shore, B. W. Coherent population transfer among quantum states of atoms and molecules. *Rev. Mod. Phys.* **70**, 1003-1006 (1998).

6. Chu, S. The manipulation of neutral particles. *Rev. Mod. Phys.* **70**, 685-706 (1998).

7. Cohen-Tannoudjii, C. N. Manipulating atoms with photons. *Rev. Mod. Phys.* **70**, 707-719 (1998).

8. Phillips, W. D. Laser cooling and trapping of neutral atoms. *Rev. Mod. Phys.* **70**, 721-741 (1998).

9. Hess, H. F. Evaporative cooling of magnetically trapped and compressed spin-polarized hydrogen. *Phys. Rev. B* **34**, 3476-3479 (1986).

10. Masuhara, N. *et al.* Evaporative cooling of spin-polarized atomic hydrogen. *Phys. Rev. Lett.* **61**, 935-938 (1988).

11. Anderson, M. H., Ensher, J. R., Matthews, M. R., Wieman, C. E. & Cornell, E. A. Observation of Bose-Einstein condensation in a dilute atomic vapor. *Science* **269**, 198-201 (1995).

12. Davis, K. B. *et al.* Bose-Einstein condensation in a gas of sodium atoms. *Phys. Rev. Lett.* **75**, 3969-3973 (1995).

13. Bradley, C. C., Sackett, C. A. & Hulet, R. G. Bose-Einstein condensation of lithium: observation of limited condensate number. *Phys. Rev. Lett.* **78**, 985-989 (1997).

14. Hau, L. V. *et al.* Near-resonant spatial images of confined Bose-Einstein condensates in a 4-Dee magnetic bottle. *Phys. Rev. A* **58**, R54-R57 (1998).

15. Hau, L. V., Golovchenko, J. A. & Burns, M. M. A new atomic beam source: The "candlestick". *Rev. Sci. Instrum.* **65**, 3746-3750 (1994).

16. Harris, S. E., Field, J. E. & Kasapi, A. Dispersive properties of electromagnetically induced transparency. *Phys. Rev. A* **46**, R29-R32 (1992).

度为 40 mW · cm^{-2} 时我们测量到一 1.3 × 10^6 rad · s^{-1} 的交流斯塔克位移。而对于测量所得的 17 m · s^{-1} 的群速度（图 4），可得非线性折射率为 0.18 cm^2 · W^{-1}。该非线性折射率是在冷铯原子中相应值的 10^6 倍[21]。

应用上文所述的可避免 |1⟩ − |2⟩ 退相率的系统（通过调谐至钠 D$_1$ 线而获得），可以对这里所使用的方法进一步发展以得出在建立电磁感应透明过程中发生的双凝聚态碰撞诱导失相率（参见文献 22 和 23）。在这种情况下，态 |3⟩ 概率幅的平方在脉冲传播过程中可以保持在 10^{-5} 以下，因此不存在凝聚态加热。由于我们装置频率稳定性的提高以及耦合强度的减小，更低的光速也是可能实现的，可能是几厘米每秒的量级，可与玻色−爱因斯坦凝聚中的音速相比。在此条件下我们预期光脉冲传播通过凝聚体时有声子被激发。通过谨慎调谐另一激光束至 |2⟩→|4⟩ 跃迁，应该有可能在单光子水平上演示光学开关[24]。最后本文指出，光束在原子云传播过程中，光脉冲在 z 方向被压缩，比率为 c/v_g。对于本实验所采用的参数，这导致脉冲的空间范围仅为 43 μm。

（崔宁 翻译；石锦卫 审稿）

17. Grobe, R., Hioe, F. T. & Eberly, J. H. Formation of shape-preserving pulses in a nonlinear adiabatically integrable system. *Phys. Rev. Lett.* **73,** 3183-3186 (1994).

18. Xiao, M., Li, Y.-Q., Jin, S.-Z. & Gea-Banacloche, J. Measurement of dispersive properties of electromagnetically induced transparency in rubidium atoms. *Phys. Rev. Lett.* **74,** 666-669 (1995).

19. Kasapi, A., Jain, M., Yin, G. Y. & Harris, S. E. Electromagnetically induced transparency: propagation dynamics. *Phys. Rev. Lett.* **74,** 2447-2450 (1995).

20. Schmidt, H. & Imamoglu, A. Giant Kerr nonlinearities obtained by electromagnetically induced transparency. *Opt. Lett.* **21,** 1936-1938 (1996).

21. Lambrecht, A., Courty, J. M., Reynaud, S. & Giacobino, E. Cold atoms: A new medium for quantum optics. *Appl. Phys. B* **60,** 129-134 (1995).

22. Hall, D. S., Matthews, M. R., Wieman, C. E. & Cornell, E. A. Measurements of relative phase in two-component Bose-Einstein condensates. *Phys. Rev. Lett.* **81,** 1543-1546 (1998).

23. Ruostekoski, J. & Walls, D. F. Coherent population trapping of Bose-Einstein condensates: detection of phase diffusion. *Eur. Phys. J. D* (submitted).

24. Harris, S. E. & Yamamoto, Y. Photon switching by quantum interference. *Phys. Rev. Lett.* **81,** 3611-3614 (1998).

Acknowledgements. We thank J. A. Golovchenko for discussions and C. Liu for experimental assistance. L.V.H. acknowledges support from the Rowland Institute for Science. S.E.H. is supported by the US Air Force Office of Scientific Research, the US Army Research Office, and the US Office of Naval Research. C.H.B. is supported by an NSF fellowship.

Correspondence and requests for materials should be addressed to L.V.H. (e-mail: hau@rowland.org).

Anticipation of Moving Stimuli by the Retina

M. J. Berry II *et al.*

Editor's Note

Our senses seem to show us the world as it is. But what we see is not just the imprint of photons hitting the retina: it is produced as our neural systems filter and highlight this incoming information. Thus it takes time to generate even the simplest neural message, creating the risk that the real world has changed by the time we perceive it—a risk that crucially applies to fast-moving objects such as predators or prey. But as this paper by Michael J. Berry at Harvard University and coworkers shows, our neural systems can predict where an object in our visual field is likely to be, giving us a better chance, say, of evading or catching it.

A flash of light evokes neural activity in the brain with a delay of 30–100 milliseconds[1], much of which is due to the slow process of visual transduction in photoreceptors[2,3]. A moving object can cover a considerable distance in this time, and should therefore be seen noticeably behind its actual location. As this conflicts with everyday experience, it has been suggested that the visual cortex uses the delayed visual data from the eye to extrapolate the trajectory of a moving object, so that it is perceived at its actual location[4-7]. Here we report that such anticipation of moving stimuli begins in the retina. A moving bar elicits a moving wave of spiking activity in the population of retinal ganglion cells. Rather than lagging behind the visual image, the population activity travels near the leading edge of the moving bar. This response is observed over a wide range of speeds and apparently compensates for the visual response latency. We show how this anticipation follows from known mechanisms of retinal processing.

BECAUSE a moving object often follows a smooth trajectory, one can extrapolate from its past position and velocity to obtain an estimate of its current location. Recent experiments on motion perception[5-7] indicate that the human brain possesses just such a mechanism: Subjects were shown a moving bar sweeping at constant velocity; a second bar was flashed briefly in alignment with the moving bar. When asked what they perceived at the time of the flash, observers reliably reported seeing the flashed bar trailing behind the moving bar. This flash lag effect has been confirmed repeatedly[8-10], and various high-level processes have been invoked to explain it, such as a time delay due to the shift of visual attention. To assess whether processing in the retina contributes to this effect we analysed the "neural image" of these two stimuli at the retinal output. We recorded simultaneously the spike trains of many ganglion cells in the isolated retina of tiger salamander or rabbit. The responses to flashed and moving bars were then analysed by plotting the firing rate in the retinal ganglion-cell population as a function of space and time.

视网膜可对运动物体的刺激进行预测

贝里等

编者按

我们的感官似乎将世界原原本本地呈现给我们。其实我们所看到的世界不仅仅是投射在视网膜上的光子印记：它是输入信息经由神经系统过滤并筛选产生的。因此即使生成最简单的神经信号都需要耗费时间，这就产生以下风险，真实的世界在我们感知到它的时候就已经改变了——这对快速移动的物体诸如捕食者和猎物都是至关重要的。但是如哈佛大学迈克尔·贝里及其同事在文章中所指出的，我们的神经系统能够预测视野中物体的可能位置，从而给我们提供更好的机会去逃避追捕或者抓住猎物。

闪光所引发的脑神经活动会有一个30~100毫秒的延迟 [1]，而延迟的大部分原因在于光感受器中迟缓的视觉传导过程 [2,3]。在这段时间里一个运动的物体会移动相当远的距离，因此当物体被看到时，已经显著落后于其实际位置。这与我们的日常体验相矛盾，有一种解释是视觉大脑皮层利用从眼睛中获取滞后的视觉数据来推测运动物体的轨迹，从而察觉其真正的位置 [4-7]。本文报道这种对运动刺激的预测始发于视网膜。移动光柱会在一群视网膜神经节细胞诱发一个随之移动的放电。这种总体的活动紧跟着移动光柱的最前缘行进，而不是滞后于视觉的影像。这种反应在很宽的速度范围内均能观察到，很明显这是在补偿视觉反应的潜伏期。我们在本文揭示这种预测是如何遵循已知视觉处理机制的。

由于运动的物体一般沿着平滑的轨迹移动，人们可以根据其过去的位置和速度推测其目前的大体位置。最近运动知觉的实验表明 [5-7]，人类大脑拥有这样一种机制：一个移动光柱匀速从受试者前掠过；第二个光柱与这个移动光柱重合，但短暂闪烁。当询问受试者在闪烁出现时觉察到什么时，他们肯定地回答看到一个闪烁的光柱尾随在移动光柱后面。这种闪烁滞后效应已经被多次证实 [8-10]，许多高级活动被援引用来解释这种现象，例如视觉注意转换引起时间的滞后。为了确定视网膜的活动是否参与了这种效应，我们分析了这两种刺激在视网膜输出中所产生的"神经影像"。我们同时记录了从虎蝾或兔分离出来的视网膜神经节细胞所产生的放电序列。通过把群体放电频率绘制为空间和时间的函数的方法，来分析视网膜神经节细胞对闪烁和移动光柱的反应。

Figure 1 illustrates the responses of individual OFF-type ganglion cells to a dark bar flashed briefly over the receptive-field centre. In both salamander (Fig. 1a) and rabbit (Fig. 1b), the cells remained silent for a latency of ~50 ms, then fired a burst of spikes that lasted another 50 ms. When the bar was swept over the retina at constant speed (Fig. 1c, d), these same cells fired for a more extended period, beginning some time before the bar reached the position at which the flash occurred, and extending for a shorter time thereafter. When the bar was swept in the opposite direction (Fig. 1e, f), it produced a very similar response, showing that these cells had no direction-selective preference.

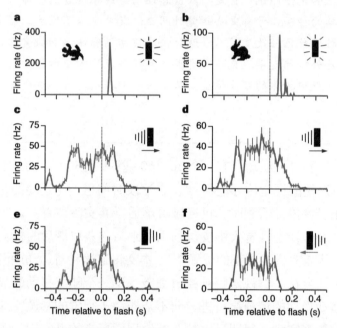

Fig. 1. Responses of two ganglion cells to flashed and moving bars. **a**, **c**, **e** Results from a "fast OFF" ganglion cell in salamander retina; **b**, **d**, **f**, results from a brisk-sustained OFF cell in rabbit retina. **a**, **b**, Firing rate as a function of time after a dark bar (90% contrast,133 μm width) was flashed for 15 ms on the receptive-field centre. **c**, **d**, Firing rate of the same two cells while the dark bar moved continuously across the retina at 0.44 mm s[-1]. At time zero, the bar was aligned with the position of the flash in **a** and **b**. **e**, **f**, As in **c**, **d**, but with the bar moving in the opposite direction. Error bars denote standard error across repeated presentations of the stimulus.

From many single-unit measurements such as these, we compiled the neural image of the visual stimulus in the population of ganglion cells. This was done by plotting the firing rate of every cell as a function of distance from the bar stimulus and interpolating these points with a smooth line (see Methods). The neural image of the flashed bar among salamander "fast OFF" cells is shown in Fig. 2a. After a latency of 40 ms, a hump of neural activity appears that increases rapidly to a peak at 60 ms, then declines and disappears at 100 ms. As might be expected, the profile is centred on the bar. It has a width on the retina of ~200 μm at half-maximum, close to the size of the receptive-field centre for these neurons[11]. The width increases somewhat during the late phase of the response, creating the impression of an outward "splash"[12].

图 1 显示了单个的 OFF 型神经节细胞对暗光柱在感受域中央短暂闪现的反应。虎螈（图 1a）和兔（图 1b）的细胞均保持大约 50 ms 的静息，然后爆发一段持续 50 ms 的放电。当光柱匀速掠过视网膜时（图 1c 和 1d），同样一群细胞放电持续时间较长，开始于光柱到达闪烁位置之前的某个时刻，然后仍持续较短的一段时间。当光柱以相反的方向扫过时（图 1e 和 1f），会产生非常相似的反应，这表明这些细胞没有方向选择的偏好性。

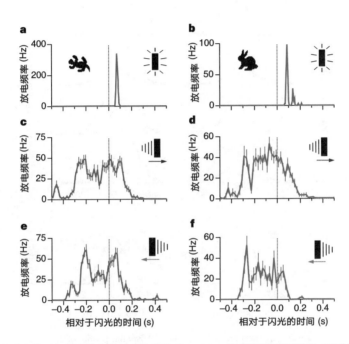

图 1. 两个神经节细胞对闪烁光柱和移动光柱的反应。**a, c, e,** 从虎螈视网膜"快速 OFF"神经节细胞中得来的结果；**b, d, f,** 从兔视网膜"持续激活 OFF"细胞得来的结果。**a, b,** 暗光柱（对比度 90%，宽度 133 μm）在感受域中央闪现时间为 15 ms，放电频率记为时间的函数。**c, d,** 当暗光柱以同样的速度 0.44 mm·s⁻¹ 持续不断地经过视网膜时，两个细胞的放电频率。在时间为 0 时，**a** 和 **b** 中移动光柱与闪烁光柱的位置相对应。**e** 和 **f** 与 **c** 和 **d** 一样，但是移动光柱向相反的方向移动。误差线表示刺激重复表现出来的标准偏差。

从许多像这样的单个细胞测量的结果中，我们记录到对一群神经节细胞进行视觉刺激而产生的神经影像数据。我们采用以下方法：将每个细胞的放电频率记为移动光柱距离的函数，并将这些点用平滑的曲线连接（参见实验方法）。图 2a 显示在虎螈"快速 OFF"细胞中闪烁光柱的神经影像。潜伏 40 ms 后，神经活动的高峰期开始出现，在 60 ms 时快速增长到峰值，然后减弱，在 100 ms 时消失。正如预测的那样，这些变化以移动光柱为中心。在峰值一半时，其在视网膜的宽度大约是 200 μm，这个数值接近于这些神经元感受域中央的大小 [11]。这个宽度在反应后期有所增加，给人一种向外"扩散"的感觉 [12]。

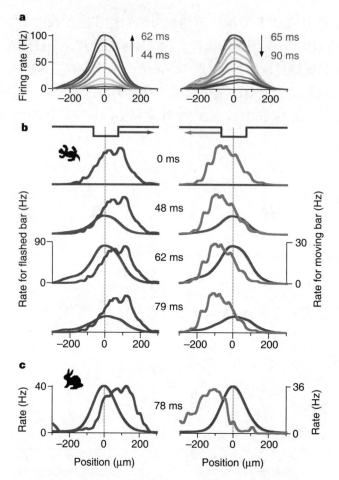

Fig. 2. Population response to flashed and moving bars. **a**, Spatial profile of firing in the population of salamander fast OFF ganglion cells in response to a flashed dark bar (90% contrast, 133 μm width, see stimulus trace in **b**) at a series of times after the flash (colour scale, 3 ms steps). **b**, Profile of the population response at four time points following a flashed bar (red, from **a**), and the same bar travelling at 0.44 mm s⁻¹ rightward (blue) and leftward (green). At time 0 ms, the moving bars were aligned with the position of the flash; at 62 ms, the flash response was maximal. Curves in **a** and **b** derived from 15 cells. **c**, As in **b**, for the population of brisk-sustained OFF cells in rabbit retina. At 78 ms, the flash response was maximal. Curves derived from six cells.

The neural image of the moving bar is shown in Fig. 2b. Again, a hump of firing activity is observed, which now sweeps over the retina along with the moving bar. If this response were subject to the same time delay as the flash, one would expect the neural image to trail behind the visual image of the bar. Instead, the hump of firing activity is clearly ahead of the centre of the bar, and the peak firing rate seems to occur near the bar's leading edge. The same response occurs when the bar moves in the opposite direction. By superposing on this the response to a flash that was aligned with the moving bar (from Fig. 2a), we find that the two neural images are clearly separated—at the time when the response to the flash peaks, the response of the moving bar is displaced ~100 μm ahead in the direction of motion. A very similar displacement between the two neural images was observed among

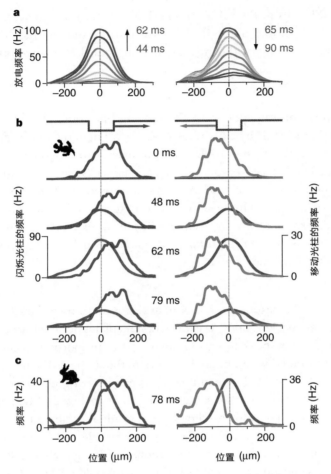

图 2. 对闪烁光柱和移动光柱的群体响应。**a**，虎螈快速 OFF 神经节细胞群体对闪烁暗光柱（对比度 90%，宽度 133 μm，同 **b** 的刺激一样）在放电之后一系列时间点上的放电空间特性（色标，3 ms 时间间距）。**b**，闪烁光柱后四个时间点上的群体反应特性（红色曲线数据来自于 **a**），同一移动光柱以 0.44 mm·s⁻¹ 速度向右移动（蓝）和向左移动（绿）。在 0 ms 时，移动光柱与闪烁光柱的位置对齐；在 62 ms 时，对闪烁光柱的反应达到最大化。**a** 和 **b** 的曲线数据来源于 15 个细胞。**c** 和 **b** 一样，是兔视网膜持续激活 OFF 细胞群体的结果。在 78 ms 时反应最大。曲线数据采集于 6 个细胞。

  图 2b 所示为移动光柱的神经影像。我们再次观察到驼峰状的放电活动，且与移动光柱掠过视网膜的过程一致。如果这种反应与闪烁光柱滞后时间相同，人们应该会看到神经影像尾随在移动光柱的视觉影像之后。相反，放电活动的峰值明显位于移动光柱中心之前，其放电频率的最高点似乎出现在靠近移动光柱最前缘的地方。当移动光柱向相反方向移动时会发生相同的反应。通过将这种反应与一个闪烁光柱（该光柱与移动光柱对齐）所引发的反应进行重叠（图 2a），我们发现两个神经影像完全分开——在对闪烁光柱的反应达峰值时，移动光柱的反应沿着运动方向向前移动大约 100 μm。在兔视网膜持续激活 OFF 细胞中（图 2c）也观察到类似的两个神

brisk-sustained OFF cells in the rabbit retina (Fig. 2c). If subsequent stages of the visual system estimate the location of the flashed bar and the moving bar by the position of these humps of neural activity, they must conclude that the moving bar is ahead of the flashed bar.

How does this apparent anticipation of the moving bar come about? One suspects that cells ahead of the bar start firing early (Figs 1c–f, 2b, c) when the bar begins to invade their receptive-field centre. The firing profile does not extend to an equal distance behind the trailing edge of the bar, perhaps because these ganglion cells have transient responses: They fire while stimulation increases as the bar invades the receptive field, not while stimulation decreases as the bar leaves it. However, we found that spatial and temporal filtering by the ganglion cell's receptive field was by itself insufficient to explain the response profiles (see below). Instead, there is another important component, which was revealed in experiments varying the intensity of the moving bar.

Figure 3a illustrates the response of this neural population to dark bars of increasing contrast relative to the background. As expected, bars of higher contrast produced stronger modulations in firing. The peak firing rate increased in proportion to contrast at first, but then appeared to saturate (Fig. 3a, inset). In addition, the shape of the neural image changed significantly with contrast. At low contrast, the peak in firing occurred behind the bar's leading edge. At high contrast—the same condition as in Fig. 2—the peak of the profile was ahead of the leading edge, followed by a more gradual decline in firing. The saturation of the peak firing rate and the shift in the response profile can be explained if the high-contrast stimulus somehow desensitizes the response of the ganglion cell after a short time delay. In that case, a ganglion cell just ahead of the bar should be strongly excited as the edge begins to enter its receptive-field centre, but then its response gain gets reduced and the firing rate declines even before the edge is half-way across. A well-known component of retinal processing that fits this description is the "contrast-gain control"[13-15].

Following ref. 16, we incorporated this aspect into a quantitative description of a ganglion cell's light response (Fig. 4). In this scenario, the retina integrates the light stimulus over space and time, with a weighting function $k(x,t)$ given by the ganglion cell's receptive field, and the resulting signal determines the neuron's firing rate. If the stimulus provides strong excitation for an extended period of time, a negative feedback loop reduces the gain at the input and consequently the response to subsequent stimulation[17]. With just four free parameters, this model produced a satisfying account of neural responses throughout the entire contrast series (Fig. 3a). It indicates that the retinal gain is modulated as much as fourfold during passage of the high-contrast bar (Fig. 3b), which pushes the response profile towards the leading edge of the bar. Without contrast-gain control the predicted profile always lagged significantly behind (Fig. 3a).

经影像之间的位置偏差。如果视觉系统的后续阶段通过这些神经活动峰值出现的位置来估计闪烁光柱束和移动光柱的位置，他们的结论一定是移动光柱位于闪烁光柱之前。

这种对移动光柱明显的预感是如何产生的呢？有人猜测当移动光柱开始进入神经节细胞的感受域中心时，在移动光柱之前的细胞提早放电（图 1c ~ f、2b 和 2c）。这种放电特性在移动光柱的后缘并未进行等距离的延长，可能因为这些神经节细胞具有瞬态响应：神经节细胞在移动光柱进入感受域、刺激增加时放电；而在移动光柱离开、感受域刺激降低时不放电。但是，我们发现，通过神经节细胞感受域对空间和时间的过滤本身不足以解释反应的特性（见下文）。相反，还有另外一种重要的组分，是通过变化移动光柱的强度发现的。

图 3a 阐明了神经细胞群体对相对背景的对比度逐渐增强的暗光柱的反应。正如所料，具有较高对比度的移动光柱对放电具有更强的调节能力。起初放电的峰值频率与对比度成正比，但是随后出现饱和（图 3a，插入图）。另外，神经影像的形状随对比度的变化而显著变化。在对比度低时，放电的峰值出现在移动光柱前缘之后。当对比度高时——与图 2 条件相同——放电的峰值在移动光柱前缘之前，随后放电逐渐减弱。如果高对比度刺激在短时间之后，从某种程度上降低了神经节细胞反应的敏感性，那么放电频率峰值的饱和与放电特性之间的偏差就可以解释了。在这种情况下，当移动光柱边缘开始进入其感受域的中心时，位于移动光柱之前的神经节细胞应该非常兴奋，但是随后其反应的增益变弱，其放电频率甚至在移动光柱边缘仅通过一半时就开始衰减。符合这种描述的一个著名的视网膜信息处理过程叫做"对比度增益控制"[13-15]。

按照参考文献 16 的方法，我们把这方面纳入神经节细胞对光反应的定量描述中（图 4）。在本情形中，视网膜将空间和时间上的光刺激与神经节细胞感受域得出的权重函数 $k(x,t)$ 相结合，最终信号决定了神经元的放电频率。如果光刺激引发的强烈兴奋超过特定的时间，负反馈环在输入端减少增益，进而降低对随后刺激的响应[17]。仅用这四个独立的参数，此模型会很好地解释贯穿整个对比系列中的神经反应（图 3a）。结果显示在高对比度移动光柱经过时视网膜增益可被调节到 4 倍（图 3b），从而将反应的曲线推向移动光柱的前缘。没有对比度增益控制，预测的曲线总是明显滞后（图 3a）。

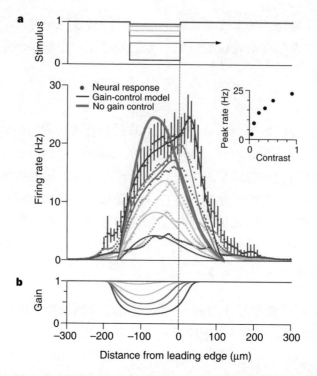

Fig. 3. Dependence of motion extrapolation on contrast. **a**, Stimulation with moving dark bars (133 μm width, 0.44 mm s^{-1} speed) of varying contrast: 5, 10, 20, 33, 50 and 90% (see top stimulus traces). Main panel shows the response profile derived from 15 salamander fast OFF ganglion cells (coloured dots), and the predicted response (coloured lines) from a model incorporating contrast-gain control (Fig. 4). Grey line shows the prediction without contrast-gain control. Inset, the peak firing rate of the response profile as a function of the contrast of the bar. Error bars denote standard error, derived from variation among ganglion cells. **b**, Gain variable g of the gain control model (Fig. 4). Model parameters: $\theta = 0$, $\alpha = 85$ Hz, $B = 45$ s^{-1}, $\tau = 170$ ms.

This explanation indicates that there will be clear limits to what stimuli can be anticipated. For example, if the bar moves fast enough to cross the receptive field before the contrast-gain control sets in, then the peak of the firing profile should lag behind the leading edge. Figure 5a explores these limits, and shows that up to speeds of about 1 mm s^{-1} on the retina, the shape of the firing profile among ganglion cells remained essentially unchanged, with a peak near or ahead of the leading edge. At higher speeds, however, the response profile began to slip significantly behind the leading edge. This basic relationship was confirmed for several different populations of ganglion cells in both rabbit and salamander (Fig. 5b). The various cell types differed in the extent of anticipation at low speeds, but all began to show a lag in the neural image at speeds of 1–2 mm s^{-1}. In particular, direction selectivity does not play a special role in motion anticipation.

370

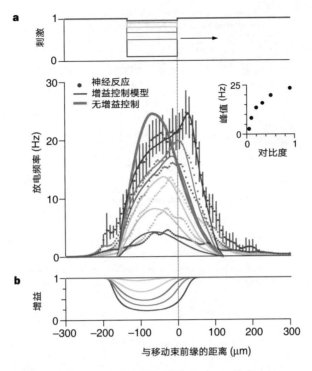

图 3. 运动推测对对比度的依赖性。**a**，移动暗光柱（宽度 133 μm，速度 0.44 mm · s⁻¹）在不同对比度：5%、10%、20%、33%、50% 和 90%（参见顶端的刺激轨迹）下的刺激。主图显示的是 15 个虎蝾快速 OFF 神经节细胞采集到的反应特性（彩色的点），和来自混合的对比度增益控制模型中的预测反应特性（彩色的线）（图 4）。灰色的线显示没有对对比度增益控制的预测。插图显示反应曲线的放电频率峰值与移动光柱对比度之间的函数关系。误差线表示从不同的神经节细胞变异中得来的标准误差。**b**，增益控制模型的增益变异 g（图 4）。模型参数：$\theta = 0$，$\alpha = 85$ Hz，$B = 45$ s⁻¹，$\tau = 170$ ms。

　　这种解释表明细胞对于何种刺激可被预测有着明确的限定。例如，如果一个光柱移动得足够快，在对比度增益控制介入之前经过感受域，那么细胞放电的峰值应该滞后于光束前缘。图 5a 探索了这种限定，结果显示在视网膜上速度高达大约 1 mm · s⁻¹，神经节细胞放电特性本质上保持不变，峰值靠近前缘或在前缘之前。然而，在更高速度下，反应特性开始明显落后于前缘。这种基本关系在几种不同群体的神经节细胞（如兔和虎蝾中）中得到验证（图 5b）。在低速情况下，不同类型细胞对运动的预测程度不同，但是所有细胞在速度为 1 ~ 2 mm · s⁻¹ 时均开始出现神经影像的滞后。尤其是方向选择性在运动的预测中没有起到特殊作用。

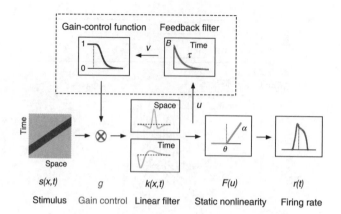

$$s(x,t) \qquad g \qquad k(x,t) \qquad F(u) \qquad r(t)$$

Stimulus   Gain control   Linear filter   Static nonlinearity   Firing rate

Fig. 4. Cascade model for a ganglion cell's light response. The stimulus $s(x,t)$ is multiplied by a gain factor g, convolved with a spatiotemporal filter $k(x,t)$, and rectified by a static nonlinear function $F(u)$ to produce the firing rate $r(t)$. The contrast-gain control mechanism (boxed region) takes the output of the linear filter u, averages it by exponential filtering in time, and uses the result v to set the gain factor g through a decreasing gain control function $g(v)$. Formally,

$$u(t) = g(v) \int_{-\infty}^{\infty} dx \int_{-\infty}^{t} dt' \, s(x,t') \, k(x,t-t')$$

$$v(t) = \int_{-\infty}^{t} dt' \, u(t') \, B \exp\left(-\frac{t-t'}{\tau}\right)$$

$$g(v) = \begin{cases} 1 & v < 0 \\ 1/(1+v^4) & v > 0 \end{cases}$$

$$F(u) = \begin{cases} 0 & u < 0 \\ \alpha(u-\theta^4) & u > 0 \end{cases}$$

The filter $k(x,t)$ is measured (see Methods), and $g(v)$ is taken from a previous, successful model of contrast-gain control in the salamander retina[17]. Thus, the model has four parameters: the threshold θ and slope α of the rectifier $F(u)$, and the amplitude B and time constant τ of the gain control filter.

In summary, we have shown that the extrapolation of a moving object's trajectory begins in the retina. In the neural image that the eye transmits to the brain, the moving object is clearly ahead of the corresponding flashed object (Fig. 2). According to a successful model for the ganglion cell's light response (Figs 3, 4), motion anticipation in these populations can be explained on the basis of the spatially extended receptive field, the biphasic temporal response and a nonlinear contrast-gain control. There are several indications that this retinal mechanism contributes strongly to human perception of moving stimuli. First, the requisite components of our model are well documented in many species. In the primate retina, a nonlinear contrast-gain control is found specifically in the M-type ganglion cells[15], neurons that feed the central pathways leading to motion perception[18]. Second, retinal motion extrapolation breaks down at speeds above 1 mm s^{-1} (Fig. 5b). This corresponds well with observations on human subjects: At retinal speeds of 0.3–0.9 mm s^{-1}, perceptual motion extrapolation appeared to compensate for the entire visual delay[5], whereas at speeds of ~4 mm s^{-1} only partial extrapolation was observed[10]. Finally, the retina anticipates high-contrast stimuli more than low-contrast stimuli (Fig. 3), a further departure from ideal extrapolation. Again, this effect has been observed in human psychophysics[9].

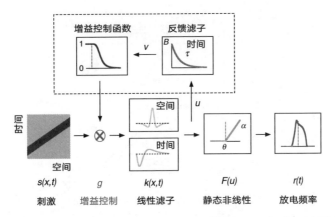

图 4. 一个神经节细胞光反应的级联模型。刺激 $s(x,t)$ 乘以增益因子 g，与时间空间滤子 $k(x,t)$ 卷积，经过静态的非线性函数 $F(u)$ 校正，产生放电频率 $r(t)$。增益控制机理（方框区域）接受线性滤子 u 的输出，通过时间的指数滤子求其平均值，通过减少增益控制函数 $g(v)$，用结果 v 设定增益因子 g。用公式表示为：

$$u(t) = g(v) \int_{-\infty}^{\infty} \mathrm{d}x \int_{-\infty}^{t} \mathrm{d}t'\, s(x,t')\, k(x,t-t')$$

$$v(t) = \int_{-\infty}^{t} \mathrm{d}t'\, u(t')\, B \exp\left(-\frac{t-t'}{\tau}\right)$$

$$g(v) = \begin{cases} 1 & v < 0 \\ 1/(1+v^4) & v > 0 \end{cases}$$

$$F(u) = \begin{cases} 0 & u < 0 \\ \alpha(u-\theta^4) & u > 0 \end{cases}$$

滤子 $k(x,t)$ 可以测量（参见方法），$g(v)$ 从先前成功的虎蝾视网膜增益控制模型中得来 [17]。因此，模型有 4 个参数：整流器 $F(u)$ 的阈值 θ 和斜率 α，以及振幅 B 和增益控制滤子的时间常数 τ。

总而言之，我们证明对运动物体轨迹的推测始于视网膜。在经眼传输至脑的神经影像中，运动物体明显先于相应的闪烁物体（图 2）。根据神经节细胞光反应的成功模型（图 3 和图 4），这些群体的运动预测可以基于如下解释：感受域空间上的延伸，时间上两阶段反应和非线性对比度增益控制。有些迹象表明这种视网膜机制极大地促成了人对运动刺激的感知。首先，我们模型中的必要组件在许多物种中得到很好的证实。在灵长类视网膜中，非线性的对比度增益控制被发现，尤其存在于 M 型神经节细胞中 [15]，满足中枢通路的神经元导致对运动的感知 [18]。第二，视网膜对运动的预测在速度大于 $1\ \mathrm{mm \cdot s^{-1}}$ 时失效（图 5b）。这与人体实验的观察结果很好地对应：视网膜速度为 $0.3 \sim 0.9\ \mathrm{mm \cdot s^{-1}}$，知觉运动的预测能够对整个视觉延迟产生补偿 [5]，当速度约为 $4\ \mathrm{mm \cdot s^{-1}}$ 时，只观察到对运动的部分预测 [10]。最后，视网膜对高对比度刺激的预测要强于对低对比度刺激的预测（图 3），这更加偏离理想推测。再一次强调，这种效应在人类精神物理学中已经观察到 [9]。

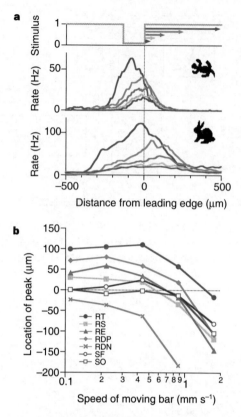

Fig. 5. Dependence of motion extrapolation on speed. **a**, Stimulation with moving dark bars (90% contrast, 133 μm width) of varying speed: 0.11, 0.22, 0.44, 0.88 and 1.76 mm s⁻¹ (see top stimulus traces). Firing profiles are plotted for salamander fast OFF ganglion cells (middle panel, 19 cells) and rabbit brisk-transient OFF cells (bottom panel, 3 cells). **b**, Motion extrapolation as a function of speed for various populations of ganglion cells. The distance between the peak in the firing-rate profile and the leading edge of the moving bar is plotted as a function of speed; positive numbers indicate that the response profile peaked ahead of the leading edge. The functional types are: salamander, fast OFF (SF, 19 cells); salamander, other OFF (SO, 16 cells); rabbit, brisk-transient OFF (RT, 3 cells); rabbit, brisk-sustained OFF (RS, 6 cells); rabbit, local edge detectors (RE, 4 cells); rabbit, ON/OFF direction-selective cells (5 cells) probed in the preferred direction (RDP) and the null direction (RDN).

In general, an animal is likely to benefit from anticipating the future position of an object, for example to pounce on it or to evade it. This is particularly urgent when the primary sensory data are delayed. In principle, this delay could be compensated anywhere within the behavioural loop, even within the motor system that executes the response. However, it is advantageous to perform the correction early, before different sensory pathways merge. For example, in many animals the retina projects directly to the tectum or the superior colliculus, where a visual map of space is overlaid with an auditory map[19,20]. Auditory transduction in hair cells incurs a much shorter delay than phototransduction[21]. If the visual and auditory images of a moving object should align on the target map, the compensation for the delay in the visual pathway must occur within the retina.

It is likely that subsequent stages of the visual system continue this process, possibly by using a similar mechanism. Within the visual cortex we can certainly find the requisite

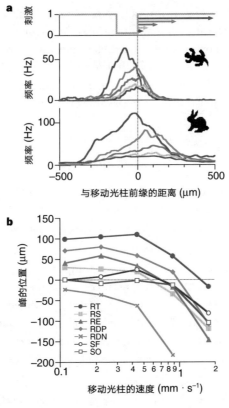

图 5. 运动推测对速度的依赖性。**a**, 移动暗光柱(对比度 90%, 宽度 133 μm)在不同速度下: 0.11、0.22、0.44、0.88 和 1.76 mm·s⁻¹(参见顶部的刺激轨迹)的刺激。放电特性是根据虎鳅快速 OFF 神经节细胞(中间图, 19 个细胞)和兔临时激活 OFF 细胞(底部图, 3 个细胞)的数据绘制而来的。**b**, 不同群体的神经节细胞速度函数的运动推测。细胞放电频率峰值和移动光柱前缘的距离被绘制为速度的函数; 正数表明反应在光柱前缘之前出现峰值。功能类型包括: 虎鳅, 快速 OFF 细胞(SF, 19 个); 虎鳅, 其他 OFF 细胞(SO, 16 个); 兔, 临时激活 OFF 细胞(RT, 3 个); 兔, 持续激活 OFF 细胞(RS, 6 个); 兔, 局部边缘探测器细胞(RE, 4 个); 兔, 在首选方向(RDP)和无效方向(RDN)探测的 ON/OFF 方向选择性细胞(5 个)。

通常, 动物可能会受益于对运动物体未来位置的预测能力, 例如用来袭击或逃脱。如果初级感官数据传输滞后, 这一点就显得尤其重要。从原理上来说, 这种滞后在行为回路中任何地方均可以补偿, 甚至在执行反应的运动系统中。然而, 在不同的感觉路径合并之前, 提前执行正确的决定是有利的。例如, 许多动物的视网膜直接映射到上表皮层或者视上丘, 在此处空间的视觉图像与听觉图像叠加[19,20]。在毛细胞中的听觉传导引发的滞后比视觉传导更短[21]。如果运动物体的视觉和听觉影像能与目标图像并列, 对视觉通路的滞后补偿一定发生在视网膜。

很可能通过相似的机理, 视觉系统随后的信息处理阶段会延续这个过程。在视觉皮层我们能够确切地发现图 4 模型中的必要组件: 例如, 兴奋输入的局部汇集,

components of the model in Fig. 4: for example, local pooling of excitatory inputs, time-delayed inhibition and mechanisms of nonlinear gain control[22,23]. More generally, there are many instances within the cortex where variables relevant to our behaviour are mapped onto two-dimensional sheets of neurons[24]. If the time course of these variables produces a smooth trajectory of neural activity on the cortical map, then a mechanism such as that described here can predict their future from past observations.

Methods

Recording. Retinae were obtained from larval tiger salamanders and Dutch belted rabbits. A piece of isolated retina was placed ganglion-cell-layer-down on a multi-electrode array, which recorded spike trains simultaneously from many ganglion cells, as described previously[11,25].

Stimulation. Visual stimuli were generated on a computer monitor and projected onto the photoreceptor layer, as described[25]. All experiments used a background of white light, with a photopic intensity of $M = 11$ mW m^{-2}. Dark bars of intensity B were presented on this background, and the contrast of a bar is defined as $C = (M - B)/M$. A screen pixel of the monitor measured 6.7 μm on the retina, and each video frame lasted 15 ms. Thus, a bar sweeping at 0.44 mm s^{-1} moved by one pixel every video frame. Flashed bars were presented for a single video frame.

Receptive fields. The spatiotemporal receptive fields of all ganglion cells were measured by reverse correlation to randomly flickering stripes[25], orientated parallel to the bars from other experiments. Each of the contiguous 13-μm wide stripes was randomly turned on or off every 30 ms. From ~60 min of recording, we computed for each ganglion cell the average stimulus sequence in the one second preceding an action potential. This reverse correlation is a measure of how the ganglion cell integrates light over space and time. Its time-reverse is the ganglion cell's linear kernel $k(x,t)$ (ref. 26), which can also be interpreted as the effect of a thin line flashed at distance x on the cell's firing rate at time t after the flash[11]. As expected, $k(x,t)$ had a "Mexican hat" spatial profile, reflecting opposite effects from centre and surround[27], and a biphasic time course (see Fig. 4 and ref. 11).

Cell types. Retinal ganglion cells appear in distinct functional types, and we took care to analyse these subpopulations separately. Salamander cells were classified based on their spatiotemporal receptive fields and on responses to uniform square-wave flashes, as described[28]. Rabbit cells were classified based on the spatiotemporal receptive field and the shape of the spike train's autocorrelation function, following the criteria of ref. 29. Direction-selective cells produced at least tenfold more spikes to one direction of the moving bar than to the opposite direction.

Population activity. The profile of population activity was evaluated along the spatial dimension perpendicular to the bars. Each cell's position was defined as the middle of its receptive field, determined by fitting a spatial gaussian to the centre lobe of the kernel $k(x,t)$. To estimate the population response to a flashed bar, the flash was repeated 75 times in each of 15 locations, separated by 33 μm. For each flash location and each ganglion cell, the firing rate following the flash

时间滞后的抑制，和非线性增益控制的机理[22,23]。通常，在皮质层有许多这样的例子，在那里与我们行为有关的可变物被映射到神经元的二维层[24]。如果这些可变物的时间进程在大脑皮层的神经活动中产生一个平稳的轨迹，那么本文描述的机制能够从过去的观察结果中预测可变物的未来。

方　法

记录　视网膜采集于幼年期的虎蝾和荷兰条纹兔。将一片分离的视网膜放置在一个多电极阵列上，放置方向为神经节细胞层朝下，按照之前所述的方法[11,25]同时从许多神经节细胞记录放电信号。

刺激　视觉刺激由计算机显示器产生，然后按照文献描述的方法投射到光感受器细胞层[25]。所有实验以白光为背景，光强为 $M = 11$ mW · m^{-2}。强度为 B 的暗光柱呈现在此背景中，暗光柱的对比度定义为 $C=(M-B)/M$。显示器单个屏幕像素在视网膜上的投射大小为 6.7 μm，每个视频帧持续的时间为 15 ms。因此，移动光柱以 0.44 mm · s^{-1} 的速度运动，每个视频帧就是一个像素。闪烁光柱在单个视频帧里出现。

感受域　所有神经节细胞的时空感受域均通过随机闪烁光柱的负相关测量[25]，方向与其他实验里的移动光柱平行。每一个连续的 13 μm 宽的条纹每隔 30 ms 随机开关一次。在大约 60 分钟的记录中，我们可以为每一个神经节细胞计算出动作电位前一秒钟之内的平均刺激序列。这种负相关是对神经节细胞如何对光进行时空整合的一种测量。其时间的逆转是神经节细胞的线性核 $k(x,t)$（文献 26），也可以解释为一条细线闪现后在 t 时间 x 距离处细胞放电频率所产生的效应[11]。正如所料，$k(x,t)$ 在空间曲线上有一个"墨西哥帽"，反映了中间和周围的相反效应[27]，以及双相的时间进程。（参见图 4 和参考文献 11）。

细胞类型　视网膜神经节细胞具有不同的功能类型，我们分别对这些亚种群进行了仔细地分析。正如前人所描述的，虎蝾的细胞是根据它们的时空感受域及其对统一的矩形波闪烁光柱的反应来区分的[28]。兔的细胞是按照参考文献 29 的标准，根据时间和空间感受域以及放电序列自相关函数的形状进行区分的。方向选择性的细胞对移动光柱移动的特定方向产生的放电至少是相反方向的 10 倍。

群体活动　根据垂直于光柱的空间维度来评价种群活动的特性。每个细胞的位置定位在其感受域中央，由空间的高斯曲线拟合核 $k(x,t)$ 的中心突起形状决定。为了估计细胞群体对闪烁光柱的反应，闪烁光柱在间隔 33 μm 的 15 个位置处每处重复 75 次。对于每一个闪烁光柱的位置和每个神经节细胞，其在闪烁光柱之后的放电频率（图 1a 和 1b）用二进制 2 ms

(Fig. 1a, b) was calculated using a time bin of 2 ms. To compose the firing profile at a given time after the flash, each ganglion cell's firing rate was plotted against the cell's position relative to the flashing line. This plot was smoothed by convolution with a gaussian of standard deviation 20 μm. To estimate the population response to a moving bar, the stimulus was repeated 50 times, and each ganglion cell's firing rate computed as for flashes, but with a time bin of 15 ms. Then the firing rate was plotted against the distance from the bar, and averaged over all cells.

(**398**, 334-338; 1999)

Michael J. Berry II, Iman H. Brivanlou, Thomas A. Jordan* & Markus Meister
Department of Molecular and Cellular Biology, Harvard University, 16 Divinity Avenue, Cambridge, Massachusetts 02138, USA
* Present address: Department of Psychiatry and Behavioral Science, Stanford University School of Medicine, Stanford, California 94305, USA.

Received 11 December 1998; accepted 2 February 1999.

References:

1. Maunsell, J. H. & Gibson, J. R. Visual response latencies in striate cortex of the macaque monkey. *J. Neurophysiol.* **68,** 1332-1344 (1992).
2. Lennie, P. The physiological basis of variations in visual latency. *Vision Res.* **21,** 815-824 (1981).
3. Schnapf, J. L., Kraft, T. W. & Baylor, D. A. Spectral sensitivity of human cone photoreceptors. *Nature* **325,** 439-441 (1987).
4. De Valois, R. L. & De Valois, K. K. Vernier acuity with stationary moving Gabors. *Vision Res.* **31,** 1619-1626 (1991).
5. Nijhawan, R. Motion extrapolation in catching. *Nature* **370,** 256-257 (1994).
6. Khurana, B. & Nijhawan, R. Extrapolation or attention shift? *Nature* **378,** 566 (1995).
7. Nijhawan, R. Visual decomposition of colour through motion extrapolation. *Nature* **386,** 66-69 (1997).
8. Baldo, M. V. & Klein, S. A. Extrapolation or attention shift? *Nature* **378,** 565-566 (1995).
9. Purushothaman, G., Patel, S. S., Bedell, H. E. & Ogmen, H. Moving ahead through differential visual latency. *Nature* **396,** 424 (1998).
10. Whitney, D. & Murakami, I. Latency difference, not spatial extrapolation. *Nature Neurosci.* **1,** 656-657 (1998).
11. Smirnakis, S. M., Berry, M. J., Warland, D. K., Bialek, W. & Meister, M. Adaptation of retinal processing to image contrast and spatial scale. *Nature* **386,** 69-73 (1997).
12. Jacobs, A. L. & Werblin, F. S. Spatiotemporal patterns at the retinal output. J. *Neurophysiol.* **80,** 447-451 (1998).
13. Shapley, R. M. & Victor, J. D. The effect of contrast on the transfer properties of cat retinal ganglion cells. *J. Physiol.* **285,** 275-298 (1978).
14. Sakai, H. M., Wang, J. L. & Naka, K. Contrast gain control in the lower vertebrate retinas. *J. Gen. Physiol.* **105,** 815-835 (1995).
15. Benardete, E. A., Kaplan, E. & Knight, B. W. Contrast gain control in the primate retina: P cells are not X-like, some M cells are. *Visual Neurosci.* **8,** 483-486 (1992).
16. Victor, J. D. The dynamics of the cat retinal X cell centre. *J. Physiol.* **386,** 219-246 (1987).
17. Crevier, D. W. & Meister, M. Synchronous period-doubling in flicker vision of salamander and man. *J. Neurophysiol.* **79,** 1869-1878 (1998).
18. Merigan, W. H. & Maunsell, J. H. How parallel are the primate visual pathways? *Annu. Rev. Neurosci.* **16,** 369-402 (1993).
19. Sparks, D. L. Translation of sensory signals into commands for control of saccadic eye movements: role of primate superior colliculus. *Physiol. Rev.* **66,** 118-171 (1986).
20. Knudsen, E. I. Auditory and visual maps of space in the optic tectum of the owl. *J. Neurosci.* **2,** 1177-1194 (1982).
21. Corey, D. P. & Hudspeth, A. J. Response latency of vertebrate hair cells. *Biophys. J.* **26,** 499-506 (1979).
22. Carandini, M., Heeger, D. J. & Movshon, J. A. Linearity and normalization in simple cells of the macaque primary visual cortex. *J. Neurosci.* **17,** 8621-8644 (1997).
23. Abbott, L. F., Varela, J. A., Sen, K. & Nelson, S. B. Synaptic depression and cortical gain control. *Science* **275,** 220-224 (1997).
24. Knudsen, E. I., du Lac, S. & Esterly, S. D. Computational maps in the brain. *Annu. Rev. Neurosci.* **10,** 41-65 (1987).
25. Meister, M., Pine, J. & Baylor, D. A. Multi-neuronal signals from the retina: acquisition and analysis. *J. Neurosci. Methods* **51,** 95-106 (1994).
26. Hunter, I. W. & Korenberg, M. J. The identification of nonlinear biological systems: Wiener and Hammerstein cascade models. *Biol. Cybern.* **55,** 135-144 (1986).
27. Rodieck, R. W. Quantitative analysis of cat retinal ganglion cell response to visual stimuli. *Vision Res.* **5,** 583-601 (1965).
28. Warland, D. K., Reinagel, P. & Meister, M. Decoding visual information from a population of retinal ganglion cells. *J. Neurophysiol.* **78,** 2336-2350 (1997).
29. Devries, S. H. & Baylor, D. A. Mosaic arrangement of ganglion cell receptive fields in rabbit retina. *J. Neurophysiol.* **78,** 2048-2060 (1997).

Acknowledgements. We thank J. Keat for assistance in generating the visual stimulus, and H. Berg and T. Holy for comments on the manuscript. This work was supported by a NRSA to M.B. and a grant from the NIH and a Presidential Faculty Fellowship to M.M.

Correspondence and requests for material should be addressed to M.J.B. (e-mail: berry@biosun.harvard.edu).

的时间计算。为了描绘闪烁光柱之后给定时间的放电特性，每个神经节细胞的放电频率根据相对于闪烁光柱线的细胞位置描绘。该曲线通过与标准偏差为 20 μm 的高斯曲线卷积变得平滑。为了估计对移动光柱的群体反应，每个刺激被重复 50 次，每个神经节细胞对闪光的放电频率都以二进制 15 ms 的时间计算。然后，对神经节细胞的放电频率以及其相对于光柱的距离进行绘图，并对所有的细胞进行平均。

（董培智 翻译；巩克瑞 审稿）

Observation of Contemporaneous Optical Radiation from a γ-ray Burst

C. Akerlof *et al.*

Editor's Note

Gamma-ray bursts are brief, extremely intense flashes of gamma rays emitted by astrophysical objects and associated with high-energy explosions. Since their discovery in the late 1960s, many have been observed, but their explanation is still debated. Although numerous visible-light objects coincident with the location of gamma-ray bursts had been seen since 1997, before this paper none had been observed while the burst was still active. Carl Akerlof and colleagues used a robotic, wide-field camera to find the optical counterpart of GRB990123 while the burst was still underway, with the observations starting just 22 seconds after the burst began. The visible emission decayed more slowly than the gamma rays. Such observations help to narrow down the identity of the emitting object.

The origin of γ-ray bursts (GRBs) has been enigmatic since their discovery[1]. The situation improved dramatically in 1997, when the rapid availability of precise coordinates[2,3] for the bursts allowed the detection of faint optical and radio afterglows—optical spectra thus obtained have demonstrated conclusively that the bursts occur at cosmological distances. But, despite efforts by several groups[4-7], optical detection has not hitherto been achieved during the brief duration of a burst. Here we report the detection of bright optical emission from GRB990123 while the burst was still in progress. Our observations begin 22 seconds after the onset of the burst and show an increase in brightness by a factor of 14 during the first 25 seconds; the brightness then declines by a factor of 100, at which point (700 seconds after the burst onset) it falls below our detection threshold. The redshift of this burst, $z \approx 1.6$ (refs 8, 9), implies a peak optical luminosity of 5×10^{49} erg s^{-1}. Optical emission from γ-ray bursts has been generally thought to take place at the shock fronts generated by interaction of the primary energy source with the surrounding medium, where the γ-rays might also be produced. The lack of a significant change in the γ-ray light curve when the optical emission develops suggests that the γ-rays are not produced at the shock front, but closer to the site of the original explosion[10].

THE Robotic Optical Transient Search Experiment (ROTSE) is a programme optimized to search for optical radiation contemporaneous with the high-energy photons of a γ-ray burst. The basis for such observations is the BATSE detector on board the Compton Gamma-Ray Observatory. Via rapid processing of the telemetry data stream, the GRB Coordinates Network[11] (GCN) can supply estimated coordinates

伽马射线暴同时的光学辐射的观测

阿克洛夫等

编者按

伽马射线暴是一类由遥远天体在短时间内发出的极强伽马射线耀发，是一种高能爆发现象。自 20 世纪 60 年代末以来，人们虽已发现并观测到多个伽马射线暴，但是对于它们起源问题的解释却是众说纷纭。尽管自 1997 年以来在很多伽马射线暴的位置处人们发现了相应的发光天体，然而在本工作之前还没有人在伽马射线暴发生的过程中找到其发光对应体。卡尔·阿克洛夫和他的同事利用一个自动的宽场相机在 GRB990123 暴的过程中，即暴开始仅 22 秒之后，找到了它的光学对应体。结果发现，可见光辐射强度的衰减明显慢于伽马射线辐射。这样的观测将有助于我们证认发出辐射的天体。

自 γ 射线暴（GRB）被发现以来，它的起源问题就一直是个谜 [1]。这个状况在 1997 年有了很大的改善，因为当时 γ 射线暴快速有效的精确定位技术 [2,3] 的发展使得探测暗弱的光学和射电余辉成为可能——而由此探测到的光谱则证实 γ 射线暴是在宇宙学距离上发生的。尽管几个团队一直在不断地努力 [4-7]，然而迄今为止，人们都没能在短暂的暴持续时间内探测到 γ 射线暴的光学辐射。这里我们报道在 GRB990123 暴期间所发出的明亮的光学辐射的探测。我们的观测在暴开始 22 秒后进行，结果显示，在最初的 25 秒内其亮度增加到 14 倍；随后其亮度下降了 99%，从此（在暴开始 700 秒之后）其亮度下降到我们的探测阈值之下。这个 γ 射线暴的红移为 $z \approx 1.6$（参考文献 8 和 9），这意味着它的光学光度的峰值为 5×10^{49} erg·s^{-1}。一般认为，γ 射线暴的光学辐射产生于主要能量源和周围介质相互作用所形成的激波波前，而 γ 射线辐射也可能在这产生。由于光学辐射发展时 γ 射线的光变曲线并没有发生明显的变化，这表明 γ 射线不是在激波波前产生，而是在更接近原始爆发位置处产生 [10]。

自动光学暂现源搜寻实验（ROTSE）是个以搜寻和 γ 射线暴的高能光子同时产生的光学辐射为目标的项目。这个观测的基础来自安装在康普顿 γ 射线天文台上的 BATSE 探测仪。通过对遥测数据流的快速处理，GRB 坐标定位网 [11]（GCN）能够在

to distant observatories within a few seconds of the burst detection. The typical error of these coordinates is 5°. A successful imaging system must match this field of view to observe the true burst location with reasonable probability.

The detection reported here was performed with ROTSE-I, a two-by-two array of 35 mm camera telephoto lenses (Canon $f/1.8$, 200 mm focal length) coupled to large-format CCD (charge-coupled device) imagers (Thomson 14 μm 2,048 × 2,048 pixels). All four cameras are co-mounted on a single rapid-slewing platform capable of pointing to any part of the sky within 3 seconds. The cameras are angled with respect to each other, so that the composite field of view is 16°×16°. This entire assembly is bolted to the roof of a communications enclosure that houses the computer control system. A motor-driven flip-away cover shields the detector from precipitation and direct sunlight. Weather sensors provide the vital information to shut down observations when storms appear, augmented by additional logic to protect the instrument in case of power loss or computer failure. The apparatus is installed at Los Alamos National Laboratory in northern New Mexico.

Since March 1998, ROTSE-I has been active for ~75% of the total available nights, with most of the outage due to poor weather. During this period, ROTSE-I has responded to a total of 53 triggers. Of these, 26 are associated with GRBs and 13 are associated with soft γ-ray repeaters (SGRs). The median response time from the burst onset to start of the first exposure is 10 seconds.

During most of the night, ROTSE-I records a sequence of sky patrol images, mapping the entire visible sky with two pairs of exposures which reach a 5σ V magnitude threshold sensitivity (m_v) of 15. These data, approximately 8 gigabytes, are archived each night for later analysis. A GCN-provided trigger message interrupts any sequence in progress and initiates the slew to the estimated GRB location. A series of exposures with graduated times of 5, 75 and 200 seconds is then begun. Early in this sequence, the platform is "jogged" by ±8° on each axis to obtain coverage of a four times larger field of view.

At 1999 January 23 09:46:56.12 UTC, an energetic burst triggered the BATSE detector. This message reached Los Alamos 4 seconds later and the first exposure began 6 seconds after this. Unfortunately, a software error prevented the data from being written to disk. The first analysable image was taken 22 seconds after the onset of the burst. The γ-ray light curve for GRB990123 was marked by an initial slow rise, so the BATSE trigger was based on relatively limited statistics. Thus the original GCN position estimate was displaced by 8.9° from subsequent localization, but the large ROTSE-I field of view was sufficient to contain the transient image. At 3.8 hours after the burst, the BeppoSAX satellite provided an X-ray localization[12] in which an optical afterglow was discovered by Odewahn et al.[13] at Mt Palomar. This BeppoSAX position enabled rapid examination of a small region of the large ROTSE-I field. A bright and rapidly varying transient was found in the ROTSE images at right ascension (RA) 15 h 25 min 30.2 s, declination (dec.) 44° 46′ 0″, in excellent agreement with the afterglow found by Odewahn et al. (RA 15 h 25 min 30.53 s, dec. 44° 46′ 0.5″). Multiple absorption lines in the spectrum of the optical afterglow

探测到暴的几秒之内向远处的天文台提供估计的位置坐标。这些坐标的典型误差为5°。一个成功的成像系统必须覆盖这么大的视场，才能在适当的概率上观测到暴的真正位置。

这里我们所报道的探测是由 ROTSE-I 完成的，它是一个由 35 mm 相机摄远镜头（佳能 $f/1.8$，焦距 200 mm）组成的 2×2 阵，并配备大幅面 CCD（电荷耦合装置）成像仪（汤姆逊 14 μm 2,048 × 2,048 像素）。所有的 4 个相机都安装在一个快速回转平台上，这使得该仪器能够在 3 秒内指向天区的任何位置。相机彼此之间具有一定的倾角，这使得其仪器的综合视场为 16° × 16°。整个装置固定在装有计算机控制系统的通讯围罩的顶上。有一个电机驱动的滑动盖用以保护探测仪不受降雨和日光直射的影响。当暴风雨来临前，气象传感仪将提供重要信息，关闭探测仪器，增强额外的保护措施，以避免功率损耗或者计算机关机对设备仪器带来的损坏。整个设备安装在新墨西哥州北部的洛斯阿拉莫斯国家实验室处。

自 1998 年 3 月开始，ROTSE-I 在约 75% 可观测的夜晚都在工作，而其他大多时候设备都是因为糟糕的天气而不得不停止。这段时期内，ROTSE-I 总共对 53 次触发作出了反应，其中 26 次和 GRB 成协，13 次和软 γ 重复暴成协。从暴开始到首次曝光开始之间的中值反应时间为 10 秒。

夜晚大部分时间内，ROTSE-I 记录一系列巡天图像：对整个可视天区进行两次成双的曝光，5σ V 波段极限星等 (m_v) 达到 15 星等。这样，每个晚上记录下约 8 千兆的数据有待后续分析。若 GCN 提供触发信息，将打断正在进行的巡天观测，并启动望远镜使其回转到所估计的 GRB 位置处。然后望远镜将开始对那块天区进行一系列 5 秒、75 秒和 200 秒的曝光。在这个观测序列早期，观测平台的每个轴都"慢跑" ±8° 以覆盖达到视场 4 倍大的天区。

在 1999 年 1 月 23 日 09:46:56.12 UTC，一个能量巨大的暴触发了 BATSE 探测仪。这个信息在 4 秒后抵达洛斯阿拉莫斯，再 6 秒后首次曝光开始。不幸的是，一个软件错误导致这次观测的数据没有被写到硬盘里。而首幅可用于分析的图像是在暴开始 22 秒后得到的。由于 GRB990123 的 γ 射线光变曲线特征为初始缓慢的增加，所以 BATSE 触发的准确度相对有限。这使得初始的 GCN 位置估计和随后的定位位置相差 8.9°，但是 ROTSE-I 的大视场已足够包含住该暂现源的像。在暴发生 3.8 小时后，BeppoSAX 卫星给出了一个 X 射线定位 [12]，而在该位置上奥德万等人 [13] 在帕洛马山发现了一个光学余辉。这个 BeppoSAX 给出的位置使得我们可以快速检验 ROTSE-I 的大视场中的一个小区域。我们在 ROTSE 图像赤经 15 h 25 min 30.2 s、赤纬 44° 46′ 0″ 处发现了一个明亮且快速变化的暂现源，其位置和奥德万等人发现的光学余辉的位置（赤经 15 h 25 min 30.53 s、赤纬 44° 46′ 0.5″）符合得很好。从光学余辉

indicate a redshift of $z > 1.6$. Dark-subtracted and flattened ROTSE-I images of the GRB field are shown in Fig. 1. Details of the light curve are shown in Table 1.

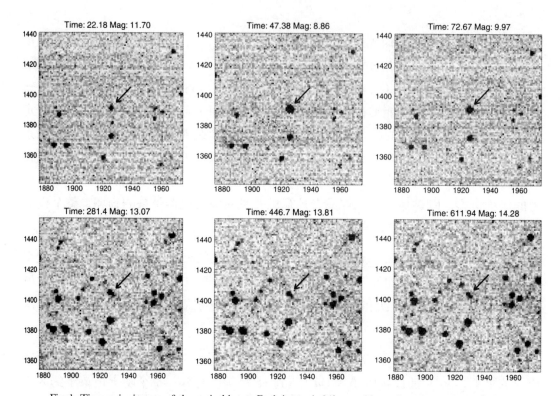

Fig. 1. Time series images of the optical burst. Each image is 24′ on a side, and represents 6×10^{-4} of the ROTSE-I field of view. The horizontal and vertical axes are the CCD pixel coordinates. The sensitivity variations are due to exposure time; the top three images are 5-s exposures, the bottom three are 75-s exposures. The optical transient (OT) is clearly detected in all images, and is indicated by the arrow. South is up, east is left. Thermal effects are removed from the images by subtracting an average dark exposure. Flat field images are generated by median averaging about 100 sky patrol (see text) images. Object catalogues are extracted from the images using SExtractor[18]. Astrometric and photometric calibrations are determined by comparison with the ~1,000 Tycho[19] stars available in each image. Residuals for stars of magnitude 8.5–9.5 are $< 1.2''$. These images are obtained with unfiltered CCDs. The optics and CCD quantum efficiency limit our sensitivity to a wavelength range between 400 and 1,100 nm. Because this wide band is non-standard, we estimate a "V equivalent" magnitude by the following calibration scheme. For each Tycho star, a "predicted ROTSE magnitude" is compared to the 2.5 pixel aperture fluxes measured for these objects to obtain a global zero point for each ROTSE-I image. For the Tycho stars, the agreement between our predicted magnitude and the measured magnitude is ± 0.15. These errors are dominated by colour variation. The zero points are determined to ± 0.02. With large pixels, we must understand the effects of crowding. (This is especially true as we follow the transient to ever fainter magnitudes.) To check the effect of such crowding, we have compared the burst location to the locations of known objects from the USNO A V2.0 catalogue[20]. The nearest object, 34″ away, is a star with R-band magnitude $R = 19.2$. More important is an $R = 14.4$ star, 42″ away. This object affects the measured magnitude of the OT only in our final detection. It can be seen in the final image to the lower right of the OT. A correction of $+0.15$ is applied to compensate for its presence. Magnitudes for the OT associated with GRB990123, measured as described, are listed in Table 1. Further information about the ROTSE-I observations is available at http://www.umich.edu/~rotse.

光谱的多条吸收线中，我们发现其红移为 $z > 1.6$。图 1 显示了扣除暗曝光并平场化的 ROTSE-I GRB 区域图像。光变曲线的细节则在表 1 中进行了展示。

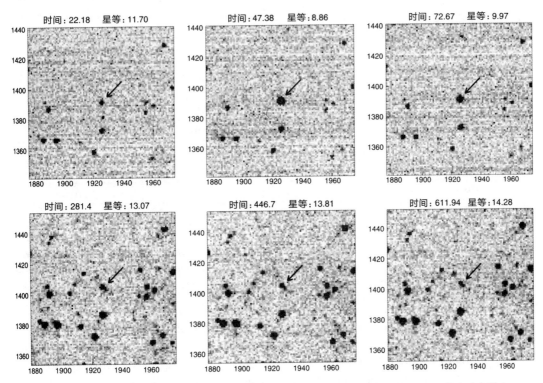

图 1. 光学暴的时间序列成像图。每幅图边长 24′，代表 6×10^{-4} 的 ROTSE-I 视场。水平轴和垂直轴为 CCD 像素坐标。灵敏度会随着曝光时间的变化而发生变化；上面三幅图曝光时间为 5 秒，下面三幅曝光时间为 75 秒。所有图上都能清楚地探测到光学暂现源（OT），以箭头标示。上边是南，左边是东。热效应已通过扣掉平均暗曝光从图像中去除。平场图像则是通过对大约 100 幅巡天图进行平均得到（见正文）。我们利用 SExtractor[18] 从图像中提取出目标源表，再通过与每幅图中存在的 1,000 颗左右的第谷恒星 [19] 进行比较来进行天体测量与测光的定标。最终，星等为 8.5 ~ 9.5 的恒星的位置残差 < 1.2″。这些图像都是通过未经滤波的 CCD 得到的。光学和 CCD 量子效率限制了我们的灵敏度，使得我们的波长范围为 400 ~ 1,100 nm。因为这个宽波带是非标准的，我们通过下面的校准程序得到一个"等效 V"星等。对每颗第谷恒星而言，我们通过比较"ROTSE 预期星等"和测量到的 2.5 个像素孔径内的流量来确定每幅 ROTSE-I 图像的整体零点。这样，对第谷恒星而言，我们的预期星等和测量星等在 ±0.15 之内基本相符，这里误差主要来自色差。零点确定精确到 ±0.02 之内。对于大的像素，我们必须考虑成团效应（对更弱星等的暂现源这个效应尤其重要）。为了检查这种成团效应，我们比较了暴的位置和 USNO A V2.0 列表 [20] 的已知天体的位置。其中距离最近的天体位于 34″ 开外，是一颗 R 波段星等为 19.2 的恒星。更为重要的是一颗 42″ 开外，$R = 14.4$ 的恒星，这个天体仅会在我们最后的观测中影响 OT 的测量星等。最后一张图中，可以在 OT 的右下方看到该天体。为了弥补这个影响，我们对 OT 的星等进行了 +0.15 的修正。最终我们将如上所述与 GRB990123 成协的 OT 的观测星等列于表 1。更多关于 ROTSE-I 的观测信息请参见 http://www.umich.edu/~rotse。

Table 1. ROTSE-I observations

Exposure start	Exposure duration	Magnitude	Camera
−7,922.08	75	< 14.8	C
22.18	5	11.70 ± 0.07	A
47.38	5	8.86 ± 0.02	A
72.67	5	9.97 ± 0.03	A
157.12	5	11.86 ± 0.13	C
281.40	75	13.07 ± 0.04	A
446.67	75	13.81 ± 0.07	A
611.94	75	14.28 ± 0.12	A
2,409.86	200	< 15.6	A
5,133.98	800	< 16.1	A

Exposure start times are listed in seconds, relative to the nominal BATSE trigger time (1999 January 23.407594 UT). Exposure durations are in seconds. Magnitudes are in the "V equivalent" system described in Fig. 1 legend. Errors include both statistical errors and systematic errors arising from zero-point calibration. They do not include errors due to variations in the unknown spectral slope of the emission. Magnitude limits are 5σ. The final limit results from co-adding the last four 200-s exposures and is quoted at the mean time of those exposures. Camera entries record the camera in which each observation was made.

By the time of the first exposure, the optical brightness of the transient had risen to $m_v = 11.7$ mag. The flux rose by a factor of 13.7 in the following 25 seconds and then began a rapid, apparently smooth, decline. This decline began precipitously, with a power-law slope of ~ -2.5 and gradually slowed to give a slope of ~ -1.5. This decline, 10 minutes after the burst, agrees well with the power-law slope found hours later in early afterglow measurements[14]. These observations cover the transition from internal burst emission to external afterglow emission. The composite light curve is shown in Fig. 2.

A number of arguments establish the association of our optical transient with the burst and the afterglow seen later. First, the statistical significance of the transient image exceeds 160σ at the peak. Second, the temporal correlation of the light curve with the GRB flux and the spatial correlations to the X-ray and afterglow positions argue strongly for a common origin. Third, the most recent previous sky patrol image was taken 130 minutes before the burst and no object is visible brighter than $m_v = 14.8$ mag. This is the most stringent limit on an optical precursor obtained to date. Searches further back in time (55 images dating to 28 September 1998) also find no signal. Finally, the "axis jogging" protocol places the transient at different pixel locations within an image and even in different cameras throughout the exposure series, eliminating the possibility of a CCD defect or internal "ghost" masquerading as a signal.

表 1. ROTSE-I 的观测

曝光开始时间	曝光持续时间	星等	照相机
−7,922.08	75	< 14.8	C
22.18	5	11.70 ± 0.07	A
47.38	5	8.86 ± 0.02	A
72.67	5	9.97 ± 0.03	A
157.12	5	11.86 ± 0.13	C
281.40	75	13.07 ± 0.04	A
446.67	75	13.81 ± 0.07	A
611.94	75	14.28 ± 0.12	A
2,409.86	200	< 15.6	A
5,133.98	800	< 16.1	A

曝光开始时间是相对于 BATSE 触发时间(1999 年 1 月 23.407594 日 UT)而言的,以秒为单位。曝光持续时间以秒为单位。星等系统为图 1 注中说明的"等效 V"系统。误差包括统计误差和零点校准引起的系统误差,不包括由于未知的辐射谱斜率变化所引起的误差。星等极限为 5σ。最后的极限星等是通过叠加最后 4 个 200 秒的曝光结果得到的,并按照这些曝光的平均时间给出。最后一列记录每次观测使用的照相机。

在首次曝光的时候,暂现源的光学亮度已经增加到了 $m_v = 11.7$ 星等。流量在接下来的 25 秒内增加到 13.7 倍,然后开始快速平滑的衰减。衰减开始很急剧,其幂律斜率约为 −2.5,然后逐渐变慢,幂律斜率变为约 −1.5。同时,暴 10 分钟后的衰减幂律和几个小时后的早期余辉测量获得的幂律斜率[14]一致。这些观测覆盖了从内部暴辐射到外部余辉辐射转换的整个过程。合成的光变曲线在图 2 中展示。

有多个论据表明我们观测到的光学暂现源和该暴及随后的余辉存在关联。首先,暂现源图像的统计显著性在峰值处超过 160σ。其次,其光变曲线和 GRB 流量在时间上的相关,以及其位置和 X 射线及余辉位置在空间上的相关,都强烈表明它们具有相同的起源。第三,最近观测的巡天图拍摄于暴前 130 分钟,而图中并没有亮度超过 $m_v = 14.8$ 星等的天体。这是迄今为止获得的对光学前兆最严格的限制。追溯以往的巡天图(至 1998 年 9 月 28 日共 55 幅图像)也没有发现信号。最后,"轴慢跑"方案使得暂现源出现在图像的不同像素位置,甚至在一系列曝光中出现在不同的照相机中,从而排除了因 CCD 缺损或内部"鬼影"产生假信号的可能性。

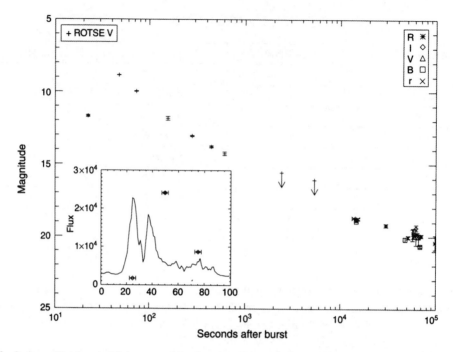

Fig. 2. A combined optical light curve. Afterglow data points are drawn from the GCN archive[21-33]. The early decay of the ROTSE-I light curve is not well fitted by a single power law. The final ROTSE limit is obtained by co-adding the final four 200-s images. The inset shows the first three ROTSE optical fluxes compared to the BATSE γ-ray light curve in the 100–300 keV energy band. The ROTSE-I fluxes are in arbitrary units. Horizontal error bars indicate periods of active observation. We note that there is no information about the optical light curve outside these intervals. Vertical error bars represent flux uncertainties. Further information about GCN is available at http://gcn.gsfc.nasa.gov/gcn.

The fluence of GRB990123 was exceptionally high (99.6 percentile of BATSE triggers; M. Briggs, personal communication), implying that such bright optical transients may be rare. Models of early optical emission suggest that optical intensity scales with γ-ray fluence[15-17]. If this is the case, ROTSE-I and similar instruments are sensitive to 50% of all GRBs. This translates to ~12 optically detected events per year. Our continuing analysis of less well-localized GRB data may therefore reveal similar transients. To date, this process has been hampered by the necessity of identifying and discarding typically 100,000 objects within the large field of view and optimizing a search strategy in the face of an unknown early time structure. The results we report here at least partially resolve the latter problem while increasing the incentive to complete a difficult analysis task. The ROTSE project is in the process of completing two 0.45-m telescopes capable of reaching 4 magnitudes deeper than ROTSE-I for the same duration exposures. If γ-ray emission in bursts is beamed but the optical emission is more isotropic, there may be many optical transients unassociated with detectable GRBs. These instruments will conduct sensitive searches for such events. We expect that ROTSE will be important in the exploration to come.

(**398**, 400-402; 1999)

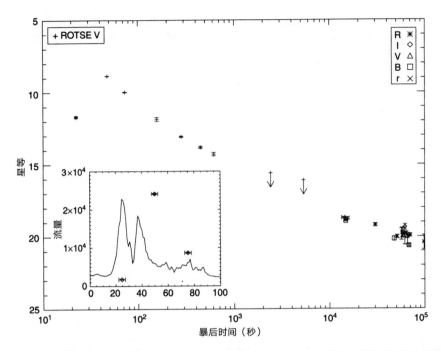

图 2. 联合的光学光变曲线。余辉数据来自 GCN 的归档数据[21-33]。ROTSE-I 光变曲线的早期衰减无法用单个幂律来进行拟合。最后的 ROTSE 极限是通过叠加最后 4 个 200 秒曝光的图像得到的。内嵌插图显示前 3 个 ROTSE 光学流量与 100～300 keV 能段的 BATSE γ 射线光变曲线的比较。ROTSE-I 流量单位是任意选取的。水平误差棒表示有效的观测时段。注意这里没有在这些时段之外的光学光变曲线信息。垂直误差棒代表流量不确定度。更多关于 GCN 的信息请参见 http://gcn.gsfc.nasa.gov/gcn。

GRB990123 的能流非常大（占据 BATSE 触发事件的 99.6%；布里格斯，个人交流），这表明这样亮的光学暂现源可能很少见。早期光学辐射的模型显示光学强度和 γ 射线能流成比例[15-17]。事实如此的话，ROTSE-I 和类似设备将对 50% 的 GRB 都非常灵敏。这意味着每年我们将可以探测到约 12 例光学事件。因此我们接下来对那些定位不是很好的 GRB 数据进行分析可能可以发现类似的暂现源。迄今为止，这个方法的困难之处在于必须证认并排除大视场内约 100,000 个天体，并且在未知早期时间结构的情况下必须最大限度地优化搜寻策略。我们这里报道的结果至少部分解决了第二个问题，并激励我们去完成这些困难的分析任务。ROTSE 项目目前正在搭建两个 0.45 米望远镜使得在相同曝光时间下的星等极限比 ROTSE-I 深 4 个星等。假如暴的 γ 射线辐射是束状的，而光学辐射更加各向同性，那么就可能有很多光学暂现源并没有成协的、能探测到的 GRB。这些仪器将对这种事件做出灵敏的搜寻。我们预期 ROTSE 将在未来的研究探索中发挥非常重要的作用。

（肖莉 翻译；黎卓 审稿）

C. Akerlof*, R. Balsano†, S. Barthelmy‡§, J. Bloch†, P. Butterworth‡∥, D. Casperson†, T. Cline‡, S. Fletcher†, F. Frontera¶, G. Gisler†, J. Heise#, J. Hills†, R. Kehoe*, B. Lee*, S. Marshall☆, T. McKay*, R. Miller†, L. Piro**, W. Priedhorsky†, J. Szymanski† & J. Wren†

* University of Michigan, Ann Arbor, Michigan 48109, USA

† Los Alamos National Laboratory, Los Alamos, New Mexico 87545, USA

‡ NASA/Goddard Space Flight Center, Greenbelt, Maryland 20771, USA

§ Universities Space Research Association, Seabrook, Maryland 20706, USA

∥ Raytheon Systems, Lanham, Maryland 20706, USA

¶ Università degli Studi di Ferrara, Ferrara, Italy

Space Research Organization, Utrecht, The Netherlands

☆ Lawrence Livermore National Laboratory, Livermore, California 94550, USA

** Instituto Astrofisica Spaziale, Rome, Italy

Received 5 February; accepted 19 February 1999.

References:

1. Klebesadel, R. W., Strong, I. B. & Olson, R. A. Observations of gamma-ray bursts of cosmic origin. *Astrophys. J.* **182**, L85-L88 (1973).

2. Piro, L. *et al.* The first X-ray localization of a γ-ray burst by BeppoSAX and its fast spectral evolution. *Astron. Astrophys.* **329**, 906-910 (1998).

3. Costa, E. *et al.* Discovery of an X-ray afterglow associated with the γ-ray burst of 28 February 1997. *Nature* **387**, 783-785 (1997).

4. Krimm, H. A., Vanderspek, R. K. & Ricker, G. R. Searches for optical counterparts of BATSE gamma-ray bursts with the Explosive Transient Camera. *Astron. Astrophys. Suppl.* **120**, 251-254 (1996).

5. Hudec, R. & Soldán, J. Ground-based optical CCD experiments for GRB and optical transient detection. *Astrophys. Space Sci.* **231**, 311-314 (1995).

6. Lee, B. *et al.* Results from Gamma-Ray Optical Counterpart Search Experiment: a real time search for gamma-ray burst optical counterparts. *Astrophys. J.* **482**, L125-L129 (1997).

7. Park, H. S. *et al.* New constraints on simultaneous optical emission from gamma-ray bursts measured by the Livermore Optical Transient Imaging System experiment. *Astrophys. J.* **490**, L21-L24 (1997).

8. Kelson, D. D., Illingworth, G. D., Franx, M., Magee, D. & van Dokkum, P. G. *IAU Circ.* No. 7096 (1999).

9. Hjorth, J. *et al. GCN Circ.* No. 219 (1999).

10. Fenimore, E. E., Ramirez-Ruiz, E., Wu, B. GRB990123: Evidence that the γ-rays come from a central engine. Preprint astro-ph9902007 at ⟨http://xxx.lanl.gov⟩ (1999).

11. Barthelmy, S. *et al.* in *Gamma-Ray Bursts: 4th Huntsville Symp.* (eds Meegan, C. A, Koskut, T. M. & Preece, R. D.) 99-103 (AIP Conf. Proc. 428, Am. Inst. Phys., College Park, 1997).

12. Piro, L. *et al. GCN Circ.* No. 199 (1999).

13. Odewahn, S. C. *et al. GCN Circ.* No. 201 (1999).

14. Bloom, J. S. *et al. GCN Circ.* No. 208 (1999).

15. Katz, J. I. Low-frequency spectra of gamma-ray bursts. *Astrophys. J.* **432**, L107-L109 (1994).

16. Mészáros, P. & Rees, M. J. Optical and long-wavelength afterglow from gamma-ray bursts. *Astrophys. J.* **476**, 232-237 (1997).

17. Sari, R. & Piran, T. The early afterglow. Preprint astro-ph/9901105 at ⟨http://xxx.lanl.gov⟩ (1999).

18. Bertin, E. & Arnouts, S. SExtractor: Software for source extraction. *Astron. Astrophys. Suppl.* **117**, 393-404 (1996).

19. Høg, E. *et al.* The Tycho reference catalogue. *Astron. Astrophys.* **335**, L65-L68 (1998).

20. Monet, D. *et al. A Catalog of Astrometric Standards* (US Naval Observatory, Washington DC, 1998).

21. Zhu, J. & Zhang, H. T. *GCN Circ.* No. 204 (1999).

22. Bloom, J. S. *et al. GCN Circ.* No. 206 (1999).

23. Gal, R. R. *et al. GCN Circ.* No. 207 (1999).

24. Sokolov, V. *et al. GCN Circ.* No. 209 (1999).

25. Ofek, E. & Leibowitz, E. M. *GCN Circ.* No. 210 (1999).

26. Garnavich, P., Jha, S., Stanek, K. & Garcia, M. *GCN Circ.* No. 215 (1999).

27. Zhu, J. *et al. GCN Circ.* No. 217 (1999).

28. Bloom, J. S. *et al. GCN Circ.* No. 218 (1999).

29. Maury, A., Boer, M. & Chaty, S. *GCN Circ.* No. 220 (1999).

30. Zhu, J. *et al. GCN Circ.* No. 226 (1999).

31. Sagar, R., Pandey, A. K., Yadav, R. K. R., Nilakshi & Mohan, V. *GCN Circ.* No. 227 (1999).

32. Masetti, N. *et al. GCN Circ.* No. 233 (1999).

33. Bloom, J. S. *et al. GCN Circ.* No. 240 (1999).

Acknowledgements. The ROTSE Collaboration thanks J. Fishman and the BATSE team for providing the data that enable the GCN localizations which made this experiment possible; and we thank the BeppoSAX team for rapid distribution of coordinates. This work was supported by NASA and the US DOE. The Los Alamos National Laboratory is operated by the University of California for the US Department of Energy (DOE). The work was performed in part under the auspices of the US DOE by Lawrence Livermore National Laboratory. BeppoSAX is a programme of the Italian Space Agency (ASI) with participation of the Dutch Space Agency (NIVR).

Correspondence and requests for materials should be addressed to C.A. (e-mail: akerlof@mich.physics.lsa.umich.edu).